D1010024

PENGUIN BOOKS

FROM ONE TO ZERO

Born in Morocco and widely traveled, Georges Ifrah has for many years been a teacher of mathematics in France, where he published many scholarly articles and where *From One to Zero* first appeared to wide acclaim.

Lowell Bair, who translated this work from the original French in consultation with the author, lives in Woodstock, New York, where he has worked full time as a translator for many years.

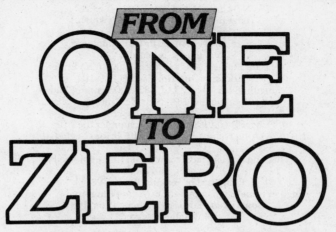

FROM ONE TO ZERO

A Universal History of Numbers

GEORGES IFRAH

Translated by Lowell Bair

PENGUIN BOOKS

PENGUIN BOOKS
Viking Penguin Inc., 40 West 23rd Street,
New York, New York 10010, U.S.A.
Penguin Books Ltd, Harmondsworth,
Middlesex, England
Penguin Books Australia Ltd, Ringwood,
Victoria, Australia
Penguin Books Canada Limited, 2801 John Street,
Markham, Ontario, Canada L3R 1B4
Penguin Books (N.Z.) Ltd, 182–190 Wairau Road,
Auckland 10, New Zealand

Originally published in French under the title
Histoire Universelle des Chiffres by Editions Seghers 1981

This English-language translation first published by
Viking Penguin Inc. 1985
Published in Penguin Books 1987
English translation copyright © Viking Penguin Inc., 1985
Copyright © Editions Seghers, Paris, 1981
All rights reserved

All illustrations, including calligraphy, are by the author.

LIBRARY OF CONGRESS CATALOGING IN PUBLICATION DATA
Ifrah, Georges.
From one to zero.
Translation of: Histoire universelle des chiffres.
Bibliography: p.
Includes index.
1. Numeration — History.   2. Numerals — History.
I. Title.
QA141.2.I3613   1987   513'.5   86-19023
ISBN 0 14 00.9919 0

Printed in the United States of America by
R. R. Donnelley & Sons Company, Harrisonburg, Virginia
Set in Trump Medieval

For you, my wife,
who were a patient witness to the joys
and anxieties which this hard labor
gave me for several years.
For you, Anna,
to whom this book and its author owe so much.

Contents

Introduction

This book began with a child's question. I was teaching mathematics at the time, and like all good teachers I tried never to let any question go unanswered, no matter how strange or naive it might seem. Intelligence is often nurtured by curiosity.

One day I was asked a question so simple that for a moment it left me speechless: "Where do figures come from? When did people learn how to count? How did numbers start?"

What *is* the origin of numbers? They are so familiar to us in our everyday lives that they give the deceptive impression of being innate, and in adulthood we are inclined to assume that they have always been part of our thoughts—like language, another tool whose use must be learned.

When I was asked where figures came from, I could say only that they were invented a long time ago, an answer that did not even disguise my ignorance. Groping my way, I tried to go back through the centuries to a time before our Hindu-Arabic and Roman numerals existed. Were there other numerals before them? If so, what were they like? And would a search for the origin of figures reveal some vestige of the brilliant invention made by the first human being who conceived the idea of counting?

I am afraid my answers that day were sketchy, incomplete, and perhaps inaccurate. I had an excuse. In the arithmetic books that were the tools of my trade, the question was not even raised. In history books it was only touched on lightly, as if there were a strange conspiracy to conceal one of humankind's most fantastic discoveries, and perhaps the most fruitful, since it has enabled us to try to measure the world, understand it a little better, and put some of its secrets to use. There were, as I later learned, a few specialized works describing what was known on the subject; I will mention them often in this book, and I owe a great deal to them. But they had been written almost exclusively for specialists and were far from being complete, despite the erudition shown in them.

My pupil's question stayed in my mind for years, to the point where I regretfully gave up teaching to devote myself entirely to a quest that many might consider as senseless as a quest for some sort of Holy Grail. Such a quest, of course, has no end. This book will take its place, perhaps a minor one, in a cohort of eminent volumes. It will not be the last, for there are still many discoveries to be made and many problems to be solved. But I like to believe that it may be the first to offer the general public a comprehensive treatment of its subject—the worldwide history of the invention of figures—written in simple language, making abundant use of illustrations, assuming no previous knowledge, and requiring at most only a little attention.

The history of that invention began long ago, though we cannot say exactly where. Unable to conceive of numbers in themselves, people did not yet know how to count. To them, numbers must have been concrete realities, inseparable from the nature of the objects considered.

In our own time, a few "primitive" peoples of Oceania, Africa, and America are still at that "zero degree" of experience with numbers. Guided solely by their natural ability to recognize concrete quantities at a glance, they can conceive, discern, and designate only a single object or a pair, and so their numerical notions are limited to "one," "two," and "many."

But there is no need to know how to count as we do in order to determine and transmit the date of a ceremony, or find out whether all the sheep, goats, and cattle taken out in the morning have come back at the end of the day. Even without the use of language, memory, or abstract thought, such things can be done by means of various devices. Some present-day "primitives," knowing only the principle of one-to-one correspondence, make notches in bone or wood for that purpose. Others use piles or rows of pebbles, seashells, bones, or sticks. Still others indicate parts of the body: fingers, toes, arm and leg joints, eyes, nose, mouth, ears, breasts, chest.

Nature has provided many models: 2 can be symbolized by a bird's wings, 3 by a clover leaf, 4 by an animal's legs, 5 by the fingers of one hand, 10 by the fingers of two hands together, and so on. Living in this world of numbers, people gradually came to grasp the abstraction of number.

And since everyone began counting on his ten fingers, most numerical systems now in existence have 10 as their base. A few eccentric cultures have chosen the base 12. The Mayas, Aztecs, and Celts, who realized that by bending down a little they could count their toes, adopted the base 20. The Sumerians, who invented the oldest known form of writing, and the Babylonians, whose invention of the zero is enough in itself to earn them a prominent place in history, used the base 60, for reasons unknown to us. To them we owe those problems of division into hours, minutes, and seconds that all our schoolchildren know and sometimes dread, and that circle strangely divided into 360 degrees. But those things already involved sophisticated calculations.

In 1937 a wolf bone more than 20,000 years old, with fifty-five notches in two rows divided into groups of five, was discovered at Vestonice, Czechoslovakia. It is one of the oldest known "counting machines." Perhaps it belonged to a skilled hunter who made a notch in a bone each time he killed an animal, using a different bone for each kind of animal: one for bears, one for bison, one for wolves. If so, he invented the rudiments of accounting, because his procedure amounted to writing figures in the simplest numerical language.

It might be thought that this was a primitive technique with no future. It was indeed primitive, but it was certainly not without a future. It has come down to us almost unaltered, one of the longest-lived inventions ever made. Not even the wheel goes back that far. Only the use of fire can rival it.

The many notches found beside pictures of animals on the walls of prehistoric caves leave no doubt that they were used as an accounting device, and the technique has scarcely changed since its beginning. From time immemorial, shepherds in the Alps and other parts of Europe have recorded the numbers of their flocks by carving vertical lines, V's, and X's in wooden tablets. In the eighteenth century, this rustic accounting system was still used in the archives of the very serious British Parliament, and in Russia and the German states it was still used for lending sums of money.

At the end of the nineteenth century, in all of rural France as well as in Indochina, notched sticks still served as accounting ledgers, written agreements, and credit records. Only ten years ago, the baker in a small French village near Dijon still cut notches into a piece of wood to keep track of the number of loaves for which his customers owed him.

The fact that this primitive technique has endured so long is all the more remarkable because it is the origin of a written numeration that we often use in addition to Hindu-Arabic numerals: older than the civilization that transmitted them to us, Roman numerals are derived from the use of notches.

Pebbles are also important in the history of arithmetic because they initiated the human race into the art of calculation. The word "calculation" itself takes us back to the remote past, since it comes from the Latin *calculus*, meaning "small stone."

In Madagascar, not so long ago, military leaders used a practical method to determine how many soldiers they had: they made them walk through a narrow passage in single file, and each time a soldier passed, a pebble was dropped on the ground. When there was a pile of ten pebbles, it was replaced with one pebble that was the beginning of a pile representing tens. The process was repeated till the tens pile contained ten pebbles, then another pile was begun for hundreds; and this continued till all the soldiers had been counted.

It would be a mistake to believe that the reasoning of those Madagascans was of a "primitive" form. The ancient Greeks and Romans, and the peoples of medieval Europe, calculated according to the same principle: they placed pebbles or counters on a counting board with columns representing units, tens, hundreds, etc. The Chinese, Japanese, and Russians still use the abacus, which differs from the ancient counting board only in its construction and the nature of its counters.

One day a group of accountants had the idea of replacing the usual pebbles with clay objects whose shapes would represent different orders of units: for example, a cone for 1, a sphere for 10, and a disk for 100. An accounting system was then developed on that basis and proved to be extremely useful. This happened in the fourth millennium B.C., in Sumer and Elam, near the Persian Gulf. The Sumerian and Elamite civilizations were already expanding economically but still had no form of writing, which meant that the limits of human memory determined how much information could be preserved, and for how long. The new system could be used not only for calculation but also for keeping many kinds of records. And its importance was all the greater because it was the origin of written accounting.

Another concrete means of counting is even older. The hand was undoubtedly the first "calculating machine." In China, India, the Soviet Union, and the Auvergne region of France, fingers are still used to perform multiplication, without the aid of any other material device. There was once a time when, using only their fingers, the Egyptians, Persians, Arabs, and Europeans were able to express numbers from 1 to 10,000 in a system similar to modern deaf-mute languages.

The method of indicating numbers with knots in string was used in an ingenious system of record keeping and administration by the Incas of South America, and is still found in western Africa, the Ryukyu Islands of Japan, the Hawaiian Islands, and the Caroline Islands.

It is striking to see how, proceeding by trial and error, peoples living at great distances from each other sometimes followed the same paths. The Egyptians, Cretans, Hittites, and Aztecs developed written numerations based on the same concept. This was also true of the Sumerians, Greeks and Romans, and the ancient peoples of southern Arabia. The Assyrians, Aramaeans, southern Indians, Ethiopians, and Chinese used systems that were, if not identical, at least similar.

It is also fascinating to see the stages of mathematical thought. The Babylonians invented the oldest known place-value numeration. (A place-value numeration is one in which the sign for the number 3, for example, has different values according to its position in a written number: units, tens, hundreds, and so on.) The Mayas, Indians, and Chinese later made the same invention, entirely independently. For centuries the Babylonians did not know the zero and they "invented" it only after a very slow process of evolution. The Chinese learned of it from Indian mathematicians. The Mayas developed it and placed it in the middle and at the end of their numerical expressions, but did not know how to use it for performing operations. Finally, the Indian zero had more or less the same possibilities as the one we know. It was transmitted to us by the Arabs at the same time as what we often call Arabic (more accurately, Hindu-Arabic) numerals, which are actually Indian numerals slightly deformed by time and travel.

We must also speak of those ingenious Phoenicians, who, in the second millennium B.C., were probably the first to discover the principle of alphabetic writing. The Greeks learned it from them and transmitted it to the whole Western world.

The Greeks, Jews, Syrians, and Arabs then had the idea of writing numbers by means of their letters. Once a number had been assigned to each letter, every word also had a numerical value and this was the basis of the mystical religious doctrine called gematria by the Cabalists and isopsephia by the Greeks and Gnostics. The Gnostics believed that by this means they could determine the formula of God's real name in order to apprehend his secrets. The same procedure led the Jewish Cabalists, and the Christian and Moslem esoterists after them, to all sorts of symbolic interpretations and various calculations for predicting the future. It enabled soothsayers to interpret dreams and make talismans. Among poets like Leonidas of Alexandria it inspired literary works of a special kind, and later, among Turkish, Persian, and North African poets, it gave rise to the art of composing chronograms.

The Greeks or Jews who put together the first alphabetic numeration certainly did not foresee that 2000 years later a Catholic theologian named Petrus Bungus would write a book in which he demonstrated to his own satisfaction that the numerical value of Martin Luther's name was 666. The meaning was clear to those versed in such calculations because, in the Book of Revelation, 666 is the number of the "beast" identified with the Antichrist. Bungus was not a pioneer in this kind of interpretation: long before him, others like him had succeeded in "proving" that 666 was the number of Emperor Nero.

Today, I could perhaps give an almost satisfactory answer to my pupil's question. His curiosity—which I hope you already share—goaded me into beginning and continuing my investigation, and I was greatly helped by luck. The luck, for example, that endows each unique human being with a unique culture. Having been born in Morocco, I knew Arabic. Being Jewish, I had learned Hebrew. Being fascinated by mathematics, I had a familiarity with figures that enabled me to pick out the rules in a complex system. And having a certain amount of dexterity, I managed to draw the illustrations that enliven this book, in a naive style for which I ask to be excused. I was also sustained, through the years, by questions from lecture audiences, by the encouragement and precious information I received from the many cooperative scientists to whom I owe so much of my knowledge, and by the demands and advice of my publishers.

The experience of drawing illustrations to accompany discussions of ancient numerical systems made me better able to see those systems from the viewpoint of the people who used them, and in one case it unexpectedly enabled me to decipher a notation whose meaning had so far remained unknown to specialists.

The most extraordinary personal discovery I made in the course of doing the research for this book—a discovery I hope I can communicate to you—is that figures, far from being only the arid symbols often denounced as instruments of our technological society, have always, since their beginning, *also* been materials for literature, a stimulus to daydreams, fantasies, and metaphysical speculation, and a means of probing the uncertain future, or at least of trying to satisfy the desire to predict it. Almost as much as words, figures have been tools for the poet as well as the accountant and the scientist. By their universality, which can be seen in the multiplicity of solutions devised for the problem of numeration, and by their history, which slowly but surely converges on the decimal place-value system that has prevailed all over the world, figures bear witness, better than the babel of languages, to the underlying unity of human culture. When we consider them, our awareness of the prodigious and fruitful diversity of societies and histories gives way to a feeling of almost absolute continuity. Though they are only one part of human history, they bind it together, sum it up, and run through it from one end to the other, like that red thread which, according to Goethe, ran through all the ropes of the British navy, so that one could not cut a piece from any of them without recognizing that it belonged to the Crown. Figures are profoundly human.

And if I have any advice to give you before you embark on this adventure of the mind, it is always to pay attention to children's questions and do your best to answer them. But they may take you very far, much farther than you ever thought you would go.

GEORGES IFRAH

PART I

Awareness of Numbers

The Origin and Discovery of Numbers

Are animals able to count?

Ingenious experiments conducted by specialists in animal behavior have shown that some species apparently have a kind of rudimentary direct perception of concrete quantities, which is not the same as the ability to count. This faculty has been called the number sense. It enables an animal to discern the difference in size between two small collections of similar elements, and in some cases to recognize that a single small collection is no longer the same after an element has been added or removed. "It does happen that a domestic animal, a dog, ape, or elephant, perceives the disappearance of an object in a restricted ensemble with which it is familiar. In many species, the mother shows by unmistakable signs that she knows that one of her little ones has been taken from her" (Lévy-Bruhl).*

A similar ability, probably more precise than in domestic animals, has been observed in birds. Experiments have shown that a goldfinch, trained to choose its food from two small piles of seeds, generally succeeds in distinguishing three from one, three from two, four from two, four from three, and six from three, but nearly always confuses five and four, seven and five, eight and six, ten and six.

Even more remarkable is the example of the crow and the magpie, which are apparently able to distinguish concrete quantities ranging from one to four. "A squire was determined to shoot a crow which made its nest in the watch-tower of his estate. Repeatedly he had tried to surprise the crow, but in vain: at the approach of man the crow would leave its nest. From a distant tree it would watchfully wait until the man had left the tower and then return to its nest. One day the squire hit upon a ruse: two men entered the tower, one remained within, the other came out and went on. But the bird was not deceived: it kept away until the man

---

*Following a quotation, a name between parentheses refers to the source of the quotation given in the bibliography.

within came out. The experiment was repeated in the succeeding days with two, three, then four men, yet without success. Finally, five men were sent: as before, all entered the tower, and one remained while the other four came out and went away. Here the crow lost count. Unable to distinguish between four and five it promptly returned to its nest" (Dantzig).

It should be noted that this number sense in animals is limited to rather obvious differences between two collections. And although animals sometimes perceive concrete quantities and differences between them, they never conceive absolute quantities because they lack the faculty of abstraction.

From this we may conclude that animals are not able to count. It seems safe to assume that counting is an exclusively human ability, closely related to the development of intelligence and involving a mental process more complex than the number sense.

The behavior of the insect known as the solitary wasp is striking: "The mother wasp lays her eggs in individual cells and provides each egg with a number of live caterpillars on which the young feed when hatched. Now, the number of victims is remarkably constant for a given species of wasp: some species provide five, others twelve, others again as high as twenty-four caterpillars per cell. But most remarkable is the case of the genus *Eumenus*, a variety in which the male is much smaller than the female. In some mysterious way the mother knows whether the egg will produce a male or a female grub and apportions the quantity of food accordingly; she does not change the species or size of the prey, but if the egg is male she supplies it with five victims, if female with ten" (Dantzig).

Impressive though this example is, however, it is not convincing evidence of an ability to count, since the wasp's behavior is connected with a basic life function and seems to be instinctive.

All sorts of rash statements have been made about circus animals by writers who often have trouble distinguishing between fact and fancy. Some of them maintain that dogs can count up to 10 or beyond, and to support this claim they cite examples of dogs able to indicate numbers by tapping their paws or barking as many times as necessary. But to anyone who has ever seen a performance by such a trained dog it should be clear that the writers who make the claim have been taken in by tricksters as skilled in the art of deception as in training their dogs.

Other writers of the same kind, either showing their own gullibility or trying to mislead their readers, have expressed amazement at the phenomenal feats of dogs, horses, and elephants supposedly able not only to add, subtract, and multiply but also to extract square and cube roots! Some have even said that a few of these clever animals had learned to recognize and use the letters of the alphabet and, when necessary, could write in order to let their masters know what they wanted. Such claims are so obviously naive that it would be pointless to argue against them. I have mentioned them only to show how far one can be led by an uncritical determination to draw parallels between human beings and animals.

Natural faculties of number perception

What was the main reason that prompted people to develop the notion of number? Was it astronomical concerns (phases of the moon, repetitive calendars of days and nights, cycle of the seasons, etc.), as some authors maintain? Or was it simply the needs of communal life? How and when did people discover, for example, that the fingers of one hand and the toes of one foot represent a single concept? How did the need for numerical calculaton become apparent to them? When did the first genuine oral numerations begin? Was the abstract concept of number developed before language? Were the cardinal and ordinal aspects of numbers grasped simultaneously or at different times? Did the number concept arise from experience, or did experience act as a catalyst in bringing out something that was already latent in the minds of our remote ancestors?

These questions cannot be answered conclusively because we have no direct information on the thought processes of early human beings, and also because they are bound up with the unsolved (and perhaps insoluble) problem of the origin of human intelligence. Our approach to such questions must be conjectural.

Even so, there are still good reasons for believing that there was a time when people did not know how to count. Not knowing how to count, however, does not imply that they had no notion of number, but only that this notion was limited to a kind of number sense, that is, to what direct perception enabled them to recognize at a glance. We may assume that, to them, the number concept was a concrete reality inseparable from objects and that it was manifested only in direct perception of physical plurality. They were probably unable to conceive of numbers in themselves, as abstractions. If so, they must not have been aware that such aggregates as a brace of hares, the wings of a bird, or the eyes, ears, arms, or legs of a person have the common characteristic of "being two."

This may be hard for us to accept, since mathematics has made such rapid and amazing progress in relatively recent times that the basic idea of number now seems childishly simple to us, but it is supported by studies of existing preliterate societies and infant behavior which show that, although "primitive"* people and young children have a natural

---

*Some writers use "primitive" subjectively, to designate members of human societies that they judge to be at a rudimentary stage of mental development, in comparison with ours. To avoid misunderstanding, I will always place that word within quotation marks.

ability to distinguish between concrete quantities, it is not much greater than that of certain animals as long as they have not become capable of grasping such concepts as one-to-one correspondence, classification, and hierarchical order.

When several similar objects have been separated, a fourteen-month-old baby can usually reassemble them into a single group. If something is missing from a small group that is familiar to him, he will immediately notice it. But his abilities are so limited that he fails to perceive numerical differences or equality in the people or objects around him as soon as their number goes beyond three or four. He cannot conceive any absolute quantity because he is not yet capable of counting, in the strict meaning of the term.

Contemporary "primitive" people also seem unable to grasp number considered in its abstract, conceptual aspect. "As a matter of fact, if a well-defined and fairly restricted group of persons or things interests the primitive ever so little, he will retain it with all its characteristics. In the representation he has of it the exact number of these persons or things is implied: it is, as it were, a quality in which this group differs from one which contained one more, or several more, and also from a group containing any lesser number. Consequently, at the moment this group is again presented to his sight, the primitive knows whether it is complete, or whether it is greater or less than before" (Lévy-Bruhl). "Primitive" people are thus affected only by a change in their visual perception, since they generally lack the abstract notion of the synthesis of distinct units.

"A rudimentary number sense, not greater in scope than that possessed by birds, was the nucleus from which our number concept grew. And there is little doubt that, left to this direct number perception, man would have advanced no further in the art of reckoning than the birds did. But through a series of remarkable circumstances man has learned to aid his exceedingly limited perception of number by an artifice which was destined to exert a tremendous influence on his future life. This artifice is counting, and it is to *counting* that we owe the extraordinary progress which we have made in expressing our universe in terms of number" (Dantzig).

Early in this century there were still peoples in Africa, Oceania, and America who could not clearly perceive or precisely express numbers greater than 4. To them, numbers beyond that point were vague, general notions related to physical plurality. It is probably significant that, as Lévy-Bruhl reports, some Oceanic tribes declined and conjugated in the singular, the dual, the trial, the quadrual, and finally the plural.

Members of the Aranda tribe in Australia had only two basic number words: *ninta* ("one") and *tara* ("two"). For "three" and "four" they said *tara-ma-ninta* ("two-and-one") and *tara-ma-tara* ("two-and-two"). Beyond *tara-ma-tara* they used a word meaning "many."

Islanders in Torres Strait, between New Guinea and Australia, had only these number words: *netat* ("one"), *neis* ("two"), *neis-netat* ("three," literally "two-one") and *neis-neis* ("four," literally "two-two"); beyond that, they used a word meaning something like "a multitude."

Among other examples of the same kind, we can mention the Indians of Tierra del Fuego, the Abipones in Paraguay, the Bushmen and Pygmies in Africa, and the Botocoudos in Brazil. (When the Botocoudos said their word for "many," they pointed to their hair, as if to say, "Beyond four, things are as countless as the hairs on my head.")

Since these people were able to use combinations of their two basic number words for the next two numbers—"two-one" for 3, "two-two" for 4—one might wonder why they did not use "two-two-one" for 5, "two-two-two" for 6, "two-two-two-one" for 7, and so on. That would be overlooking the fact that to express 3 and 4, numbers that they can recognize by direct perception, people at this stage "simply *pair* 1 and 2, then 2 and 2; for them, these are still pairs, though for us they are whole numbers when we designate them as 3 and 4. Being able to conceive, recognize, and name only a single element or a pair of elements, how could these people, on their own, express 5 as 2 + 2 + 1, or 6 as 2 + 2 + 2, when each of those expressions would contain three elements?" (Gerschel 1).

It would be a mistake to believe that we could do much better than those "primitive" people *if we let ourselves be guided only by our natural ability to recognize numbers at a glance.* "In every practical case where civilized man is called upon to discern number, he is consciously or unconsciously aiding his direct number sense with such artifices as symmetric pattern reading, mental grouping or counting. *Counting* especially has become such an integral part of our mental equipment that psychological tests on our number perception are fraught with great difficulties" (Dantzig).

How far can we go if we look at a group of similar elements and try to ascertain their number, using only our direct number sense, without counting? We can easily distinguish one, two, three, or even four elements; but after four, our ability to discern concrete quantities usually stops (fig. 1–1). "From five on, everything becomes blurred. Are there five or six steps in that staircase? Fourteen or fifteen bars in that gate? We must count them to know. It can be taken as a fact of broad significance that the sequence of numbers came to a stop at four: when a numerical notation consists in representing a number by as many aligned similar marks as there are elements in the collection that one proposes to count, one must stop at IIII, because it is impossible to 'read' at a glance a row of five marks, IIIII; or six, IIIIII; or seven, IIIIIII; or more" (Gerschel 1).

In the language of the ancient Romans, "only the first four number words are declined; beginning with five, they have neither declension nor gender. Similarly, only the first four months of the Roman year have real names: Martius, Aprilis, etc. Beginning with the fifth, Quinctilis, they

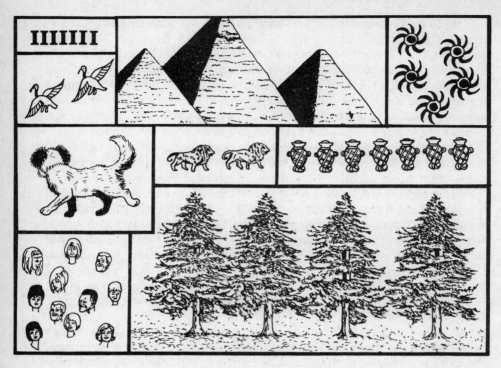

Fig. 1–1. Our direct number sense enables us to know at a glance if a collection contains one, two, three, or four elements, but beyond four we usually have to count the elements, or use such artifices as mental grouping or symmetric pattern reading, because our direct perception of plurality is no longer adequate.

are only 'serial numbers,' up to the last month of the original year, December. It has also been noted that the given names of Roman children were varied, up to and including the fourth child, but that parents simply named succeeding children Quintus, Sextus, . . . Octavius, . . . Decimus, whereas Quartus, for example, was never used. The Romans' capacity for numerical individualization stopped at four: up to that point, they gave normally constituted names, with ordinary characteristics, to numbers, months and children; beyond it, there was imprecision" (Gerschel 2).

Can one define a quantity without knowing how to count?

As we have just seen, our direct perception of number very seldom goes beyond four, so for greater quantities our mind no longer relies solely on the natural number sense: it uses the artifice of abstract counting, which is known to every "civilized" person.

Are we to conclude that, without having learned to count, the human mind is incapable of learning any numerical procedures at all? Even though someone who does not know how to count will, of course, lack the concepts that we call "one," "two," "three," "four," "five," "six," etc., is it legitimate to infer that such a person can never learn or invent a number technique that will enable him, in certain circumstances, to identify concrete quantities?

The answers to those questions are negative. There are excellent reasons, which we are about to examine, to support the conjecture that *for centuries people used many numbers without being able to conceive them abstractly.*

Anthropological studies show that several contemporary "primitive" peoples, who are not yet prepared for the abstract concept of number, have number techniques that can be called "concrete" in comparison with ours; up to a certain point, they can achieve the same results by using material objects (pebbles, shells, bones, fruit, dried animal droppings, sticks, notched wood or bone).

How can these methods be used by people unable to count or conceive number? If such people see a row of notches on a bone, or a pile of pebbles, are they any better off than if they were looking at a group of persons, animals, or objects?

Their methods are much less powerful than ours (and sometimes much more complicated), but they can be used, for example, to determine whether as many sheep came back in the evening as went out in the morning. To do that, there is no need to understand the artifice of counting.

Let us imagine a shepherd, unable to count, who has a flock of sheep that he keeps in a cave at night. There are fifty-five of them, but he has no understanding of what "the number 55" means. He knows only that he has "many" sheep. But he would like to be sure that all of them come back every evening. One day he has an idea. He sits down at the entrance of his cave and has the sheep go into it one by one. Each time one of them passes in front of him, he makes a notch in a bone. When all the sheep have passed, he has made exactly fifty-five notches, without knowing the arithmetical meaning of that number. From now on, he has his sheep go into the cave one by one every evening. Each time one of them passes he puts his finger on a notch, starting at one end of the bone, and if his finger reaches the last notch he is reassured because he knows that his whole flock has safely returned.

In this connection, a discovery made by an American archaeological team in 1928, and analyzed in 1959 by A. L. Oppenheim, is of great interest. In the ruins of the palace of Nuzi, a Mesopotamian city dating from about the fifteenth century B.C., in the Kirkuk region southwest of Mosul, Iraq, the archaeologists found an egg-shaped clay envelope (fig. 1–2) bearing a cuneiform inscription translated as follows:

Objects concerning sheep and goats	4 male lambs
21 ewes that have lambed	6 female goats that have kidded
6 female lambs	1 male goat
8 adult rams	[2] kids

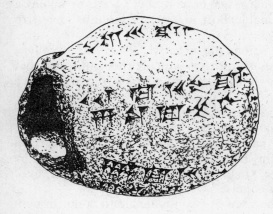

Fig. 1–2. Clay envelope discovered in the ruins of the ancient Mesopotamian city of Nuzi. Harvard Semitic Museum, Cambridge, Mass.

Total: forty-eight animals. When the envelope was opened, it was found to contain forty-eight clay balls, which were later lost.

The archaeologists would probably not have attached much importance to this discovery if an incident had not revealed the original function of the clay balls: "A servant of the expedition was sent to the market to buy chickens. When he returned, the chickens were mistakenly put into the poultry yard without having been counted. Since the servant was completely uneducated and did not know how to count, he was unable to say how many chickens he had bought. He could not have been accurately reimbursed for the purchase if he had not presented a number of pebbles that he had set aside: one for each chicken, he explained" (Guitel 1).

Without knowing it, the uneducated servant had repeated a procedure used by equally uneducated shepherds who lived in that region 3500 years earlier. The clay envelope had belonged to an accountant (who, unlike the shepherds, knew how to write). The shepherds had gone to see him before taking their master's flock to pasture. He made as many clay balls as there were animals in the flock and put them into the envelope, which was then closed, covered with a cuneiform inscription describing the composition of the flock, and marked with the owner's seal. If things had happened normally, when the shepherds returned the envelope would have been broken open (we do not know why it remained intact) and the number of sheep and goats would have been checked by matching them with the clay balls. There could have been no dispute, because two records had been kept: the clay balls for the shepherds, the seal and inscription for the owner.

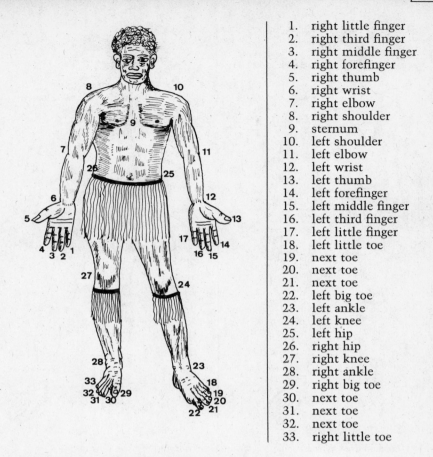

1.	right little finger
2.	right third finger
3.	right middle finger
4.	right forefinger
5.	right thumb
6.	right wrist
7.	right elbow
8.	right shoulder
9.	sternum
10.	left shoulder
11.	left elbow
12.	left wrist
13.	left thumb
14.	left forefinger
15.	left middle finger
16.	left third finger
17.	left little finger
18.	left little toe
19.	next toe
20.	next toe
21.	next toe
22.	left big toe
23.	left ankle
24.	left knee
25.	left hip
26.	right hip
27.	right knee
28.	right ankle
29.	right big toe
30.	next toe
31.	next toe
32.	next toe
33.	right little toe

Fig. 1–3. Counting method used by Torres Strait islanders.

In the nineteenth century, some Torres Strait islanders "counted visually" in the following way (fig. 1–3). On the right side of the body, they touched each of the five fingers, the wrist, the elbow, and the shoulder; then they touched the sternum and, on the left side of the body, the fingers, wrist, elbow, and shoulder. This gave a total of 17. If that was not enough, they added the toes, ankles, knees, and hips on both sides, which gave 16 more, or 33 in all. For numbers greater than 33, they used small sticks.

Other Torres Strait islanders used variants of the same technique for reaching numbers up to 29 in some cases, 19 in others. Similar techniques were used by the Papuans and Elemas of New Guinea (fig. 1–4) and peoples in various parts of Africa, Oceania, and America.

These methods give us a plausible idea of how our remote ancestors

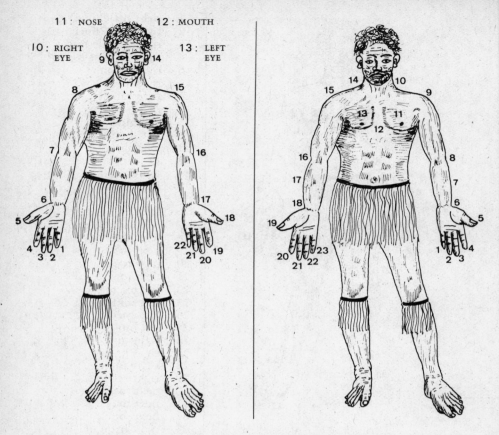

Fig. 1–4. Method used by the Papuans (left) and Elemas (right) of New Guinea.

must have used parts of the body as a material basis for the number concept, and they are probably at the origin of the slow, gradual progress of the human mind toward the abstraction of counting. They enable the people who still use them to satisfy concrete needs. But these people, of course, do not view them in the light of our number concept. They associate certain quantities with the movements of touching various parts of the body in a fixed order, and if necessary they can refer to each quantity again by repeating the series of movements that led up to it. In a commercial transaction, for example, a man may remember at which part of his body a number of objects ended, and by repeating the operation, beginning with his left little finger (if that is the starting point of his system), he can reproduce the number in question whenever he wants.

To form a more specific idea of such practices, let us imagine a "primitive" tribe whose council of elders has convened to decide on the day and the month when many tribes will gather to perform a religious ceremony in common. It will be in several "moons" because time is needed to notify the other tribes so they can be at the right place on the right day.

After a long discussion, the council schedules the ceremony for what we, with our sophisticated number concept, would call "the tenth day of the seventh moon," starting from the first day of the next moon. But the members of the tribe are unable to conceive abstract number and the notions contained in our terms "tenth" and "seventh." They have only concrete techniques based on parts of the body and other material objects.

How will they go about expressing and remembering the date they have just decided on? Let us suppose that they use the technique illustrated in figure 1–3, which enables them to reach numbers up to 17 by successively touching the fingers of the right hand, beginning with the little finger, then the right wrist, elbow, and shoulder, then the sternum, and finally the left fingers, wrist, elbow, and shoulder. They can then express the date by saying something like this: "Moon, right elbow; day, left shoulder."

So as not to forget this important date, the chief of the tribe uses some sort of durable coloring substance to mark his own right elbow and left shoulder; he may, for example, draw a line on his left shoulder to indicate the day of the ceremony, and a circle on his right elbow for the "rank" of the corresponding moon. He then tells some of his subordinates to do the same in order to take the message to the other tribes.

To make sure they will recognize the date when it arrives, the tribe members combine observation of the next new moon with one of the ingenious procedures handed down to them by tradition and originally developed by their ancestors after generations of trial and error.

On the first day of the next moon after the meeting of the council, the chief of the tribe takes one of the bones with thirty notches in it that he uses whenever he needs to consider the days of a single moon in their order of succession. Then he ties a string around the first notch of this "lunar calendar." The next day, he ties another string around the second notch, and so on till the end of the lunar month (fig. 1–5). When he has tied the thirtieth string, he unties it and all the others, and to show that one moon has passed he draws a circle on the little finger of his right hand.

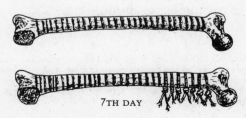

Fig. 1–5.   7TH DAY

At the beginning of the next moon he again ties a string around the first notch of the bone, and continues in this way till the end of the moon, when he draws a circle on the third finger of his right hand. This time, however, he stops at the twenty-ninth notch, since his ancestors observed long ago that the lunar month alternates between twenty-nine and thirty days.

He continues this procedure, alternating between twenty-nine and thirty days, until he marks his right elbow with a circle. He now knows that there are no more moons to count. He also knows that the tribe must soon leave for the place of the ceremony because there are only ten more days—or rather, from his viewpoint, the time to reach his left shoulder (fig. 1–3)—before the date that has been agreed on.

This plausible reconstruction (several elements of which are found among the Australian aborigines, for example) shows that such techniques make it possible to reach relatively large numbers when parts of the body, touched in a fixed order, are associated with other objects: knotted strings, sticks, pebbles, notched bones, etc.

Rudimentary "accounting" techniques

When people who have not learned abstract counting want to verify a previously known total, they use methods based on a concept that is highly important in modern mathematics: one-to-one correspondence, also known as bijection or matching.* Two collections are matched if they are both exhausted when each element of one has been assigned to an element of the other (fig. 1–6).

If all the seats in a theater are occupied and no one is standing, each seat corresponds to one and only one member of the audience, and conversely. The seats and the audience are matched. We can describe the situation by saying that there are as many seats as people, and so, using the concept of matching, we have no need to count in order to know that both collections have the same number of elements. Matching therefore suggests an abstract notion expressing a common property of the two collections, a notion *entirely independent of the nature of their elements.*

It is precisely when two collections are no longer distinguished according to the nature of their elements that matching becomes capable of playing a significant part in the development of numeration. "The transition from relative number to absolute is not difficult. It is necessary

---

*By the age of two, a child seems able to grasp the principle of one-to-one correspondence. If he is given as many chairs as dolls, for example, he will probably associate one doll with each chair. Playing in this way, he will be matching the objects of one collection (the dolls) with those of another (the chairs). If, however, he is given more chairs than dolls, or vice versa, he will probably be perplexed after a certain time; he will have discovered that matching is impossible in this case.

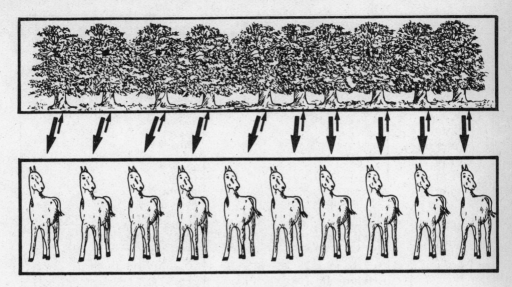

Fig. 1–6. Matching the elements of one collection with those of another.

only to create *model collections*, each typifying a possible collection. Estimating any given collection is then reduced to the selection among available models of one which can be matched with the given collection member by member" (Dantzig). Thus twenty pebbles, twenty sticks, or twenty notches (or the collection of ten fingers and ten toes) can be made to correspond to twenty people, twenty horses, twenty sheep, or twenty measures of wheat. Objects that differ in nature but are equal in quantity are given the same numerical interpretation, which means that matching is a "concrete measure" of *quantity*, independent of the *quality* of the objects involved.

Expressing number orally and by gestures

As we have seen, some "primitive" peoples use visual number techniques rather than true oral numerations. In a conversation about a trade transaction or a message concerning the date of a ceremony, for example, they do not use number words in the strict sense of the term. Instead, they name certain parts of the body in a fixed order and touch them at the same time. But does this not really amount to the same thing as using an ordered sequence of number words? In trying to answer that question, it will be useful to examine some anthropological studies.

The first example is taken from the report of the Cambridge Expedition to Torres Strait. People in villages along the Musa River, in the northwestern part of what was formerly British New Guinea, used the

NUMBERS	CORRESPONDING GESTURES	WORDS ASSOCIATED WITH THESE GESTURES
1	right little finger	*anusi*
2	right third finger	*doro*
3	right middle finger	*doro*
4	right forefinger	*doro*
5	right thumb	*ubei*
6	right wrist	*tama*
7	right elbow	*unubo*
8	right shoulder	*visa*
9	right ear	*denoro*
10	right eye	*diti*
11	left eye	*diti*
12	nose	*medo*
13	mouth	*bee*
14	left ear	*denoro*
15	left shoulder	*visa*
16	left elbow	*unubo*
17	left wrist	*tama*
18	left thumb	*ubei*
19	left forefinger	*doro*
20	left middle finger	*doro*
21	left third finger	*doro*
22	left little finger	*anusi*

TABLE 1–1.

system of touching parts of the body: first the fingers of the right hand, beginning with the little finger, then the right wrist, elbow, shoulder, ear, and eye, then the left eye, the nose, the mouth, the left ear, the left shoulder, and so on down to the left little finger. Each gesture was accompanied by a Papuan word, as shown in table 1–1.

The words used were simply names of parts of the body, not true number words. Most of them designated more than one part of the body and were associated with several different numbers. *Doro*, for example, designated the third finger, middle finger, or forefinger of either hand, and was associated with 2, 3, 4, 19, 20, and 21. It would have been impossible to know which of those six numbers it indicated if, when it was spoken, the act of touching one of those six fingers had not specified its meaning.

In other parts of New Guinea, the system shown in table 1–2 was used.

Here, the word *ano*, meaning either the right or the left side of the neck, was associated with both 10 and 14, which shows that it was not a genuine number word. Yet the system did not cause confusion, because the words were spoken as parts of the body were touched in a fixed order.

In such systems of naming parts of the body, the accompanying gestures are essential. Counting does not require any kind of oral expression. It can be done entirely by means of gestures with preestablished meanings.

All this makes it seem likely that numerical gestures preceded oral

NUMBERS	CORRESPONDING GESTURES	WORDS ASSOCIATED WITH THESE GESTURES
1	left little finger	*monou*
2	left third finger	*reere*
3	left middle finger	*kaupu*
4	left forefinger	*moreere*
5	left thumb	*aira*
6	left wrist	*ankora*
7	between the left wrist and the left elbow	*mirika mako*
8	left elbow	*na*
9	left shoulder	*ara*
10	left side of the neck	*ano*
11	left breast	*ame*
12	sternum	*unkari*
13	right breast	*amenekai*
14	right side of the neck	*ano*
etc.	etc.	. . . .

TABLE 1–2.

expressions of numbers and that our present number words had their remote origin in names of parts of the body that were used in visual, concrete number techniques.

In some languages the original meanings of the number words are still discernible and they often reveal a relation to such techniques. Here, for example, are the number words of the Bugilai in New Guinea, with their original meanings:

1: *tarangesa*, left hand: little finger   6: *gaben*, wrist
2: *meta kina*, next finger   7: *trankgimbe*, elbow
3: *guigimeta kina*, middle finger   8: *podei*, shoulder
4: *topea*, forefinger   9: *ngama*, left breast
5: *manda*, thumb   10: *dala*, right breast

The Lengua Indians of Paraguay have words for the first three numbers that are evidently unrelated to parts of the body. "*Thlama*, 'one,' and *anit*, 'two,' are apparently root words; the rest appear to depend upon them, and on the hands. *Antanthlama*, for 'three,' appears to be made by these two words joined (3 = 2 + 1)" (Hawtrey). But for numbers 4 to 20 the Lenguas use expressions that can be translated as follows (fig. 1–7):

4: two sides alike
5: one hand
6: arrived at the other hand, one
7: arrived at the other hand, two
8: arrived at the other hand, three
9: arrived at the other hand, two sides alike
10: finished the hands
11: arrived at the foot, one
12: arrived at the foot, two
13: arrived at the foot, three
14: arrived at the foot, two sides alike
15: finished the foot
16: arrived at the other foot, one
17: arrived at the other foot, two
18: arrived at the other foot, three
19: arrived at the other foot, two sides alike
20: finished the feet

Fig. 1–7. Method for counting up to 20, using the fingers and toes as references.

The Zuñis also have number words of this type:

1: *töp-in-te*, taken to start with
2: *kwil-li*, raised with the one before
3: *ha'-i*, dividing finger
4: *a-wi-te*, all fingers raised except one
5: *öp-te*, notched off
6: *to-pa-lï-k'ya*, another added to what is already counted
7: *kwil-li-lï-k'ya*, two brought and held up with the rest
8: *ha-i-lï-k'ya*, three brought and held up with the rest
9: *ten-a-lï-k'ya*, all but one held up with the rest
10: *äs-tem-'thla*, all of the fingers
11: *äs-tem-'thla-to-pa-yä' thi-to-na*, all fingers and one more held up

"With twelve, thirteen, and so on, to nineteen, the process was but a repetition of this; but when twenty was reached, [the word was] *kwil-li-k'yën-äs-tem-'thla* . . . that is, 'two times all of the fingers' " (Cushing).

On the basis of these examples and many others like them, we can plausibly reconstruct the course that various systems of oral number expression must have followed in evolving toward true number words.

First stage. The notion of number is limited to what can be perceived at a glance. To people at this stage, number is a concrete reality inseparable from the nature of the objects directly related to it. (The people of the Fiji and Solomon islands, for example, have specific words for certain collections of objects: *bola* for 100 boats, *koro* for 100 coconuts, *salavo* for 1000 coconuts, and others.)

To solve the problem of dealing with quantities greater than four, they develop concrete procedures, based on the principle of one-to-one correspondence, that they can use up to a certain point. Among them are techniques involving the fingers and other parts of the body, which provide simple and readily available model collections. And it is these model collections that are expressed in language when the corresponding gestures are made.

Second stage. Names of parts of the body used in this way are not true number words. But "such numeration may unconsciously become half-abstract and half-concrete, as the names (especially the first five) gradually bring before the mind a fainter representation of the parts of the body and a stronger idea of a certain number which tends to separate itself and become applicable to any object whatever" (Lévy-Bruhl).

Third stage. Finally, "once the *number word* has been created and adopted, it becomes as good a model as the object it originally represented. The necessity of discriminating between the name of the borrowed object and the number symbol itself would then naturally tend to bring about a change in sound, until in course of time the very connection between the two is lost to memory. As man learns to rely more and more on his language, the sounds supersede the images for which they stood, and the originally concrete models take the abstract form of number words. Mem-

	GREEK	LATIN	ITALIAN	FRENCH	SPANISH	PORTU-GUESE	RUMANIAN	SANSKRIT (INDIA)	RUSSIAN
1	hén	unus	uno	un	uno	um	uno	éka	odjn
2	dúo	duo	due	deux	dos	dois	doi	dvi	dva
3	treîs	tres	tre	trois	tres	tres	trei	tri	tri
4	téttares	quattuor	quattro	quatre	cuatro	quatro	patru	tchatur	tchetjre
5	pénte	quinque	cinque	cinq	cinco	cinco	cinci	pañcha	pjat'
6	héx	sex	sei	six	seis	seis	shase	sas	chest'
7	heptá	septem	sette	sept	siete	sete	shapte	sapta	sem'
8	októ	octo	otto	huit	ocho	oito	opt	asta	vosem'
9	ennéa	novem	nove	neuf	nueve	noue	noue	nava	devjat'
10	déka	decem	dieci	dix	diez	dez	zece	daça	desjat'

	CZECH	BALTIC	GOTHIC	HIGH GERMAN OLD	HIGH GERMAN MIDDLE	HIGH GERMAN MODERN	OLD LOW GERMAN	ANGLO-SAXON	ENGLISH
1	jeden	vienes	ains	ein	eins	eins	en	an	one
2	dva	du	twa	zwene	zwene	zwei	twene	twegen	two
3	tri	trys	preis	dri	drei	drei	thria	pri	three
4	tchtyri	keturi	fidwor	vier	vier	vier	fiuwar	feower	four
5	pét	penki	fimf	fünf	fünf	fünf	fif	fíf	five
6	shest	sheshi	saíhs	sehs	sehs	sechs	sehs	six	six
7	sedm	septyni	sibun	siben	siben	sieben	sibun	seofou	seven
8	osm	ashtuoni	ahtaú	ahte	ahte	achte	ahto	eahta	eight
9	devét	devyni	niun	niun	niun	neun	nigun	nigon	nine
10	deset	deshimt	taíhun	zehan	zehen	zehn	tehan	tyn	ten

	DUTCH	OLD NORSE	ICELANDIC	DANISH	SWEDISH	IRISH	WELSH	BRETON
1	een	einn	einn	en	en	oin	un	eun
2	twee	tveir	tveir	to	twa	da	dau	diou
3	drie	prir	prir	tre	tre	tri	tri	tri
4	vier	fjorer	fjorir	fire	fyra	cethir	petwar	pevar
5	vijf	fimm	fimm	fem	fem	coic	pimp	pemp
6	zes	sex	sex	seks	sex	se	chwe	chouech
7	zeven	siau	sjö	syv	sju	secht	seith	seiz
8	acht	atta	atta	otte	åtta	ocht	wyth	eiz
9	negon	nio	niu	ni	nio	noi	naw	nao
10	tien	tio	tiu	ti	tio	deich	dec	dek

TABLE 1-3.

ory and habit lend concreteness to these abstract forms, and so mere words become measures of plurality" (Dantzig).

Table 1–3 shows the first ten number words in some of the Indo-European languages. Although the original meanings of these words have been lost, it may well be that in the distant past they were the names of parts of the body used in number techniques like those we have seen above.

Counting: a human faculty

Decisive progress in the art of abstract counting requires that the whole numbers be arranged in a sequence in which each number, after 1, is obtained by adding 1 to the number before it, according to what is known as the principle of recursion (fig. 1–8).

Let us now take the case of a young child. Until he has developed to the stage of being able to grasp the principle of recursion, he can recognize the persistence of a given quantity only by means of one-to-one correspondence. But when he has reached that stage—which occurs between the ages of three and four, according to Jean Piaget—he will soon be capable of learning to count and calculate. The essential feature of his development is predominance of the abstract number concept over the almost exclusively perceptual aspect of collections. He first learns to count up to 10, using his fingers, then to extend his number sequence as he comes to conceive the abstraction of number.

In a little book on the childhood of his sons, Georges Duhamel shows how Bernard, called Baba, more or less understood the natural sequence of numbers even before he knew their names:

"It was hard at first. Baba did the best he could.

" 'I came to get some candy,' he said. 'Give me some for everybody.'

" 'How many pieces?'

" 'One, one and one.'

"That was clear, but it was not real arithmetic. Then he learned to count on his fingers. When he was asked how old Maryse or Robert was, he usually showed the right number of fingers, using one hand and then the other. Then things became more complicated.

" 'How old is Jacqueline?'

"After thinking for a second, he answered, 'For Jacqueline, it takes a little toe!' "

Once whole numbers have been arranged in the natural sequence, they make possible a new ability that has vastly important consequences: counting.

Counting the objects in a collection means assigning to each of them a symbol (a word, a gesture, or a graphic sign) corresponding to a number in the natural sequence, and continuing till the collection is exhausted

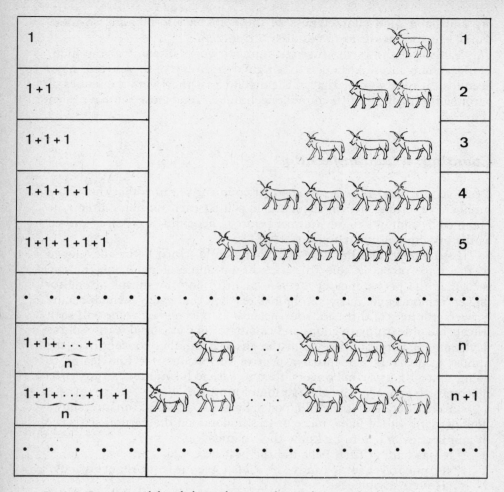

Fig. 1–8. Generation of the whole numbers according to the principle of recursion.

(fig. 1–9). Each symbol represents the ordinal number of the object to which it is assigned, and the ordinal number of the last object is the number of elements in the collection. The number obtained in this way is, of course, entirely independent of the order in which the objects are counted: whether counting begins with one object or another, the result will always be the same.

Let us consider, for example, a box containing several balls. We take out a ball at random and assign the number 1 to it. (This is the *first* ball taken from the box.) We take out another ball, still at random, and assign the number 2 to it. We continue till there are no balls left in the box. In taking out the last one, we assign to it a specific number in the sequence.

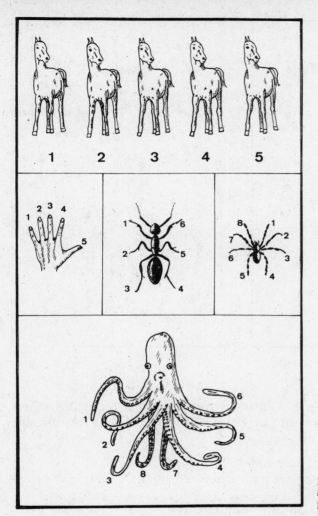

Fig. 1–9. Counting, which enables us to pass from concrete plurality to abstract number.

If that number is 20, we say that there are twenty balls, and so by counting we have changed a vague piece of information (several balls) into a precise one.

Now let us consider a group of dots scattered at random (fig. 1–10). To find out how many there are, we begin by connecting them with a line to make sure we do not overlook any of them. The dots now form what can be called a chain. We assign a number to each dot, beginning at one end of the chain. The last number is the total number of dots.

Thus, through the concepts of sequence and counting, the confused, heterogeneous, and imprecise notion of concrete plurality is transformed, in our minds, into the abstract, homogeneous notion of "absolute quantity."

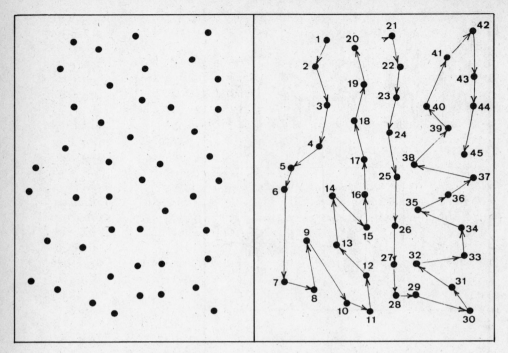

Fig. 1–10. Counting a "cloud" of dots.

From the foregoing it is clear that people cannot count the objects in a collection (using "count" as we have defined it) unless they are able to do all three of the following:

1. Assign a rank to each object.
2. Associate each object with the memory of all those considered before it.
3. Convert succession into simultaneity.

The number concept, which might seem elementary at first sight, is now revealed as being much more complicated. An anecdote told by a man named Bourdin, and reported by R. Balmès, illustrates this point: " 'I once knew someone,' said Bourdin, 'who heard the clock strike four one day, just as he was falling asleep, and he counted like this: "One, one, one, one." Then, when he realized how ridiculous that was, he said, "The clock has gone crazy: it struck one o'clock four times!" ' "

The notion of whole number has two complementary aspects: cardinal, based only on correspondence, and ordinal, requiring both correspondence and succession.

The difference between them can be illustrated by a simple example.

The month of January has thirty-one days. Here, the whole number 31 is the number of elements in the collection of days called January. It is therefore a cardinal number. But in the expression "January 31," 31 is not a cardinal number. It designates the thirty-first day of January and is therefore more accurately written as "31st January," although "January 31" is accepted by usage and causes no confusion. In "January 31," then, 31 is an ordinal number.

"We have learned to pass with such facility from cardinal to ordinal number that the two aspects appear to us as one. To determine the plurality of a collection, that is, its cardinal number, we do not bother anymore to find a model collection with which we can match it—we *count it*. And to the fact that we have learned to identify the two aspects of number is due our progress in mathematics. . . . The operations of arithmetic are based on the tacit assumption that *we can always pass from any number to its successor*, and this is the essence of the ordinal concept.

"And so matching by itself is incapable of creating an art of reckoning. Without our ability to arrange things in ordered succession little progress could have been made. Correspondence and succession, the two principles that permeate all mathematics—nay, all realms of exact thought—are woven into the very fabric of our number system" (Dantzig).

Symbolization of number

To assimilate, remember, differentiate, and combine the various whole numbers, people were gradually led, through untold centuries, to assign a symbol to each of them, a mental process that made it possible to replace an operation on things with a corresponding operation on numerical symbols. This shows that "number does not come from things, but from the laws of thought working on things" (Balmès). In other words, we change concrete things into mental objects.

Several methods of symbolizing number have been used in the course of human development:

Concrete numerations: material objects of all sorts (pebbles, shells, bones, sticks, clay objects of different shapes and sizes), notches in bone or wood, knotted strings, intuitive or conventional gestures (use of the fingers, toes, or other parts of the body).

Oral numerations, which may use words of three different kinds:

1. Concrete terms directly implying the notion of number: "sun," "moon," or "penis," for example, to designate 1; "eyes," "breasts," or "bird's wings" for 2; "clover leaf" for 3; "an animal's legs" for 4; "fingers of one hand" for 5, etc.

2. Words reflecting the use of gestures in a concrete number technique involving parts of the body: "little finger" for 1; "third finger" for 2; "middle finger" for 3; "forefinger" for 4, "thumb" for 5, etc.

3. Words showing no traces of any initial concrete meaning: "one," "two," "three," "four," "five."

Written numerations: graphic signs of various kinds (drawn, painted, or engraved marks, notches, representational signs, letters of the alphabet, signs with no direct visual reference), each called a figure or numeral, taking these terms in the broad sense of "any graphic numerical sign." (The words "figure" and "number" must not be confused. A number corresponds to a concept that has quantity as one of its aspects; a figure, or numeral, is a graphic sign that represents a number but is not identical with it.)

Figure 1–11 illustrates this division into concrete, oral, and written numerations, without indicating any order of precedence among the different kinds of symbols.

Ten fingers for counting

In the past, fingers were often used as a material support for the number concept, and sometimes they still serve that purpose. It is important that the hand, or both hands together, can be regarded not only as a total but also as a natural succession of collections of fingers (one finger, two fingers, . . . five fingers, . . . ten fingers). With the hand, the ideas of cardinal and ordinal number become intuitive. It provides the simplest and most natural succession of model collections that human beings have. so to speak, at hand.

The most elementary procedure for counting with the aid of the fingers consists in assigning to each finger a numerical value taken in order from the sequence of whole numbers, beginning with 1. We sometimes use it to stress an idea, and most children use it in learning to count.

There are several variants of this procedure:

1. All the fingers are bent down, then the left thumb is raised for 1, the left forefinger is raised for 2, and so on to the right little finger for 10 (fig.1–12, line A).

2. All the fingers are extended, then the left little finger is bent down for 1, the left third finger is bent down for 2, and so on to the right thumb (fig. 1–12, line B) or the right little finger (fig. 1–12, line C).

3. All the fingers are bent down and then raised one by one, but beginning with the left or right forefinger, rather than with the thumb or the little finger (fig. 1–12, line D).

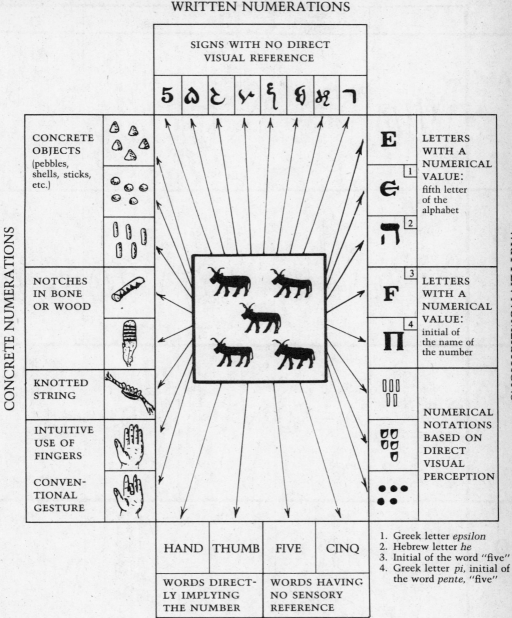

Fig. 1–11. Various symbols assigned to a single number, in this case 5.

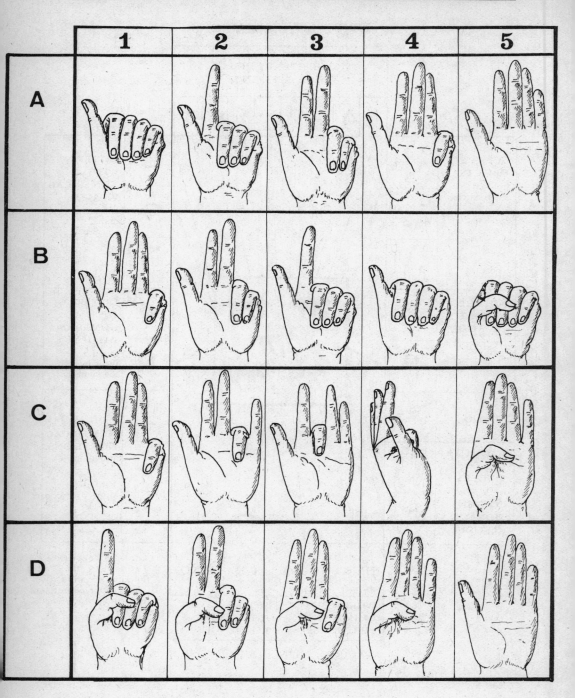

Fig. 1–12. Variants of elementary finger counting.

6	7	8	9	10	
					A
					B
					C
					D

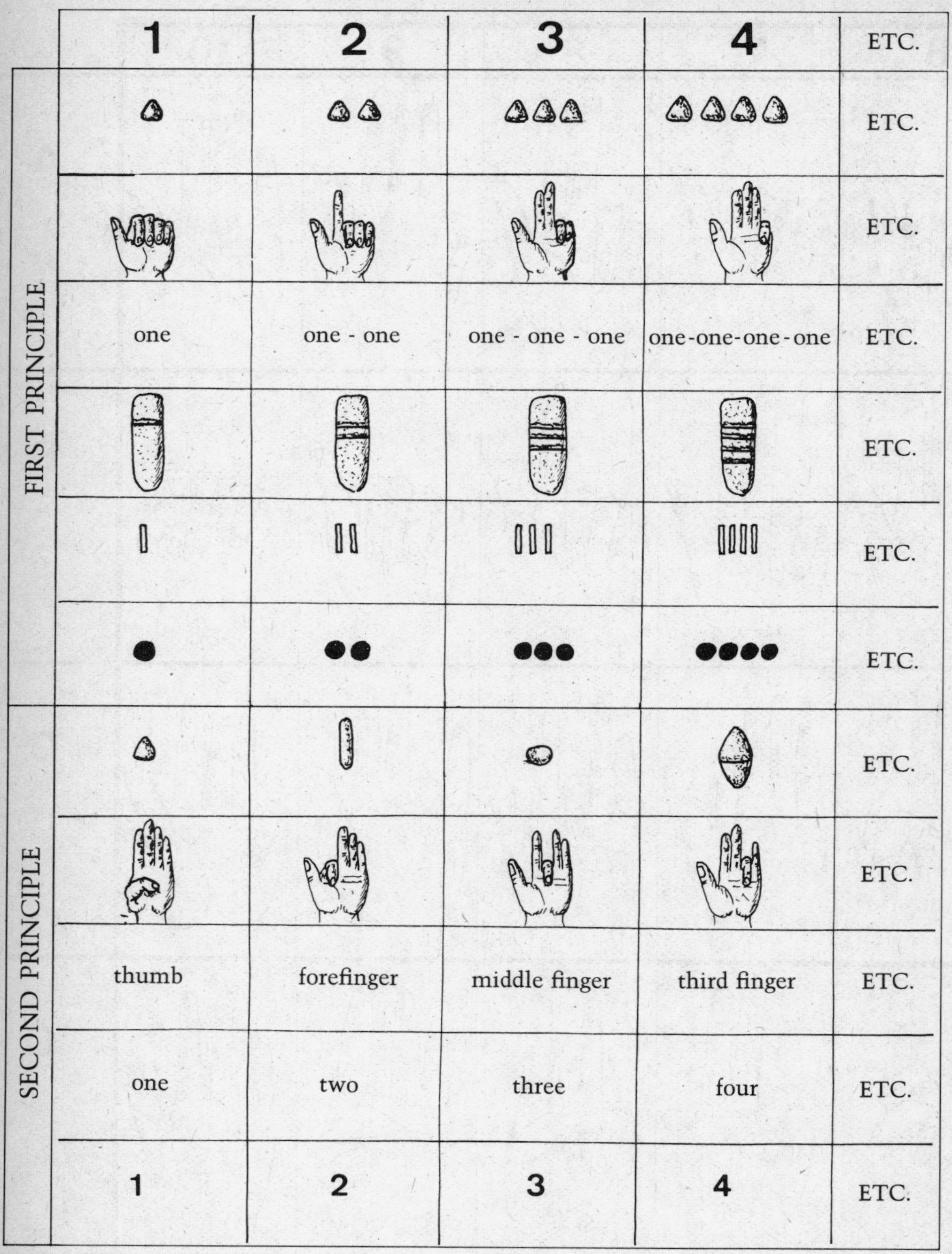

1	2	3	4	ETC.
△	△△	△△△	△△△△	ETC.
				ETC.
one	one - one	one - one - one	one-one-one-one	ETC.
				ETC.
I	II	III	IIII	ETC.
●	●●	●●●	●●●●	ETC.
△		○	◇	ETC.
				ETC.
thumb	forefinger	middle finger	third finger	ETC.
one	two	three	four	ETC.
1	2	3	4	ETC.

FIRST PRINCIPLE (rows 1–6), SECOND PRINCIPLE (rows 7–11)

Fig. 2–1. The "cardinal" and "ordinal" procedures for representing whole numbers.

2

The Principle
of the Base

When people feel a need to symbolize the notion of number, they can choose between two different procedures. One, which can be called "cardinal," consists in adopting a standard symbol for 1 and using it as many times as there are units in the number being considered. The other, which can be called "ordinal," consists in assigning to the consecutive whole numbers, beginning with 1, distinct symbols unrelated to each other (fig. 2–1).

Simple though it may be, the first procedure cannot lead very far because it requires continual repetition of the standard symbol. The second procedure also poses a problem: it seems to require continual invention of new numerical symbols. The fact that ways of surmounting both difficulties have been found is a tribute to the ingenuity of the human mind.

Using one or the other of those two basic procedures, people have developed convenient methods of designating—concretely, orally, or graphically—many numbers with relatively few symbols. To do this, a "scale" of symbols must be adopted in order to classify increasingly large numbers and thus avoid impractically great efforts of memory and representation. In the light of this preliminary observation, we can examine some of the solutions that have been devised through the centuries.

Decimal systems

The system of oral numeration now most widely used can be described as follows: its base is 10; each number from 1 to 10, and each power of 10, has an independent, individual name; the names of other numbers are combinations of the names of numbers 1 to 10 and the powers of 10, formed according to the additive or multiplicative principle. That is a theoretical description. Different languages depart from it in varying degrees. In English, for example, "ten thousand" is a combination of "ten" and "thousand" rather than an independent name, and "eleven" has no apparent relation to "ten" and "one" (though in fact it is related to "one").

The formation of number words according to the theoretical model is shown in table 2–1.

1	one	11	ten-one	30	three-ten	400	four hundred
2	two	12	ten-two	40	four-ten	500	five hundred
3	three	13	ten-three	50	five-ten	600	six hundred
4	four	14	ten-four	60	six-ten	700	seven hundred
5	five	15	ten-five	70	seven-ten	800	eight hundred
6	six	16	ten-six	80	eight-ten	900	nine hundred
7	seven	17	ten-seven	90	nine-ten	1000	thousand
8	eight	18	ten-eight	100	hundred	2000	two thousand
9	nine	19	ten-nine	200	two hundred	3000	three thousand
10	ten	20	two-ten	300	three hundred	....	

TABLE 2–1.

The Chinese oral numeration corresponds to the theoretical model with no irregularities, as shown in table 2–2, where the number words are given in Joseph Needham's transcriptions.

1	i	11	shih-i	21	erh-shih-i	100	pai
2	erh	12	shih-erh	22	erh-shih-erh	200	erh-pai
3	san	13	shih-san	23	erh-shih-san	300	san-pai
4	ssu	14	shih-ssu	....		1000	chhien
5	wu	15	shih-wu	30	san-shih	2000	erh-chhien
6	liu	16	shih-liu	40	ssu-shih	3000	san-chhien
7	chhi	17	shih-chhi	50	wu-shih	10,000	wan
8	pa	18	shih-pa	60	liu-shih	20,000	erh-wan
9	chiu	19	shih-chiu	70	ts'i-shih		
10	shih	20	erh-shih	80	pa-shih		

53,781: "wu wan san chhien chhi pai pa shih i"
$5 \times 10,000 \quad 3 \times 1000 \quad 7 \times 100 \quad 8 \times 10 \quad 1$

TABLE 2–2.

In most of the Mongolian, Semitic, and Indo-European languages (tables 1–3, 2–3), the systems of oral numeration have a decimal base and conform, at least approximately, to the theoretical model described above.

The almost universal adoption of the base 10 was undoubtedly caused by the fact that we happen to have ten fingers, since people first learned to count on their fingers. If we had six fingers on each hand, our numeration would be duodecimal, that is, its base would be 12.

To understand this better, let us imagine a tribe whose members have been temporarily forbidden to speak (because of a religious observance, for example) and have a flock of sheep. To evaluate the number

	LATIN	ITALIAN	FRENCH	SPANISH	RUMANIAN
11	undecim	un-dici	on-ze	on-ce	un spree zece
12	duodecim	do-dici	dou-ze	do-ce	doi spree zece
13	tredecim	tre-dici	trei-ze	tre-ce	trei spree zece
14	quattuor-decim	quattor-dici	quator-ze	cator-ce	patru spree zece
15	quindecim	quin-dici	quin-ze	quin-ce	cinci spree zece
16	sedecim	se-dici	sei-ze	diez-y-seiz	shase spree zece
17	septendecim	diciasette	dix-sept	diez-y-siete	shapte spree zece
18	(octodecim)	diciotto	dix-huit	diez-y-ocho	opt spree zece
19	(undeviginti)	dicianove	dix-neuf	diez-y-nueve	noua spree zece
20	viginti	venti	vingt	veinte	doua-zeci
30	triginta	trenta	trente	treinta	trei-zeci
40	quadraginta	quaranta	quarante	cuar-enta	patru-zeci
50	quinquaginta	cinquanta	cinquante	cincu-enta	cinci-zeci
60	sexaginta	sessanta	soixante	ses-enta	shase-zeci
70	septuaginta	settanta	soixante-dix	set-enta	shapte-zeci
80	octoginta	ottanta	quatre-vingts	och-enta	opt-zeci
90	nonaginta	novanta	quatre-vingt-dix	nov-enta	noua-zeci
100	centum	cento	cent	ciento	o suta
1000	mille	mille	mille	mil	o mie

	GOTHIC	OLD HIGH GERMAN	MODERN HIGH GERMAN	ANGLO-SAXON	ENGLISH
11	ain-lif	einlif	elf	endleofan	eleven
12	twa-lif	zwelif	zwölf	twelf	twelve
20	twai-tigius	zwein-zug	zwan-zig	twen-tig	twenty
30	preo-tigius	driz-zug	drei-ßig	pri-tig	thir-ty
40	fidwor-tigius	fior-zug	vier-zig	feower-tig	for-ty
50	fimf-tigius	finf-zug	fünf-zig	fif-tig	fif-ty
60	saihs	sehs-zug	sech-zig	six-tig	six-ty
70	sibunt-ehund	sibun-zo	sieb-zig	hund-seofontig	seven-ty
80	ahtaut-ehund	ahto-zo	acht-zig	hund-eahtatig	eigh-ty
90	niunt-ehund	niun-zo	neun-zig	hund-nigontig	nine-ty
100	taihun-taihund	zehan-zo	hundert	hund-teontig	hundred
1000	pusundi	dusunt ; tusent	tausend	pusund	thousand

TABLE 2–3.

Fig. 2–2.

MAN A ——→

MAN B ——→

MAN C ——→

of sheep in the flock, the chief of the tribe has devised the following procedure, illustrated by figure 2–2.

Three men, whom we will call A, B, and C, watch the sheep pass one by one. A raises his first finger when the first sheep passes, his second finger when the second sheep passes, and so on to the tenth sheep. Then B, who has been watching A, raises his first finger and A lowers all his fingers. When the eleventh sheep passes, A again raises his first finger, and he continues in this way till the twentieth sheep has passed. Meanwhile B has been keeping his first finger raised. He raises his second one when he sees A raise his tenth finger and then lower all his fingers. When the hundredth sheep has passed, C, who has been watching B's hands, raises his first finger while A and B lower all their fingers. C keeps his first finger raised till the two-hundredth sheep has passed, then he raises his second finger. And so on.

When 627 sheep, for example, have passed, the situation will be as follows (figs. 2–2, 2–3): A has seven fingers raised, B has two, and C has six. A's raised fingers represent units, B's represent tens, and C's represent hundreds.

The process has been carried out in groups of ten, without a single word having been spoken, and it is therefore each man's ten fingers that have imposed the base 10, rather than the base 12, for example.

But although the base 10 is anatomically convenient, it has few advantages from a mathematical or practical viewpoint. Any other number in the vicinity of 10 (except perhaps 9) would have done just as well and probably better.

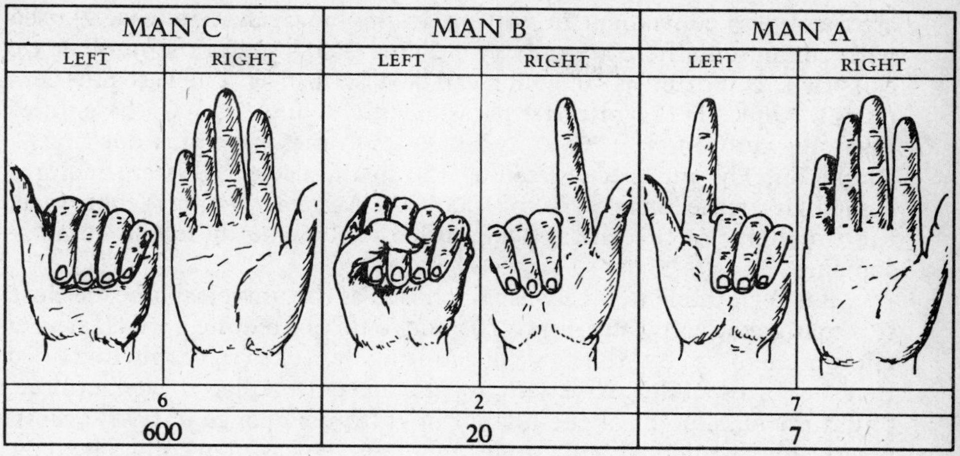

MAN C		MAN B		MAN A	
LEFT	RIGHT	LEFT	RIGHT	LEFT	RIGHT
6		2		7	
600		20		7	

Fig. 2–3.

"Indeed, if the choice of a base were left to a group of experts, we should probably witness a conflict between the practical man, who would insist on a base with the greatest number of divisors, such as *twelve*, and the mathematician, who would want a prime number, such as *seven* or *eleven*, for a base. As a matter of fact, late in the eighteenth century the great naturalist Buffon proposed that the duodecimal system (base 12) be universally adopted. He pointed to the fact that 12 has four divisors, while 10 has only two, and maintained that throughout the ages this inadequacy of our decimal system has been so keenly felt that, in spite of 10 being the universal base, most measures had 12 secondary units.

"On the other hand the great mathematician Lagrange claimed that a prime base is far more advantageous. He pointed to the fact that with a prime base every systematic fraction would be irreducible and would therefore represent the number in a unique way. In our present numeration, for instance, the decimal fraction .36 stands really for three fractions: 36/100, 18/50 and 9/25. Such an ambiguity would be completely eliminated if a prime base, such as eleven, were adopted.

"But whether the enlightened group to whom we would entrust the selection of a base decided on a prime or a composite base, we may rest assured that the number *ten* would not even be considered, for it is neither prime nor has it a sufficient number of divisors" (Dantzig).

But the base 10 does have a significant advantage over large bases such as 20 or 60, for example, since it falls within a range that does not strain human memory: in a decimal system few number words are required, and the multiplication table can be learned by heart without great difficulty.

The base 10 also has an advantage over small bases like 2 or 3, which are much less convenient in writing. In the binary system (base 2) used with computers, there are only two digits, 0 and 1, and the number we call "two" is written as 10, which can be described as "one base plus zero units"; "three" is 11 ("one base plus one unit"), "four" is 100 ("base times base plus zero units"), "five" is 101 ("base times base plus one unit"), and so on. The number written as 128 in the decimal system requires eight digits in the binary system: 10000000. As far as writing is concerned, the simplicity of the binary system does not make up for its cumbersomeness.

It is regrettable that 12 was not chosen as the universal base, because it is mathematically superior to 10 and would not require a much greater effort of memory. But the habit of counting by tens is so deeply ingrained that the corresponding base will probably never be replaced. Furthermore, "from the standpoint of the history of culture a change of base, even if practicable, would be highly undesirable. As long as man counts by tens, his ten fingers will remind him of the human origin of this most important phase of his mental life" (Dantzig).

Traces of the base 5

For the needs of everyday life, some merchants in the Indian state of Maharashtra (Bombay region) still use an interesting finger technique. A man employing this technique counts the first five units by successively raising the five fingers of his left hand. He then raises his right thumb and continues counting up to 10 with the fingers of his left hand. When all the fingers of his left hand are raised, he holds up his right forefinger to mark this second group of five. He proceeds in this way, counting successive groups of five with his left hand, till all the fingers of his right hand are raised. He has now counted to 25 (fig. 2–4) and can go on to 30 by once again counting on the fingers of his left hand; and if that is not enough, he can repeat the whole procedure and count as far as 60.

The distinctive feature of this technique is its grouping by fives. Each left finger stands for one unit, while each right finger marks a group of five units, which means that this is a remarkable example of a *concrete numeration of base 5*.

In a quinary (base 5) oral numeration there are independent names for the first five whole numbers and the powers of 5. Combinations of these names are used for the other numbers. In the example of such a system shown in table 2–4, *dan* and *mim* are words invented to designate 5^2 (25) and 5^3 (125), respectively.

While there are very few quinary systems existing in a pure state, it is possible to find a considerable number of oral numerations that, though not quinary, have obvious traces of grouping by fives, based on the five

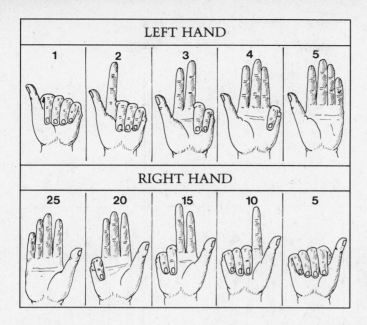

LEFT HAND

1	2	3	4	5

RIGHT HAND

25	20	15	10	5

Fig. 2–4. A quinary (base 5) finger technique used by some merchants in the Bombay region of India.

1	one	18	three-five-and-three
2	two	19	three-five-and-four
3	three	20	four-five
4	four	21	four-five-and-one
5	five	22	four-five-and-two
6	five-and-one	23	four-five-and-three
7	five-and-two	24	four-five-and-four
8	five-and-three	25	*dan*
9	five-and-four	26	*dan*-one
10	two-five	27	*dan*-two
11	two-five-and-one	30	*dan*-five
12	two-five-and-two	31	*dan*-five-and-one
13	two-five-and-three	35	*dan*-two-five
14	two-five-and-four	41	*dan*-three-five-and-one
15	three-five	125	*mim*
16	three-five-and-one	136	*mim*-two-five-and-one
17	three-five-and-two		. . . .

TABLE 2–4.

fingers of one hand (fig. 2–5). This is true of the number words in the Api language of the New Hebrides, as shown in table 2–5.

Another example of an oral numeration bearing traces of the base 5 comes from the language of the Aztecs, in which the first nine numbers are expressed as shown in table 2–6.

In the ancient language of Sumer (southern Mesopotamia), the words for certain numbers kept traces of an earlier use of the base 5.

Some linguists believe that most oral numerations probably had an

1	*tai*	
2	*lua*	
3	*tolu*	
4	*vari*	
5	*luna*	"hand"
6	*otai*	"other one"
7	*olua*	"other two"
8	*otolu*	"other three"
9	*ovari*	"other four"
10	*lua luna*	"two hands"

TABLE 2–5.

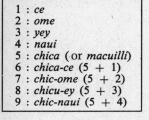

1 : *ce*
2 : *ome*
3 : *yey*
4 : *naui*
5 : *chica* (or *macuilli*)
6 : *chica-ce* (5 + 1)
7 : *chic-ome* (5 + 2)
8 : *chicu-ey* (5 + 3)
9 : *chic-naui* (5 + 4)

TABLE 2–6.

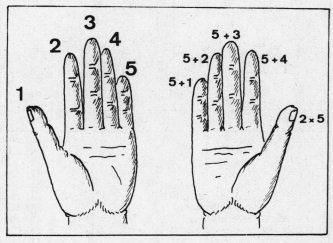

Fig. 2–5. Use of the base 5, as represented by the fingers of one hand, to express numbers 6 to 10.

early quinary stage and that the Indo-European words for the first ten numbers must have been originally formed during that stage; they cite the example of the Persian word *panche*, "five," which has the same root as *panchá*, "hand." But this is only a hypothesis.

Twenty fingers and toes for counting

Instead of counting by tens, as most peoples have done and still do, certain peoples, such as the Celts, Mayas, and Aztecs, developed the habit of counting by twenties at an early stage of their history.

The Mayan calendar had "months" of twenty days each and could be used for calculating cycles of twenty years *(katuns)*, 400 years *(baktuns)*, and 8000 years *(pictuns)*. The end of each *katun* was celebrated by erecting a commemorative monument. And, as we will see, the Mayas had an oral numeration that assigned an individual name to each power of 20.

The Aztecs also counted by twenties and powers of 20. The various products that government officials collected from subjugated peoples were usually quantified in multiples of 20, 400 (20^2), or 8000 (20^3). "To cite some examples [from the Codex Mendoza, an Aztec document], twice a year Toluca had to give 400 loads of cotton, 400 loads of decorated *ixtle* cloaks, 1200 loads of rolls of white *ixtle* cloth. . . . Quahuacan gave 3600 beams and planks four times a year and, twice a year, 800 loads of cotton and 800 loads of *ixtle* cloth. . . . Quauhnahuac contributed to the imperial treasury by paying twice yearly 3200 loads of cotton cloaks, 400 loads of loincloths, 400 loads of feminine garments, 2000 ceramic vases, 8000 reams of 'paper.' . . . Tlalcozauhtitlan supplied only 800 loads of cotton cloth, 200 jars of honey, one luxurious costume and 20 bowls of *tecozauitl*, a kind of light yellow earth which the elegant ladies of Mexico City used as a facial cosmetic. Tuxtepec, besides handing over cloth and garments to the *calpixque* (imperial functionaries), also supplied 16,000 balls of rubber, 24,000 bundles of parrot feathers, 80 bundles of *quetzal* feathers" (Soustelle 1).

The Aztec oral numeration was strictly vigesimal (base 20), as shown in table 2–7.

TABLE 2–7.

1	*ce*	11	*matlactli-on-ce* (10 + 1)	
2	*ome*	12	*matlactli-on-ome* (10 + 2)	
3	*yey*	13	*matlactli-on-yey* (10 + 3)	
4	*naui*	14	*matlactli-on-naui* (10 + 4)	
5	*chica* or *macuilli*	15	*caxtulli*	
6	*chica-ce* (5 + 1)	16	*caxtulli-on-ce* (15 + 1)	
7	*chic-ome* (5 + 2)	17	*caxtulli-on-ome* (15 + 2)	
8	*chicu-ey* (5 + 3)	18	*caxtulli-on-yey* (15 + 3)	
9	*chic-naui* (5 + 4)	19	*caxtuli-on-naui* (15 + 4)	
10	*matlactli*	20	*cem-poualli* (1 × 20)	

30	*cem-poualli-on-matlactli*	(20 + 10)
40	*ome-poualli*	(2 × 20)
50	*ome-poualli-on-matlactli*	(2 × 20 + 10)

100	*macuil-poualli*	(5 × 20)
200	*matlactli-poualli*	(10 × 20)
300	*caxtulli-poualli*	(15 × 20)
400	*cen-tzuntli*	(1 × 400)
800	*ome-tzuntli*	(2 × 400)
1200	*yey-tzuntli*	(3 × 400)

8000	*cen-xiquipilli*	(1 × 8000)

THREE CELTIC LANGUAGES

	IRISH		WELSH		BRETON	
1	oin		un		eun	
2	da		dau		diou	
3	tri		tri		tri	
4	cethir		petwar		pevar	
5	coic		pimp		pemp	
6	se		chwe		chouech	
7	secht		seith		seiz	
8	ocht		wyth		eiz	
9	noi		naw		nao	
10	deich		dec; deg		dek	
11	oin deec	1 + 10	un ar dec		unnek	1 + 10
12	da deec	2 + 10	dou ar dec		daou-zek	2 + 10
13	tri deec	3 + 10	tri ar dec		tri-zek	3 + 10
14	cethir deec	4 + 10	petwar ar dec		pevar-zek	4 + 10
15	coic deec	5 + 10	hymthec	15	pem-zek	5 + 10
16	se deec	6 + 10	un ar hymthec	1 + 15	choue-zek	6 + 10
17	secht deec	7 + 10	dou ar hymthec	2 + 15	seit-zek	7 + 10
18	ocht deec	8 + 10	tri ar hymthec (¹)	3 + 15	eiz-zek (²)	8 + 10
19	noi deec	9 + 10	pedwar ar hymthec	4 + 15	daou-zek	9 + 10
20	fiche	20	ugeint	20	ugent	20
30	deich ar fiche	10 + 20	dec ar ugeint	10 + 20	tregont	
40	da fiche	2 × 20	de-ugeint	2 × 20	daou-ugent	2 × 20
50	deich ar da fiche	10 + (2 × 20)	dec ar de-ugeint	10 + (2 × 20)	hanter-kant	½×100
60	tri fiche	3 × 20	tri-ugeint	3 × 20	tri-ugent	3 × 20
70	deich ar tri fiche	10 + (3 × 20)	dec ar tri-ugeint	10 + (3 × 20)	dek ha tri-ugent	10 + (3 × 20)
80	ceithri fiche	4 × 20	pedwar-ugeint	4 × 20	pevar-ugent	4 × 20
90	deich ar ceithri fiche	10 + (4 × 20)	dec ar pedwar-ugeint	10 + (4 × 20)	dek ha pevar-ugent	10 + (4 × 20)
100	cet		cant		kant	
1000	mile		mil		mil	

¹ or "deu naw" $\dfrac{}{2 \times 9}$ ² or "tri-(ch)ouech" $\dfrac{}{3 \times 6}$

TABLE 2–8.

Examination of the Aztec words for numbers 1 to 20 shows that the first five can be associated with the fingers of one hand, the next five with the fingers of the other hand, the third group of five with the toes of one foot, and the fourth group with the toes of the other foot.

It is clear that the vigesimal system originated in the habit of counting on the ten fingers and ten toes. In Europe, the Celtic languages (table 2–8) are the only ones still used in which counting is done by twenties, but many languages seem to bear traces of an ancient vigesimal stage. English, for example, has "score" (20), "two score" (40), "three score" (60), etc. Shakespeare often used those terms: ". . . as easy as a cannon will shoot point-blank twelve score" (*The Merry Wives of Windsor*, act 3, scene 2); "I'll procure this fat rogue a charge of foot; and I know his death will be a march of twelve-score" (*First Part of King Henry the Fourth*, act 2, scene 4).

The French and Latin words for 20 (*vingt* and *viginti*) are vestiges of a vigesimal tradition, since they are obviously independent of the words for 2 (*deux* and *duo*) and 10 (*dix* and *decim*). In Old French, terms similar to the modern *quatre-vingts* (80, literally "four-twenties") were often used. For 60, 120, and 140, for example, *trois-vingts* ("three-twenties"), *six-vingts* ("six-twenties") and *sept-vingts* ("seven-twenties") were common. A corps of 220 policemen in Paris was called the Corps des Onze-Vingts ("Corps of the Eleven-Twenties"). A Paris hospital built in the thirteenth century to house 300 blind veterans still has its original name: Hôpital des Quinze-Vingts ("Hospital of the Fifteen-Twenties"). In *Le Bourgeois Gentilhomme*, act 3, scene 4, Molière has one of his characters say *six-vingts* ("six-twenties") for 120.

The complexity of a vigesimal numeration can be seen in the Mayan dialect spoken in the Yucatán Peninsula of Central America. (It is worth noting that the Mayan language and its dialects are still spoken in Mexico— in the states of Yucatán, Campeche, Tabasco, and Chiapas, and the Quintana Roo territory—El Salvador, the western part of Honduras, and nearly all of Guatemala.) The words for numbers 1 to 19 in this dialect are shown in table 2–9.

TABLE 2–9.

1 : *hun*	11 : *buluc*	
2 : *ca*	12 : *lahca*	(*lahun* + *ca* = 10 + 2)
3 : *ox*	13 : *ox-lahun*	(3 + 10)
4 : *can*	14 : *can-lahun*	(4 + 10)
5 : *ho*	15 : *ho-lahun*	(5 + 10)
6 : *uac*	16 : *uac-lahun*	(6 + 10)
7 : *uuc*	17 : *uuc-lahun*	(7 + 10)
8 : *uaxac*	18 : *uaxac-lahun*	(8 + 10)
9 : *bolon*	19 : *bolon-lahun*	(9 + 10)
10 : *lahun*		

Up to 10, the numbers have independent names; from there to 19, they have compound names containing the word for 10, with the exception of 11, which is *baluc*, not *hun-lahun* ("one-ten"), as might be expected. Other Mayan dialects, however, did use *hun-lahun* for 11; perhaps the inhabitants of the Yucatán Peninsula felt that this term might be misinterpreted as meaning "one ten," in the sense of "one group of ten," and therefore invented another word to avoid ambiguity.

The words for numbers 20 to 39 are shown in table 2–10, and those for some of the numbers above 39 in table 2–11.

The words for numbers 21 to 39 are composed, with two exceptions, by inserting the original prefix *tu* between the word for "score" and the words for the corresponding units. The exceptions are 30, which is "ten-two-twenty" instead of "ten [after the] twentieth," and 35, which is "fifteen-two-twenty" instead of "fifteen [after the] twentieth."

What follows is a scenario that I have imagined in an attempt to explain those two exceptions.

Let us first travel in space, to a region of Mexico near the Guatemalan border. Now let us go back a few thousand years in time. We are in a village of Indians whose descendants will later belong to the prestigious civilization of the Mayas.

In preparation for a military expedition, they are about to evaluate the number of their warriors. Several men have lined up to serve as a "calculating machine" (fig. 2–6).

TABLE 2–10.

20	*hun kal,* "one score"	
21	*hun tu-kal*	one [after the] twentieth
22	*ca tu-kal*	two [after the] twentieth
23	*ox tu-kal*	three [after the] twentieth
24	*can tu-kal*	four [after the] twentieth
25	*ho tu-kal*	five [after the] twentieth
26	*uac tu-kal*	six [after the] twentieth
27	*uuc tu-kal*	seven [after the] twentieth
28	*uaxac tu-kal*	eight [after the] twentieth
29	*bolon tu-kal*	nine [after the] twentieth
30	*lahun ca kal*	ten-two-twenty
31	*buluc tu-kal*	eleven [after the] twentieth
32	*lahca tu-kal*	twelve [after the] twentieth
33	*ox-lahun tu-kal*	thirteen [after the] twentieth
34	*can-lahun tu-kal*	fourteen [after the] twentieth
35	*holhu ca kal*	fifteen-two-twenty
36	*uac-lahun tu-kal*	sixteen [after the] twentieth
37	*uuc-lahun tu-kal*	seventeen [after the] twentieth
38	*uaxac-lahun tu-kal*	eighteen [after the] twentieth
39	*bolon-lahun tu-kal*	nineteen [after the] twentieth

40	*ca kal,* "two score"	
41	*hun tu-y-ox-kal*	one—third score
42	*ca tu-y-ox-kal*	two—third score
43	*ox tu-y-ox-kal*	three—third score
44	*can tu-y-ox-kal*	four—third score
......		....
58	*uaxac-lahun tu-y-ox-kal*	eighteen—third score
59	*bolon-lahun tu-y-ox-kal*	nineteen—third score
60	***ox kal,*** **"three score"**	
61	*hun tu-y-can kal*	one—fourth score
62	*ca tu-y-can kal*	two—fourth score
......		....
78	*uaxac-lahun tu-y-can-kal*	eighteen—fourth score
79	*bolon-lahun tu-y-can-kal*	nineteen—fourth score
80	***can kal,*** **"four score"**	
81	*hun tu-y-ho-kal*	one—fifth score
82	*ca tu-y-ho-kal*	two—fifth score
......		....
98	*uaxac-labun tu-y-ho-kal*	eighteen—fifth score
99	*bolon-lahun tu-y-ho-kal*	nineteen—fifth score
100	***ho kal,*** **"five score"**	
....		
400	***hun bak,*** = "one four-hundred" (20^2)	
8 000	***hun pic,*** = "one eight-thousand" (20^3)	
160,000	***hun calab,*** = "one hundred-sixty-thousand" (20^4)	

TABLE 2–11.

An "accountant" touches the first finger of the first man when the first warrior passes, the second finger of the same man for the second warrior, and so on to the tenth warrior. Then he touches the first toe of the same man and continues till he touches the tenth toe, which corresponds to the twentieth warrior. For the next twenty warriors, he touches the second man's fingers and toes in the same way. He then goes on to the third man, and he can continue in this way, with as many men as necessary, till the last warrior has passed.

If there are fifty-three warriors, he will touch the third toe of the first foot of the third man when the last of them passes. He will then announce the result by saying something like: "There are warriors as far as three toes on the first foot of the third man." The corresponding number can also be expressed as "two hands and three toes of the third man," or "ten and three of the third score."

Following this procedure, number words can be formed rather simply, as shown in table 2–12.

Fig. 2–6.

We can now see the reason for the irregularities in the Mayan numeration used in the Yucatán Peninsula:

1. The words for numbers 21 to 39, except for 30 and 35, are formed according to Model A in table 2–12:

    21: *hun tu-kal*, "one [after the] twentieth," or "one [after the] first score."

    22: *ca tu-kal*, "two [after the] twentieth," or "two [after the] first score."

    39: *bolon-lahun tu-kal*, "nine and ten [after the] twentieth," or "nine and ten [after the] first score."

2. The words for numbers 41 to 59, 61 to 79, etc., are formed according to Model B in table 2–12:

    41: *hun tu-y-ox-kal*, "one [of the] third score."

    42: *ca tu-y-ox-kal*, "two [of the] third score."

    59: *bolon-lahun tu-y-ox-kal*, "nine and ten [of the] third score."

    60: *ox-kal*, "three score."

    61: *hun tu-y-can-kal*, "one [of the] fourth score."

    62: *ca tu-y-can-kal*, "two [of the] fourth score."

MODEL A				MODEL B
	1 one		11 eleven	
	2 two		12 twelve	
	3 three		13 thirteen	
	4 four		14 fourteen	
	5 five		15 fifteen	
	6 six		16 sixteen	
	7 seven		17 seventeen	
	8 eight		18 eighteen	
	9 nine		19 nineteen	
	10 ten		20 one man	

21	one after the first man	one of the second man	21
22	two after the first man	two of the second man	22
23	three after the first man	three of the second man	23
24	four after the first man	four of the second man	24
25	five after the first man	five of the second man	25
26	six after the first man	six of the second man	26
27	seven after the first man	seven of the second man	27
28	eight after the first man	eight of the second man	28
29	nine after the first man	nine of the second man	29
30	ten after the first man	ten of the second man	30
31	ten-one after the first man	ten-one of the second man	31
32	ten-two after the first man	ten-two of the second man	32
33	ten-three after the first man	ten-three of the second man	33
34	ten-four after the first man	ten-four of the second man	34
35	ten-five after the first man	ten-five of the second man	35
36	ten-six after the first man	ten-six of the second man	36
37	ten-seven after the first man	ten-seven of the second man	37
38	ten-eight after the first man	ten-eight of the second man	38
39	ten-nine after the first man	ten-nine of the second man	39
40	two men	two men	40
41	one after the second man	one of the third man	41
42	two after the second man	two of the third man	42
43	three after the second man	three of the third man	43
. . . .			
51	ten-one after the second man	ten-one of the third man	51
52	ten-two after the second man	ten-two of the third man	52
53	ten-three after the second man	ten-three of the third man	53
. . . .			
59	ten-nine after the second man	ten-nine of the third man	59
60	three men	three men	60
61	one after the third man	one of the fourth man	61
62	two after the third man	two of the fourth man	62
. . . .			
79	nineteen after the third man	ten-nine of the fourth man	79
80	four men	four men	80

TABLE 2–12.

79: *bolon-lahun tu-y-can-kal,* "nine and ten [of the] fourth score."
80: *can kal,* "four score."

3. Finally, the words for numbers 30 and 35 are formed according to Model B in table 2–12:

30: *lahun ca kal*, "ten-two-twenty," or "ten [of the] second score."
35: *holhu ca kal*, "fifteen-two-twenty," or "fifteen [of the] second score."

The Eskimos of Greenland, the Tamanas of Venezuela, and the Ainus of Japan are three of the many other peoples who count by scores. For 53, for example, the Greenland Eskimos use the expression *inup pinga-jugsane arkanek-pingasut*, "of the third man, three on the first foot," which is analogous to the expressions used by the Tamanas and the Mayas. For the same number, the Ainus use *wan-re wan-e-re-hotne*, "ten and three of the third score."

The sexagesimal system

As the base of a numeration, 60 is a very large number that taxes the memory, but we still use a sexagesimal system for measuring time (minutes and seconds), arcs, and angles (degrees, minutes, and seconds). If we are asked, for example, to set a precision watch at 9:08:43 A.M., we know that the time in question is nine hours, eight minutes, and forty-three seconds after midnight, a duration that can be expressed as:

$$(9 \times 60^2) + (8 \times 60) + 43 = 32,923 \text{ seconds.}$$

And when a naval officer gives his men the latitude of a place by saying, "We're at 25:36:07 north," they know that the place is $25°36'7'' = (25 \times 60^2) + (36 \times 60) + 7 = 92,167''$ north of the equator.

"With the Greeks, and then the Arabs, the sexagesimal system was a scholarly numerical system used by astronomers. . . . From the time of the Greeks onward, with rare and late exceptions, it was used only for expressing fractions. But earlier, in Babylonia, it was used for expressing whole numbers as well as fractions. It was a complete numerical system, employed by Babylonian mathematicians and astronomers. At the beginning of Assyriology, examples of it were found by the first decipherers: Hincks (1792–1866), on an astronomical tablet from the excavations at Nineveh; Rawlinson (1810–1895), on a mathematical tablet obtained by Loftus at Senkereh (Larsa). Since then, excavations in Babylonia, notably those by Sarzec at Tello, have shown that before it became a scholarly system (that is, one used only for 'scientific' texts), the sexagesimal system had been commonly and exclusively used by the Sumerians, the predecessors of the Babylonians" (Thureau-Dangin).

This system, which proceeds by successive powers of 60, has a major drawback resulting from the high value of its base. Theoretically, its only units are 1, 60, 60^2, 60^3, etc., and using this base requires knowledge of sixty different words for numbers 1 to 60. But the numerical differences between those units are so great that, in practice, an auxiliary unit is needed to reduce the effort of memory. The Sumerians introduced such a unit, taking 10 as an intermediate level between the sexagesimal units of their numerical system.

Disregarding certain variants, the Sumerian words for numbers 1 to 10 are shown in table 2–13.

1 : *gesh* (or *ash*, or *dish*)	2 : *min* 3 : *esh* 4 : *limmu*	5 : *iá* 6 : *àsh* 7 : *imin*	8 : *ussu* 9 : *ilimmu* 10 : *u*

TABLE 2–13.

According to Thureau-Dangin, the Sumerian words for 1 *(gesh)* and 2 *(min)* originally meant "man, male, penis" and "woman," respectively. *Esh*, for 3, still according to Thureau-Dangin, had the additional meaning of "plurality" and was even used as a verbal suffix marking the plural. (The same idea is found in a practice of the Hittites and the ancient Egyptians that consisted in writing a single hieroglyphic sign three times, or adding three dots or vertical lines to it, not to indicate three of the things represented by the sign but to indicate the plural.)

The forms of certain Sumerian number words show traces of an earlier quinary system:

$$7 = imin = i + min = i(á) + min = 5 + 2$$
$$9 = ilimmu = i + limmu = i(á) + limmu = 5 + 4$$

And although the derivation is less obvious than in those two cases, the word for 6 may have a similar origin:

$$6 = àsh = à + sh = (i)á + ash = 5 + 1 (?)$$

The Sumerian system next gave a name to each multiple of 10 lower than or equal to 60, and thus had a decimal form up to that point (table 2–14).

10 : *u* 20 : *nish* 30 : *ushu*	40 : *nishmin* or *nimin* or *nin* 50 : *ninnû* 60 : *gesh* or *geshta*

TABLE 2–14.

Actually, some of these words are compound. The word for 30 is formed by combining the words for 3 and 10:

$$30 = ushu = esh.u = 3 \times 10$$

The word for 40 comes from the words for 20 and 2:

$$40 = nishmin = nish.min = 2 \times 20$$

(*Nimin* and *nin*, two other words for 40, are shortened forms of *nishmin*.) The word for 50 is also compound:

$$50 = ninnû (= nimnu?) = nin + u = 40 + 10$$

Thus the words for 20, 40, and 50 are, to use Thureau-Dangin's expression, a kind of "vigesimal island in the Sumerian system."

The word for 60 *(gesh)* is the same as the word for 1, "no doubt because the Sumerians regarded 60 as the great unit" (Thureau-Dangin). To avoid confusion, 60 was sometimes called *geshta*.

With 60, another level is reached in the Sumerian oral numeration: numbers 60 to 600 are expressed by treating 60 as a new unit (table 2–15).

60 : *gesh*	360 : *gesh-àsh* (60 × 6)
120 : *gesh-min* (60 × 2)	420 : *gesh-imin* (60 × 7)
180 : *gesh-esh* (60 × 3)	480 : *gesh-ussu* (60 × 8)
240 : *gesh-limmu* (60 × 4)	540 : *gesh-ilimmu* (60 × 9)
300 : *gesh-iá* (60 × 5)	600 : *gesh-u* (60 × 10)

TABLE 2–15.

The next level is reached with 600, for despite the fact that the word for this number *(gesh-u)* is composed of *gesh* (60) and *u* (10), it acts as a new unit in expressing numbers 600 to 3600 (60^2) (table 2–16).

600 : *gesh-u*	2400 : *gesh-u-limmu* (600 × 4)
1200 : *gesh-u-min* (600 × 2)	3000 : *gesh-u-iá* (600 × 5)
1800 : *gesh-u-esh* (600 × 3)	3600 : *shàr* (independent name)

TABLE 2–16.

The number 3600 has an independent name and also acts as a new unit, since it marks yet another level (table 2–17).

3600 : *shàr*	21,600 : *shàr-àsh* (3600 × 6)
7200 : *shàr-min* (3600 × 2)	25,200 : *shàr-imin* (3600 × 7)
10,800 : *shàr-esh* (3600 × 3)	28,800 : *shàr-ussu* (3600 × 8)
14,400 : *shàr-limmu* (3600 × 4)	32,400 : *shàr-ilimmu* (3600 × 9)
18,000 : *shàr-iá* (3600 × 5)	36,000 : *shàr-u* (3600 × 10)

TABLE 2–17.

With numbers 36,000, 216,000, 2,160,000, and 12,960,000, still more levels appear (table 2–18).

36,000 : *shàr-u* (= 60^2 × 10) 72,000 : *shàr-u-min* (36,000 × 2)	2,160,000 : *shàr-gal-u* (= 60^3 × 10) 4,320,000 : *shàr-gal-u-min* (2,160,000 × 2)
180,000 : *shàr-u-iá* (36,000 × 5)	10,800 000 : *shàr-gal-u-iá* (2,160,000 × 5)
216,000 : *shàr-gal* (= 60^3) big 3600 432,000 : *shàr-gal-min* (216,000 × 2)	12,960,000 : *shàr-gal-shu-nu-tag* (= 60^4) (unit higher than the big *shàr*)
1,944,000 : *shàr-gal-ilimmu* (216,000 × 9)	

TABLE 2–18.

To sum up, the Sumerian oral numeration used 60 as a base and had consecutive levels built alternately on the auxiliary bases 6 and 10 (table 2–19). That is, the system involved a kind of compromise between two divisors of the base acting as auxiliary bases (table 2–20).

VALUES	NAMES	MATHEMATICAL STRUCTURES	
1	*gesh*	1	1
10	*u*	10	10
60	*gesh*	60	10·6
600	*gesh-u*	60×10	10·6·10
3600	*shàr*	60^2	10·6·10·6
36,000	*shàr-u*	$60^2 \times 10$	10·6·10·6·10
216,000	*shàr-gal*	60^3	10·6·10·6·10·6
2,160,000	*shàr-gal-u*	$60^3 \times 10$	10·6·10·6·10·6·10
12,960,000	*shàr-gal shu-nu-tag*	60^4	10·6·10·6·10·6·10·6

TABLE 2–19.

The origin of the sexagesimal system

The fact that people first learned to count on their ten fingers accounts for the predominance of the decimal system. The less common vigesimal system originated in the practice of counting on ten fingers and ten toes. But it is hard to imagine how the sexagesimal system came to be adopted by the Sumerians (and the duodecimal system by other peoples). Thureau-Dangin has examined the various explanations that have been suggested; none of them seems conclusive.

Units	Tens	60s	600s	3600s	36,000s
1 gesh	10 u	60 gesh	600 gesh-u	3600 shàr	36,000 shàr-u
2 min	20 nish	120 gesh-min (60 × 2)	1200 gesh-u-min (600 × 2)	7200 shàr-min (3600 × 2)	72,000 shàr-u-min (36,000 × 2)
3 esh	30 ushu	180 gesh-esh (60 × 3)	1800 gesh-u-esh (600 × 3)	10,800 shàr-esh (3600 × 3)	108,000 shàr-u-esh (36,000 × 3)
4 limmu	40 nimin	240 gesh-limmu (60 × 4)	2400 gesh-u-limmu (600 × 4)	14,400 shàr-limmu (3600 × 4)	144,000 shàr-u-limmu (36,000 × 4)
5 iá	50 ninnû	300 gesh-iá (60 × 5)	3000 gesh-u-iá (600 × 5)	18,000 shàr-iá (3600 × 5)	180,000 shàr-u-iá (36,000 × 5)
6 àsh		360 gesh-àsh (60 × 6)		21,600 shàr-àsh (3600 × 6)	
7 imin		420 gesh-imin (60 × 7)		25,200 shàr-imin (3600 × 7)	
8 ussu		480 gesh-ussu (60 × 8)		28,800 shàr-ussu (3600 × 8)	
9 ilimmu		540 gesh-ilimmu (60 × 9)		32,400 shàr-ilimmu (3600 × 9)	

TABLE 2–20.

Theon of Alexandria, a fourth-century commentator on Ptolemy, believed that 60 was chosen because it is the lowest number with a great many divisors. This opinion was also expressed fourteen centuries later by the English mathematician John Wallis (1616–1703) and in 1910 by Löffler, who maintained that the sexagesimal system must have arisen in priestly schools where it was recognized that 60 has the first six whole numbers as factors.

Another theory, expressed by the Venetian Formaleoni in 1789, and by Moritz Cantor (1829–1920) in 1880, attributes a "natural" origin to the sexagesimal system: the number of days in a year, rounded off to 360, resulted in the division of the circle into 360 degrees, and because the chord of a sextant (the sixth part of a circle) is equal to the radius, this same number engendered the division of the circle into six parts, which gave a privileged status to 60.

In 1899, Ferdinand Lehmann-Haupt (1861–1938) believed he had found the origin of the sexagesimal system in the relation between the Babylonian hour (equal to two of our hours) and the apparent diameter of the sun, expressed in units of time that were each equal to two of our minutes.

But in 1904 Kewitsch, a German Assyriologist, raised a valid objection to these theories when he stated that neither astronomy nor geometry could account for a system of numeration, which led Moritz Cantor to abandon the hypothesis he had proposed in 1880. Kewitsch maintained that the choice of base 60 must have arisen from contact between two peoples, one using the decimal system and the other a system that had been built on the number 6 as the result of a special method of finger counting. Thureau-Dangin rejects this idea because "the existence of a numeration of base 6 is a postulate with no historical foundation."

More recently, in 1927, Otto Neugebauer advanced the hypothesis that the choice of base 60 came from metrology (the science of measurement), but Thureau-Dangin believes that this is "a proposition which cannot be correct if it refers to the measurement system, properly speaking," since it seems "certain that the sexagesimal system penetrated metrology only because it already existed in numeration."

PART II

Concrete Counting

Fig. 3–1. Numerical finger signs shown in an Egyptian monument of the Old Kingdom (fifth dynasty, twenty-sixth century B.C.).

The First Calculating Machine: The Hand

Several methods of counting and calculating on the fingers, which we are presently going to examine, were used by many civilizations in ancient times, and some of them are still in use today.

Archaeological evidence going back as far as the Old Kingdom shows that a method of finger counting was commonly used in ancient Egyptian administrative affairs (figs. 3–1, 3–2). And references in the writings of Greek authors from Aristophanes (448?–380? B.C.) to Plutarch (A.D. 46?–120?) show that finger counting was used in ancient Greece and Persia.

Following an old tradition, the Romans also used finger-counting techniques in their everyday affairs. As direct evidence of this, we have a large number of Roman *tesserae*, tokens made of bone or ivory, each representing a certain sum of money, which tax collectors gave to tax-payers as receipts. Most of them have a number expressed in Roman numerals on one side and, on the other, a picture showing a finger sign for that same number (fig. 3–3). The numerical values appearing on *tesserae* used in the provinces of the empire apparently never went beyond 15.

The philosopher Lucius Annaeus Seneca (4? B.C.–A.D. 65) wrote in one of his letters: "Avarice taught me to count and to put my fingers at the disposal of my passion." In his *Historia Naturalis*, Pliny the Elder (A.D. 24–79) reported that the statue of the two-faced god Janus dedicated by King Numa showed the number of days in a year on its fingers. (Janus, one of the foremost gods in the Roman pantheon, was the god of the year, the four seasons, age, and time. Our "January" comes from the Latin Januarius, the name of the first month of the Roman year, which was dedicated to Janus.) And in his *Apologeticus*, Tertullian (A.D. c. 150–220) wrote: "But meanwhile one must sit surrounded by a multitude of papers, gesticulating with one's fingers to express numbers."

A passage in the first book of *Institutio Oratoria*, by the Roman rhetorician Quintillian (A.D. 35–95) shows that finger counting was commonly used by orators: "Knowledge of numbers is necessary not only to the orator, but to anyone who possesses even the first elements of writing. It is often used at the bar, and a lawyer who hesitates over a product, or

Fig. 3–2. Finger counting in ancient Egypt, New Kingdom: portion of a wall painting in the tomb (at Thebes) of Prince Menna, who lived during the reign of Thutmose IV, late fifteenth century B.C. Six scribes are overseeing four workers measuring grain and using containers to move it from one pile to another. Seated on one of the piles, the chief scribe is reckoning on his fingers and dictating the results to the three scribes on the left, who are recording them on wooden tablets and will later copy them on papyruses to be kept in the pharaoh's archives.

Fig. 3–3. Two Roman tokens from the first century A.D. The one on the left shows the number 9 represented by a finger sign on one side and by Roman numerals on the other. The one on the right shows a man representing the number 15 in the same system of finger signs.

merely shows uncertainty or awkwardness in counting with his fingers, immediately gives a bad opinion of his ability."

Among the peoples who, like the ancient Egyptians, Greeks, and Romans, had traditions of counting and calculating on the fingers were the Aztecs (fig. 3–4), Chinese, Indochinese, Indians, Persians, Turks, and Arabs, as well as the Coptic Christians of Egypt and the peoples of medieval Europe (figs. 3–5, 3–6, 3–9).

Fig. 3–4. Aztec finger counting. Detail from a painting by Diego Rivera, National Palace, Mexico City.

Fig. 3–5. Boethius (480?–524?), a Roman philosopher and mathematician, using a system of finger counting. Drawing by the author, after a painting attributed to Justus of Ghent (fifteenth century).

Fig. 3–6. Numerical finger signs shown in a twelfth-century Spanish manuscript.

Some of these traditions have continued down to the present in parts of Asia. Others continued in the West till relatively recent times: scarcely more than 400 years ago the use of one finger-counting technique (to which we will return) was so common among learned Europeans that arithmetic books had to contain detailed explanations of it if they were to be considered complete (figs. 3–7, 3–8).

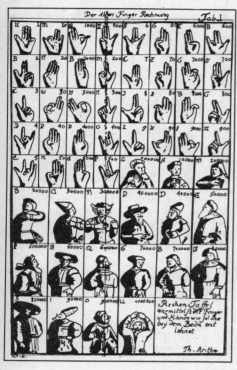

Fig. 3–7. Finger counting in *Theatrum Arithmetico-Geometricum*, by Jacob Leupold, published in Germany in 1727.

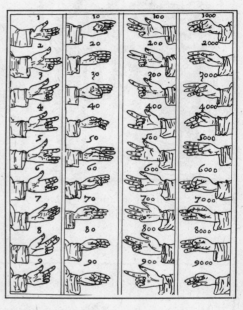

Fig. 3–8. Finger counting in *Summa de Arithmetica, Geometrica, Proportioni e Proportionalita*, by Luca Pacioli, published in Venice in 1494.

We will now review some of the finger-counting systems once used in the East and the West. They show a certain understanding of the principle of the base, and they will give us an idea of how different peoples were able to extend the numerical possibilities of the fingers. Some details of these systems are important in history and anthropology because they reveal contacts and influences that are not obvious at first sight.

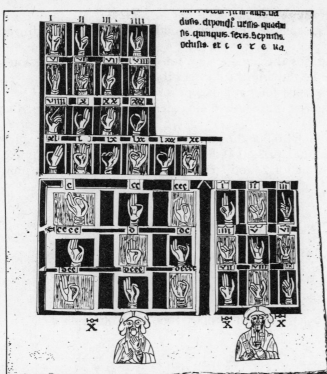

Fig. 3–9. Finger counting in a Spanish codex written on parchment, dating from about 1210. Lisbon Public Library.

A curious way of bargaining

The ancient tradition we will examine here seems to have been still rather widespread in Asia during the first half of this century.

It concerns a method of finger counting used by Oriental merchants and their customers. The eighteenth-century German traveler Carsten Niebuhr (1733–1815) gives this account of it: "I was afraid to have someone buy something for me in that way, because it allows an agent to deceive the person for whom he is making a purchase, even in his presence. The two parties in the transaction indicate prices asked or offered by touching each other's fingers or knuckles. In doing this, they hide their hands beneath their robes, not to make a mystery of the method they use, but to keep onlookers in ignorance of their bargaining."

In this system, numbers are indicated as follows (fig. 3–10):

For 1, one of the bargainers grasps the other's forefinger.

For 2, he grasps the forefinger and middle finger together.

For 3, he grasps the forefinger, middle finger, and third finger together.

For 4, he grasps the hand, except for the thumb.

For 5, he grasps the whole hand.

For 6, he twice grasps the forefinger, middle finger, and third finger together (2 × 3).

For 7, he first grasps the hand except for the thumb, then the fore-finger, middle finger, and third finger together (4 + 3).

For 8, he twice grasps the hand except for the thumb (2 × 4).

For 9, he first grasps the whole hand, then the hand except for the thumb (5 + 4).

For 10, 100, 1000, or 10,000, he grasps his partner's forefinger, as for 1; for 20, 200, 2000, or 20,000, he grasps his forefinger and middle finger together, as for 2, and so on. This cannot cause confusion,

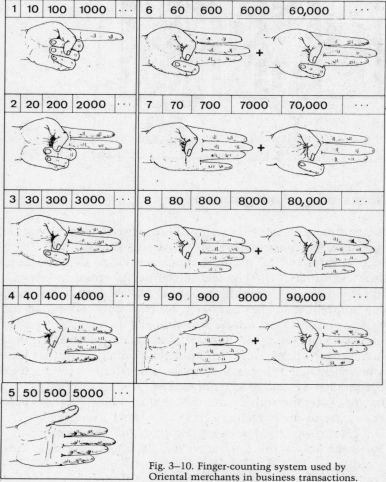

Fig. 3–10. Finger-counting system used by Oriental merchants in business transactions.

of course, because the two bargainers know the general price range in advance: if the seller wants about 400 monetary units for his merchandise, he will make sure the buyer understands that the price is in the hundreds before they begin their bargaining.

Niebuhr does not say whether he personally witnessed such bargaining, but J. G. Lemoine gathered descriptions of it as it was practiced in the early part of this century on the island of Bahrain, in the Persian Gulf, an area known for its pearl fisheries as well as its petroleum. He talked with pearl dealers who had repeatedly gone to Bahrain and used the method in question with natives of the island. Here is his summary of what they told him: "The two bargainers sit facing each other, each holding the other's right hand. With their left hands they hold a piece of cloth over their joined right hands, so that all their gestures will be hidden. The bargaining, including the disagreements that inevitably arise, takes place in silence. According to those who have seen it, the psychology of the procedure is extremely interesting because impassivity is the rule and the slightest sign of emotion is interpreted to the disadvantage of the man who shows it."

Similar bargaining systems have been reported in Syria, Iraq, Arabia, India, China, Mongolia, and Algeria. Here, for example, is how bargaining was still carried on in China early in this century, as described by P.-J. Dols, who observed life in the province of Kansu:

"The buyer puts his hands in the seller's sleeves. As they talk, he takes hold of the seller's forefinger, which means that he is offering 10, 100, or 1000.

" 'No,' says the seller.

"The buyer then takes the seller's forefinger and middle finger together.

" 'Yes,' replies the seller.

"The bargain is concluded; the object has been sold for a price of 20 or 200. The forefinger, middle finger, and third finger taken together mean 30, 300, or 3000. All fingers except the thumb: 40, 400, or 4000. The whole hand: 50, 500, or 5000. The thumb and the little finger together: 60. The buyer's thumb in the seller's palm means 70. The thumb and the forefinger together indicate 80. When the buyer touches the seller's forefinger with his thumb and forefinger together, he is offering 90."

Counting on the bones and joints of the fingers

Other systems of finger counting, also very old, use the bones of the fingers or their corresponding joints. In a system that is common in India, Indochina, and southern China, each finger bone counts as a unit, beginning with the lower bone of the little finger and going to the upper bone of the thumb. Both hands are used, each being touched with a finger of the other (fig. 3–11). Numbers 1 to 14 can be counted on one hand, 1 to 28 on both. In China, the procedure is slightly different: counting usually begins with the upper bone of the little finger and ends with the lower bone of the thumb.

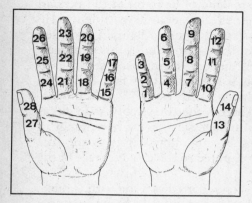

Fig. 3–11. Finger-counting technique common in India, China, and Indochina, using the fourteen finger bones of each hand.

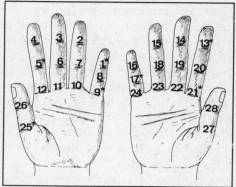

Fig. 3–12. Method used by the Venerable Bede to consider the twenty-eight years of the solar cycle in the Julian calendar, with its leap years, each of which is indicated here by an asterisk.

A Chinese from Canton has described a practical application of this method: to keep track of her menstrual cycle, his mother tied a string around one of her twenty-eight finger bones each day, beginning with the upper bone of her left little finger and going to the lower bone of her right thumb, so that if her menstruation was irregular she could determine how many days late or early it was, in relation to her normal cycle.

In his *De Ratione Temporum*, the Venerable Bede (673?–735), an English monk who had a considerable influence on the intellectual development of medieval Europe, explained a method of finger counting similar to the one described above. It concerns computations relating to the solar year and uses the lunar and solar cycles of the Julian calendar, with its leap years. Its purpose was to determine the date of Easter, which at that time was a subject of controversy between the Church of Rome and the churches of Ireland.

To count the twenty-eight years of the solar cycle, Bede used the twenty-eight joints (or bones) of both hands and began with a leap year (fig. 3–12). Starting at the upper joint of the left little finger, he counted in descending rows of joints till he came to the twelfth year of the solar cycle, represented by the lower joint of the left forefinger. He then continued counting on the right hand in the same way, except that he began with the upper joint of the forefinger. Finally, the count was completed with the thumbs, going from the lower joint of the left one to the upper joint of the right one.

To count the nineteen years of the lunar cycle (the period of time after which the phases of the moon return to the same dates), Bede used the fourteen joints and five fingernails of the left hand. He began with the base of the thumb and ended with the nail of the little finger, moving upward along each finger (fig 3–13).

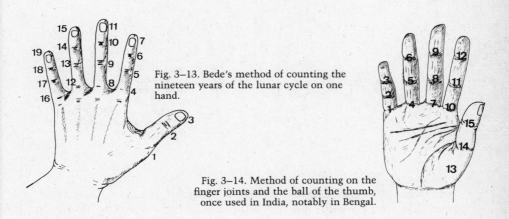

Fig. 3–13. Bede's method of counting the nineteen years of the lunar cycle on one hand.

Fig. 3–14. Method of counting on the finger joints and the ball of the thumb, once used in India, notably in Bengal.

Another method of finger counting was used for a long time in northeastern India and is said to be still used in Bengal and the Dacca region of Bangladesh. It was reported by European authors in the seventeenth and eighteenth centuries, notably by the French traveler Jean Baptiste Tavernier in his *Voyages en Turquie, en Perse et aux Indes* (1679). It consists in counting on the finger joints and the ball of the thumb, as shown in figure 3–14, which makes it possible to reach 15 on one hand. N. B. Halhed, an eighteenth-century English author, wrote that this system gave rise to a practice among Indian merchants: the buyer and the seller joined hands under a cloth and negotiated prices by touching finger joints.

The system is found nearly everywhere in the Islamic countries of Asia and North Africa. But in these countries it is related to a religious practice: Moslems traditionally use it for enumerating the ninety-nine attributes of Allah and counting the supererogatory blessings (*subḥa*) that are said after obligatory prayer. (A prophetic *hadith* says that "Allah has ninety-nine names, one hundred minus one. He who knows them all will enter paradise.")

In this religious practice, a Moslem counts as follows. He successively touches each finger joint and the ball of the thumb on each hand,

beginning with the lower joint of the left little finger and ending with
the upper joint of the left thumb, which brings him to 15 (fig. 3–15). He
reaches 30 by proceeding in the same way with his right hand. He reaches
33 by touching the tips of the little finger, third finger, and middle finger
on his right hand, or by returning to the three joints of his right forefinger.
And finally he reaches 99 by repeating the whole procedure twice.

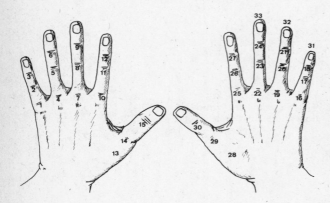

Fig. 3–15. How Moslems use the
finger joints and balls of the
thumbs to enumerate the ninety-
nine (3 × 33) attributes of Allah
and count the supererogatory
blessings said after obligatory
prayer.

Moslems use this method of counting when a rosary is not available.
It is very old, and probably preceded the use of the rosary (fig. 3–16). In
several traditional stories the Prophet proscribes the use of beads or pebbles for reciting litanies and recommends counting prayers or praises of
God on the fingers.

Fig. 3–16. Use of the rosary (*subḥa* or *sebḥa* in
Arabic) by Moslems to enumerate the ninety-
nine attributes of Allah or count the supererogatory blessings. Pilgrims and dervishes habitu-
ally carry rosaries. These are composed of
strung beads, made of wood, mother-of-pearl, or
bone, which are slipped between the fingers.
They are often divided into three groups separated by two larger beads and a much larger
one which serves as a handle. The number of
beads is usually 100 (33 + 33 + 33 + 1), but
this may vary.

There is one method of finger counting, evidently still used in India,
Indochina, Pakistan, Afghanistan, Iran, Turkey, Iraq, Syria, and Egypt,
that allows the hand to be considered as a sequence of numbers from 1
to 12. It may explain the rather general tendency, in the past, to take 12
as a secondary counting unit, along with 10. (In ancient times, 12 was
often used as a base in divisions of time and units of measurement, as
well as commercial counts, of which our dozen and gross are surviving
examples.) But that is only a conjecture.

The method consists in counting on the bones (or sometimes the joints) of four fingers on one hand, using the thumb of the same hand (fig. 3–17). The four fingers have a total of twelve bones. (The thumb is excluded, of course, because the operation is performed by means of it.) The same procedure can be repeated on the other hand to count up to 24.

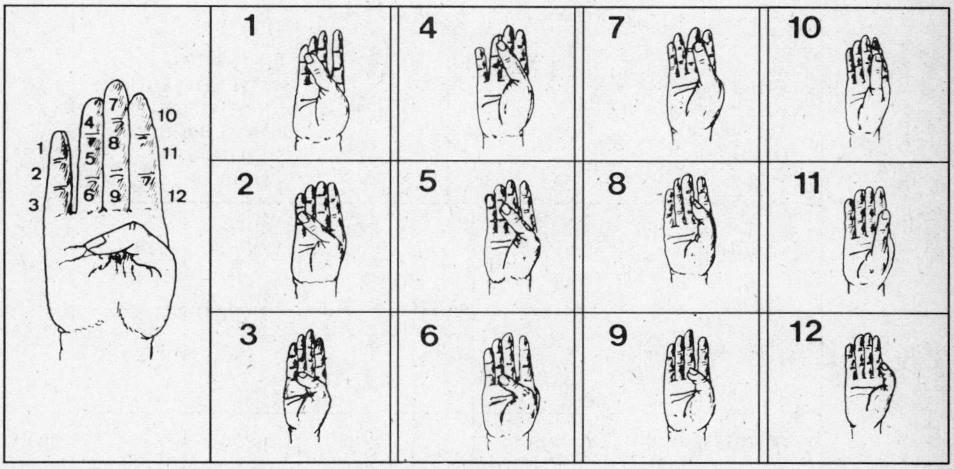

Fig. 3–17. Duodecimal finger-counting method used in India, Indochina, Pakistan, Afghanistan, Iran, Turkey, Iraq, Syria, and Egypt.

Was it this system that led the ancient Egyptians to divide day and night into twelve "hours" of light and twelve of darkness ("hours" whose length changed with seasonal variations in the duration of day and night)? Was it what led the Sumerians, and the Assyrians and Babylonians after them, to divide the cycle of day and night into twelve equal parts (called *danna*, each equivalent to two of our hours), to adopt for the ecliptic and the circle a division into twelve *beru* (30° each), and to give the number 12, as well as its divisors and multiples, a preponderant place in their various measurements? And finally (to take only one more of the many examples that could be chosen), was it what led the Romans to divide the *as* (an arithmetical and monetary unit, besides being a unit of weight) into twelve parts called *unciae?* These questions cannot be answered with certainty.

A variant of the system we have just discussed begins in the same way, by counting numbers 1 to 12 on one hand, but then counting numbers up to 60 on both hands (fig. 3–18). The procedure is as follows. The first twelve numbers are counted on the finger bones of the right hand, using the right thumb. Then the little finger of the left hand is bent down.

Numbers 13 to 24 are counted, again on the right hand. Then the third finger of the left hand is bent down. The process is repeated till all five fingers of the left hand are bent down, indicating 60.

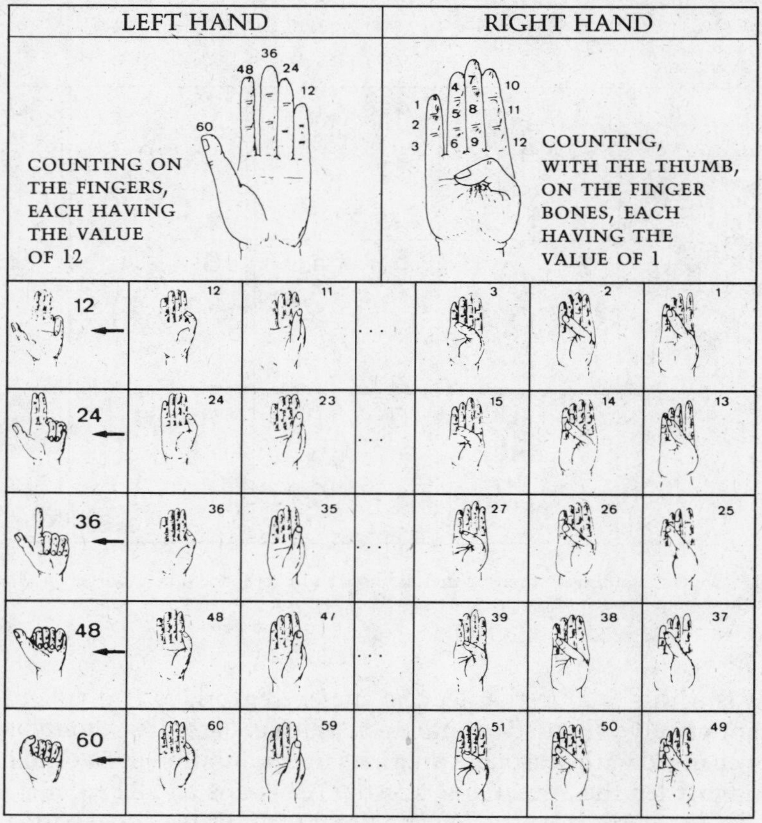

LEFT HAND				RIGHT HAND		
COUNTING ON THE FINGERS, EACH HAVING THE VALUE OF 12				COUNTING, WITH THE THUMB, ON THE FINGER BONES, EACH HAVING THE VALUE OF 1		
12 ←	12	11	…	3	2	1
24 ←	24	23	…	15	14	13
36 ←	36	35	…	27	26	25
48 ←	48	47	…	39	38	37
60 ←	60	59	…	51	50	49

Fig. 3–18. Sexagesimal finger-counting method still used in Turkey, Iraq, Iran, India, and Indochina.

This system is still used in parts of Turkey, Iraq, Iran, India, and Indochina. It is therefore not impossible that the Sumerians (and the Babylonians after them) knew it and were influenced by it to the point where, besides 10, they adopted 60 as their great counting unit and 12 (with its divisors 2, 3, 4, and 6) as an auxiliary unit relating to certain divisions. This, however, does not solve the problem of the origin of the sexagesimal system. It is only a hypothesis; it is based on anthropological reports and the fact that the tradition of counting by sixties was probably not the result of any arithmetical, geometrical, or astronomical considerations, but to the best of my knowledge there is no historical evidence to support it.

A more extensive and mathematically more interesting system of counting on the finger joints was used in China as late as the nineteenth century. In it, each joint is divided into three parts: left, right, and center (fig. 3–19). Each finger is then associated with the nine units of a decimal order, corresponding to the nine divisions of its joints: the little finger with the first nine units, the third finger with tens, the middle finger with hundreds, and so on. With this system, it is possible to count up to 100,000 on one hand.

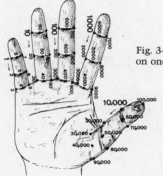

Fig. 3–19. System used in China for counting up to 100,000 on one hand.

Morra, a finger game

Morra, a game that has been played in various parts of the world since ancient times, is derived more or less directly from the habit of counting on the fingers. It is a simple game, usually played by two people.

The two players face each other, each holding up a closed fist. At a given signal, they both hold up as many fingers as they wish, and announce a number from 1 to 10. (The game can also be played with each player using both hands, in which case the number announced must be between 1 and 20.) A player who announces a number equal to the total number of fingers held up by both players scores a point. If player A, for example, shows three fingers and says "Five," and player B shows two fingers and says "Six," player A scores a point because the total number of fingers shown is 3 + 2 = 5. The game involves not only chance but also the players' intuition, observation, and quick reactions.

Morra is still quite popular in Italy and is sometimes played in southwestern France, the Basque region of Spain, Portugal, and North Africa (at least in Morocco). I myself played it with childhood friends in Marrakesh, as a form of drawing lots to see who would be chosen for some purpose or other. We stood face to face. One of the two players held his hands behind his back, then showed one hand with a certain number of fingers extended while his opponent simultaneously announced a number from 1 to 5. If the "guesser" said the right number, he won; if not, the other player won.

In China and Mongolia, a form of morra has been known a long time. P. Perni, writing in the nineteenth century, reported that it was still in vogue among the Chinese at that time. "If the guests are all friends, the host suggests a game of *tsin hoa kiuen.* If the suggestion is accepted, politeness requires the host to begin the game with one of his guests, then yield his place to another guest soon afterward. Each time a player loses, he must drink a cup of tea."

The Chinese sometimes make the game more difficult by a rule stipulating that instead of simply calling out a number a player must think of a saying that contains it, such as "A bird in the hand is worth *two* in the bush."

During the Renaissance, particularly in France and Italy, morra was in great favor with servants, who often played it to pass the time when they had no work to do. Rabelais wrote in *Pantagruel,* "The pages were flicking their fingers in a game of morra," and Malherbe wrote in one of his letters, ". . . dawdling along the way like footmen playing morra when they have been sent for wine."

Fig. 3–20. The game of morra, shown in one of the frescoes in the Villa Farnesina, Rome.

Fifteen centuries earlier the game was known in Rome, where it was often played by the common people, under the name of *micatio* or *micare e digitis* (fig. 3–20). Cicero mentions a saying applied to someone completely trustworthy: "You could play *micatio* with him in the dark." This saying, which Cicero describes as "timeworn," shows how old and well known morra was among the Romans. "Sometimes, when two people had a dispute, they would agree to settle the matter by a game of morra, as today we flip a coin or draw straws. This procedure was even used in commercial transactions, when buyer and seller could not come to terms otherwise. A fourth-century inscription (*Corpus Inscriptionum Latinarum*, VI, 1770) has preserved an edict of the Prefect of Rome forbidding it in public markets" (Lafaye).

Fig. 3–21. Morra in ancient Greece: (*left*) Painted vase in the Lambert collection, Paris; (*right*) Painted vase in the Munich Museum.

Morra was also played in ancient Greece, as is shown by certain monuments and vases (fig. 3–21). According to legend, it was invented by Helen of Troy to play with her lover Paris.

The ancient Egyptians had a traditional game similar to morra going back at least to the Middle Kingdom. We know this chiefly from the two tomb paintings reproduced in figure 3–22.

Fig. 3–22. Morra in two Egyptian tomb paintings: (*top*) Tomb no. 15 (Middle Kingdom) at Beni Hasan; (*bottom*) Theban tomb no. 36 (tomb of Aba, Twenty-sixth Dynasty).

The first of these is from a tomb at Beni Hasan and dates from the Middle Kingdom (twenty-first to seventeenth centuries B.C.). It shows two scenes, in each of which two men are squatting face to face. In the scene on the left, one player is holding both hands in front of his opponent's eyes, with one hand hiding the extended fingers of the other, while his opponent holds out his clenched fist. In the scene on the right, two players are shown in the same situation, except that the first one is holding his hands lower. Jean Yoyotte has given me this translation of the hieroglyphs above the two scenes:

Left: "Showing (or giving) *íp* on the forehead."

Right: "Showing (or giving) *íp* on the hand."

Since the Egyptian word *íp* means "count" or "calculate," the game being played is undoubtedly morra.

The second painting is from Thebes in the time of King Psamtik (seventh century B.C.), and according to Jean Leclant it was copied from a Middle Kingdom original. It also shows four men facing each other in pairs, but in this case both players in each pair are holding out a certain number of fingers.

And finally morra, known as *moukhāraja* ("that which causes to come out"), was still played in its traditional form at the beginning of this century in isolated rural areas of Arabia, Syria, and Iraq. Father Anastasius of Baghdad reported that the game was also played in other forms, and for purposes other than amusement: *mouqāra'a*, for dividing things between people; *mousāhama*, for sharing an inheritance or distributing profits in a partnership; *mounāhada*, for dividing booty.

Elements of finger computation

In the past, people used their fingers not only for counting but also for computing—that is, performing all sorts of arithmetical operations.

As an example, here is an old and ingenious method of multiplying numbers between 5 and 10 on the fingers without having to know the multiplication table for any numbers higher than 5, except for the easy case of 10 (fig. 3–23). To multiply 7 by 8, let us say, bend as many fingers on one hand as there are excess units in 7 over 5 (two fingers) and hold the other three fingers raised, then on the other hand indicate the excess of 8 over 5 in the same way (three fingers) and hold the other two fingers raised. The product is obtained by multiplying the total number of fingers raised (2 + 3 = 5) by 10 and adding the result to the product obtained by multiplying the number of raised fingers on one hand by the number of raised fingers on the other: $7 \times 8 = (2 + 3) \times 10 + 3 \times 2 = 50 + 6 = 56$.

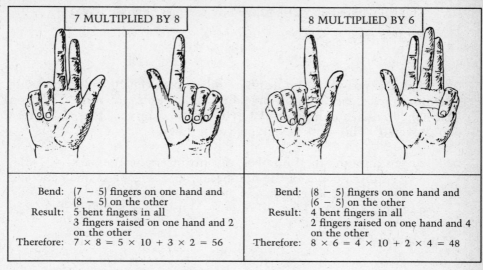

7 MULTIPLIED BY 8	8 MULTIPLIED BY 6

Bend:	(7 − 5) fingers on one hand and (8 − 5) on the other	Bend:	(8 − 5) fingers on one hand and (6 − 5) on the other
Result:	5 bent fingers in all 3 fingers raised on one hand and 2 on the other	Result:	4 bent fingers in all 2 fingers raised on one hand and 4 on the other
Therefore:	7 × 8 = 5 × 10 + 3 × 2 = 56	Therefore:	8 × 6 = 4 × 10 + 2 × 4 = 48

Fig. 3–23. Method of multiplying on the fingers.

Similarly, to multiply 8 by 6, indicate the excess of 8 over 5 on one hand and the excess of 6 over 5 on the other; that is, bend $8 - 5 = 3$ fingers on one hand and $6 - 5 = 1$ on the other. Multiply the total number of bent fingers (4) by 10 and add the result to the product of the raised fingers $(2 \times 4) : 8 \times 6 = (3 + 1) \times 10 + 2 \times 4 = 48$.

Traces of this kind of procedure are still found in India, Iran, Iraq, Syria, Serbia, Bessarabia, Wallachia, Auvergne, and North Africa. It is described by the Persian author Behā' ad-Din al'Amuli (1547–1622), whose influence was considerable in Persia and India, and by the fifteenth-century French mathematician Nicolas Chuquet in his *Triparty en la science des Nombres*.

Here is a mathematical proof of the procedure, for readers who are not baffled by elementary algebra. Let x and y be two whole numbers between 5 and 10. On one hand, bend as many fingers as there are excess units in x over 5; and on the other, as many fingers as there excess units in y over 5. The total number of fingers bent is therefore:

$$R = (x - 5) + (y - 5) = x + y - 10$$

The number of fingers raised on the first hand is equal to:

$$a = 5 - (x - 5) = 10 - x$$

and the number of fingers raised on the other hand is equal to:

$$b = 5 - (y - 5) = 10 - y$$

The rule stated above is proved by the fact that we have:

$$10R + ab = 10(x + y - 10) + (10 - x)(10 - y)$$
$$= (10x + 10y - 100) + (100 - 10x - 10y + xy)$$
$$= xy$$

It is also possible to multiply numbers between 10 and 15 by using a similar procedure. To multiply 14 by 13, for example, bend fingers on one hand to indicate the excess of 14 over 10 and, on the other, the excess of 13 over 10. The result is obtained as follows:

a. Multiply the total number of bent fingers by 10: $10 \times (4 + 3)$.
b. Add the product of the number of bent fingers on one hand and the number of bent fingers on the other: 4×3.
c. Add 100: $14 \times 13 = 10 \times (4 + 3) + 4 \times 3 + 100 = 182$.

Mathematical proof:

$$10[(x - 10) + (y - 10)] + (x - 10)(y - 10) + 100 = xy$$

An elaborate system of finger counting

In conclusion, here is a finger numeration more elaborate than those above. It was used by European peoples from ancient times to the end of the Middle Ages, and also in the Near East, where it seems to have persisted longer.

Its techniques, somewhat similar to the finger language of deaf-mutes, makes it possible to represent numbers from 1 to 10,000 by means of finger signs.

Two descriptions, one Occidental and the other Oriental, will show us how it works. The first one, written in Latin in the seventh century, is from the first chapter, *De Computo vel Loquela Digitorum* ("On Calculating and Speaking with the Fingers"), of *De Temporum Ratione* ("On the Division of Time") by the Venerable Bede. The second one is from *Farhangi Jihangiri*, a sixteenth-century Persian dictionary. These two texts show how units and tens were represented with the fingers of one hand (the left in the Occident, the right in the Orient), and hundreds and thousands with those of the other. It is striking to see the close similarities between two systems described in two widely separated parts of the world, one nearly a thousand years after the other.

OCCIDENTAL DESCRIPTION
Venerable Bede, seventh century

ORIENTAL DESCRIPTION
Persian dictionary, sixteenth century

A. UNITS (fig. 3–24A)

When you say "one," bend the left little finger and touch the middle line of the palm with it.

For the number 1, lower the little finger.

When you say "two," bend the third finger to the same place.

For the number 2, lower the third finger beside the little finger.

When you say "three," bend the middle finger in the same way.

For the number 3, lower the middle finger to the other two.

When you say "four," raise the little finger.

For the number 4, raise the little finger, with the other fingers remaining in the same position.

When you say "five," raise the third finger.

For the number 5, also raise the third finger.

When you say "six," raise the middle finger and bend the third finger down to the middle of the palm.

For the number 6, raise the middle finger and lower the third finger so that its tip touches the middle of the palm.

When you say "seven," touch the base of the palm with the little finger and hold up all the other fingers.

For the number 7, raise the third finger and lower the little finger so that its tip is close to the wrist.

When you say "eight," bend the third finger in the same way.

For the number 8, do the same with the third finger.

When you say "nine," bend the shameless [middle] finger in the same way.

For the number 9, do the same with the middle finger.

B. TENS (fig. 3–24B)

When you say "ten," place the nail of the forefinger on the middle of the thumb.

For 10, the nail of the right forefinger is placed on the upper joint of the thumb, so that the space between the two fingers resembles a circle.

With "twenty," place the tip of the thumb between the forefinger and the middle finger.

For 20, the thumbnail is placed against the lower bone of the forefinger on the side of the middle finger, so that the tip of the thumb seems to be pressed between the bases of the forefinger and the middle finger, though the middle finger plays no part in indicating the number, since variations in the position of this finger must continue to represent other numbers. The number 20 is expressed solely by joining the thumbnail to the side of the lower forefinger bone.

With "thirty," join your forefinger and thumb in a tender embrace.

For 30, the thumb is held upright and the tip of the forefinger is placed on its nail, so that the thumb and forefinger resemble a bow with its string. If, to facilitate this position, one feels a need to curve the thumb, the position will still indicate the number in question and no confusion will result.

With "forty," place the thumb on the side or back of the forefinger, with both fingers extended.

For 40, the inside of the end of the thumb is placed on the back of the lower bone of the forefinger, so that no space is left between the thumb and the edge of the palm.

With "fifty," bend the thumb toward the palm with its upper end lowered, like the Greek letter Γ.

For 50, the forefinger is held upright while the thumb is bent and placed against the palm under the forefinger.

With "sixty," keep the thumb bent as before and place the forefinger over it, just below the nail.

For 60, the thumb is held bent and the inside of the second bone of the forefinger is placed over the thumbnail.

With "seventy," place the thumbnail against the middle bone of the forefinger.

For 70, the thumb is held upright and the inside part of the first or second bone of the forefinger is pressed against the tip of the thumb in such a way that the thumbnail remains entirely uncovered.

With "eighty," place the thumbnail against the middle bone of the forefinger, as above, and bend the end of the forefinger over it.

For 80, the thumb is held upright and its end is pressed against the outside of the first joint of the forefinger.

With "ninety," place the nail of the forefinger on the base of the thumb.

For 90, the nail of the forefinger is placed on the lower joint of the thumb, just as for 10 it is placed on the upper joint.

C. HUNDREDS AND THOUSANDS (fig. 3–24C and D)

When you say "one hundred," do with the right hand as for "ten" with the left. "Two hundred" is on the right hand as "twenty" is on the left. "Three hundred" is on the right hand as "thirty" is on the left. "Four hundred" is on the right hand as "forty" is on the left; and so on to "nine hundred."

When you say "one thousand," do with the right hand as for "one" with the left. "Two thousand" is on the right hand as "two" is on the left. "Three thousand" is on the right hand as "three" is on the left; and so on to "nine thousand."

If one clearly has in mind these eighteen signs, that is, the nine combinations of the little, third, and middle fingers and the nine combinations of the thumb and forefinger, it is easy to understand that the signs made with the right hand for the units from 1 to 9 can be made with the left hand for the thousands from 1000 to 9000, and that the signs made with the right hand for the tens from 10 to 90 can be made with the left hand for the hundreds from 100 to 900. Thus, with the fingers of both hands, one can count from 1 to 9999.

Up to 9000, the only noteworthy difference between the two texts is for the number 80. From 10,000 onward, however, they diverge considerably.

In the Persian text, 10,000 is indicated by pressing the upper joint of the forefinger against the upper joint of the thumb, so that the two nails are opposite each other and their upper ends are on the same level.

In Bede's system, ten thousands are indicated as follows. For 10,000, place the back of the left hand against the middle of the chest, with the fingers pointing toward the neck. For 20,000, place the left hand on the chest with the fingers outspread. For 30,000, hold up the left hand with the palm to the right and point the thumb toward the cartilage in the

A. UNITS

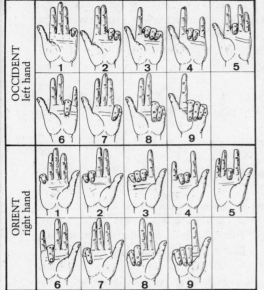

B. TENS

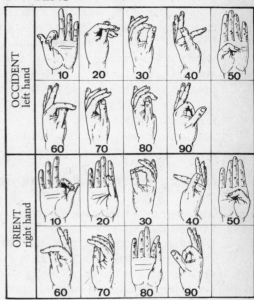

C. HUNDREDS

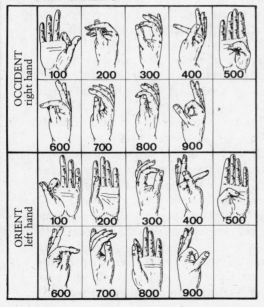

D. THOUSANDS

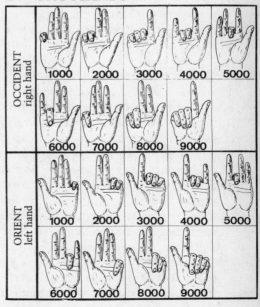

Fig. 3–24.

middle of the chest. For 40,000, place the left hand, turned downward, over the navel. For 50,000, with the left hand in the same position, point to the navel with the thumb. For 60,000, touch the left thigh with the left hand, still turned downward. For 70,000, touch the same place, but with the hand turned palm up. For 80,000, grasp the thigh with the hand. For 90,000, place the left hand on the small of the back, with the thumb pointing toward the genitals.

Bede then explains that the same signs on the right side of the body indicate numbers from 100,000 to 900,000, and that for 1,000,000 both hands are joined with their fingers interlaced.

History recounted on the fingers

The system of finger counting that we have just examined was widespread in both the Orient and the Occident. There are allusions to it in the writings of several Greek and Latin writers, such as Plutarch, who attributes these words to Orontes, son-in-law of King Artaxerxes of Persia: "Just as, in calculating, fingers sometimes have a value of ten thousand and sometimes of only one, the favorites of kings may be either everything or almost nothing."

The Roman satirist Juvenal (A.D. c. 55–135) says of Nestor, the legendary king of Pylos (who was supposed to have lived to twice the age of a crow), "Lucky Nestor, who, having passed a hundred, now counts the years with his right hand," an allusion to the fact that the Romans counted units and tens on the left hand and hundreds and thousands on the right.

Apuleius (A.D. c. 125–170), a philosopher and rhetorician from the Roman province of Numidia (modern Algeria and Tunisia), married a rich widow named Aemilia Pudentilla and was accused of having used magic to win her favor. He defended himself before Proconsul Claudius Maximus in the presence of Emilianus, his main accuser, who had unkindly said that Aemilia was sixty years old when she was actually only forty. Here, according to Apuleius himself, is how he addressed his accuser: "How dare you, Emilianus, increase the real number [of Aemilia Pudentilla's age] by half, or even a third? If you had said 'thirty' for 'ten,' it might have been thought that your mistake came from holding your fingers open when you should have held them curved. But forty is the easiest number to indicate, since it is expressed with the hand open" (fig. 3–25).

Saint Jerome (340?–420), a Latin philologist who lived in the time of Saint Augustine, wrote: "Thirty corresponds to marriage, for the conjunction of the fingers as though in a sweet kiss represents the husband and the wife. Sixty corresponds to widows, for they are in a situation of anguish and tribulation that weighs down on them. . . . And the gesture

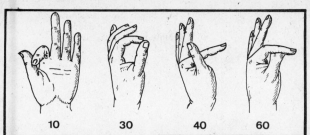

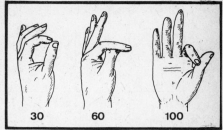

Fig. 3–25.   Fig. 3–26.

for a hundred (reader, please pay close attention), transferred from the left hand to the right, on the same fingers . . . expresses on the right hand the crown of virginity" (fig. 3–26).

One of the many reasons for the popularity of this finger numeration was its secret, even mysterious nature. The Venerable Bede gives an example of how the system can be used for silent communication. After establishing a correspondence between the letters of the Roman alphabet and the first twenty whole numbers (fig. 3–27), he says that in order to tell a friend *"Caute age!"* ("Be careful!") in the presence of indiscreet or dangerous people, one can make the number signs corresponding to the letters of the words (the signs are shown in fig. 3–28):

3	1	20	19	5	1	7	5
C	A	U	T	E	A	G	E

In Islamic regions, this kind of finger counting was used as extensively as in the West, and tradition seems to attest its use in ancient times. As early as the first few centuries of the Hegira, Arab and Persian poets subtly alluded to a person's lack of generosity by saying that his hand made 93, since the sign for that number was made with a closed hand (fig. 3–29), a common symbol of avarice.

The poet Yahya Ibn Nawfal al-Yamani (seventh century) wrote: "Ninety followed by three, which a stingy man represents by a clenched fist ready to strike, is not more niggardly than your gifts, Yazid."

According to one of his commentators, the following lines by Al-Farazdaq (died 728) refer to the number sign for 30 made by joining the tips of the thumb and forefinger, as shown in figure 3–30: "We strike the chief of every tribe while your father picks lice off himself behind his she-ass. His fingers, forming a number near his testicles, crush lice in the lowest situation in which the humblest man can be."

The poet Khaqani (1106–1200): "What is this pitched battle between Rustem and Bahram? What anger, what dissensions agitate those two descendants of illustrious families? On their 90, they struggle night and day to see which of the two armies will have the number 20."

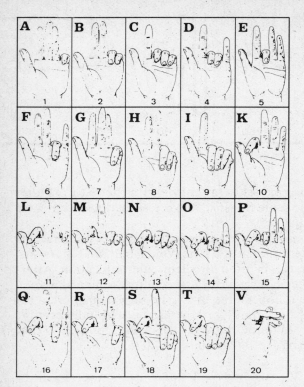

Fig. 3–27. Finger alphabet composed by the Venerable Bede, using the number signs shown in figure 3–24.

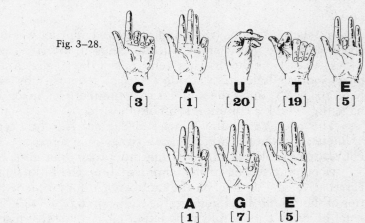

Fig. 3–28.

C [3] A [1] U [20] T [19] E [5]

A [1] G [7] E [5]

93

Fig. 3–29. In the Arab-Persian finger numeration, the number 93 was indicated by touching the lower joint of the thumb with the forefinger (for 90) and lowering the other three fingers (for 3), which resulted in a closed hand.

Fig. 3–30.

The meaning of this passage was obvious to people of the time, but it is obscure to twentieth-century readers not familiar with the ancient method of finger counting alluded to here. If, however, we examine the finger sign for 90 (fig. 3–31), we realize that "their 90" is probably the poet's way of designating the anus and, by extension, the buttocks. As for the expression "to have the number 20 (on someone)" it is evidently, as J. G. Lemoine has suggested, a pejorative designation of the sex act (which, it seems, is called "making the thumb" in Persian) and by extension, in this military context, means "to defeat." So the last sentence of the passage can be read as: "On their buttocks [their 90], they struggle night and day to see which of the two armies will defeat the other [will have the number 20]."

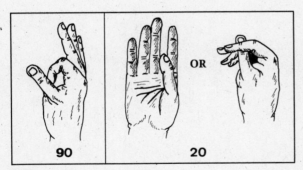

Fig. 3–31.

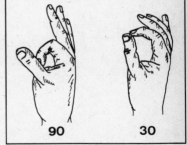

Fig. 3–32.

Here is another example of this use of 90. Ahmad al-Barbir al-Tarabulusi writes in one of his commentaries that to make his pupils remember the finger signs for 30 and 90 he could not resist saying to them, "In an epigram against an adolescent boy [named Khalid], a poet showed great subtlety in saying, 'Khalid left with a fortune of 90 dirhams, and when he came back he had only a third of it left.'"

Behind its apparent innocence, this was a malicious attack implying that the boy engaged in sodomy. "The poet means," explained Ahmad al-Barbir, "that the Khalid in question was 'narrow' when he left and 'wide' when he came back" (fig. 3–32).

Such roundabout allusions to the system of finger signs show that it must have been in common use, since otherwise readers could not have been expected to understand them.

4

Notching

The notched bones left to us by prehistoric people who lived more than 20,000 years ago are probably among the oldest objects that have served as a material support for the notion of number (fig. 4–1). The people who notched those bones must have used them for keeping account of something.

The oldest known notched bones have been discovered in western Europe. They date from the Aurignacian culture, 30,000 to 20,000 years ago, and more or less correspond to the first appearance of the Cro-Magnons. The notches on them are most likely numerical marks of some sort, but their exact purpose is hard to determine. Some specialists believe they must have been related to marking phases of the moon, but not enough examples have been studied to confirm or invalidate this theory. Others believe they were used for community record keeping, notably for indicating the number of animals killed by hunters.

This elementary form of accounting has come down to us almost unaltered through thousands of years. Not so long ago it was still used in France by bakers in rural areas when they sold bread on credit. The system involved two round or flat pieces of wood called *tailles* (tallies); the baker kept one and the customer took the other. Both tallies were notched each time the customer received a loaf of bread. Payment was made at fixed intervals, once a week, for example. No dispute was possible, since the number, location, and size of the notches on one tally were the same as those on the other. If the baker had added a notch or the customer had removed one, comparison of the two tallies would have shown the fraud (fig. 4–2).

In a book entitled *Michel Rondet*, André Philippe has given us a description of this old French system of "credit cards." The scene takes place in 1869, near Saint-Étienne: "Each woman held out a piece of wood about eight inches long, marked with a file. The pieces were all different; some were simply sections of a branch, others had been planed flat. The baker had duplicates of them, strung on a leather thong. He took the piece with the customer's name inscribed on it and placed her piece beside

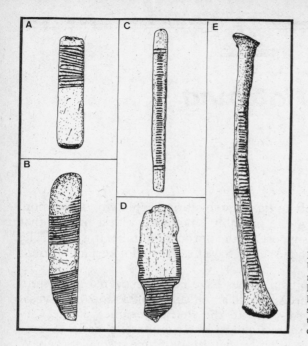

Fig. 4–1. Notched bones from the Upper Paleolithic: A and C, Aurignacian (30,000 to 20,000 B.C.); B and D, Aurignacian, from the Külna cave in Moravia, Czechoslovakia; E, Magdalenian (19,000 to 12,000 B.C.), from the Pekarna cave in Moravia.

Fig. 4–2. Tally sticks of the kind that were used by French bakers in small country towns.

it; the notches matched exactly, with the Roman numerals I, V, and X representing the weight of the bread sold."

And René Jouglet recounts a scene taking place around the turn of the century, in the Hainaut region: "The baker went from door to door in his cart. When he called out to a housewife she brought her tally, a flat, narrow piece of wood about as long as the blade of a chisel. The baker had one just like it. He put the two pieces together and, with a saw, cut a notch in their edges for each six-pound loaf of bread he delivered. This made it easy to check the account. Since the number of notches was the same on both pieces, the housewife could not remove any from both of them, and the baker could not add any."

The use of notched sticks as an instrument of credit was by no means limited to France. Its trace lingers, for example, in the English word "tally," which can mean "notch," "notched stick," "account," "to reckon," or "to agree"; and in Britain a tallyman is a merchant who sells on credit.

The system was used by the Cheremisses and Chuvashes of eighteenth-century Russia: "For lending money, they used the double tally. A tally was split lengthwise and notches representing the transaction were cut into both halves, then each partner received his half-tally. The two halves could be brought together again to make sure that no notches had been fraudulently added or cut away. In the case we are considering, when crosses or vertical lines had been cut to show the number of monetary units lent, the debtor and the creditor each put his mark on one half of the tally; this mark was chosen once and for all to be used in circumstances where we would use a signature. Then they exchanged their tallies, which to them were as valid as our strongest written commitments are to us" (Gerschel 3).

Dr. Jules Harmand wrote of the Khas Bolovens in nineteenth-century Indochina: "In their commercial transactions they use a system similar to the one used by bakers: they cut notches into two flat sticks and each party keeps one of them. But their version of the system is much more complicated, and it is hard to see how they avoid confusion. They record everything: the seller or sellers, the buyer or buyers, the witnesses, the delivery date, the nature of the objects and the price for which they are sold" (Harmand 1).

These lines by Homeyer sum up the matter: "In the past, tallies took the place of our accounting ledgers, with their debit and credit entries, and marks replaced the written names of the people concerned, particularly the debtors." And it was for this special purpose, he adds, that the use of tallies continued into modern times.

"With its strictly material nature, the tally was used for counting because it retained information better and longer than the fingers could do when the successive partial results of a sizeable operation had to be preserved. That was its main use, and the reason for its creation. Once it had been put into practice, it proved to be suitable for other applications. . . . It could not only preserve partial results till the final result had been obtained—and this takes us back to the remote time when people still knew nothing of arithmetic—but it could also preserve the final result. It was in this latter function, this new application, that it continued into our time, but it then served an *economic* purpose, and not simply an arithmetical one" (Gerschel 3).

This no doubt explains why the system was still being used in the Court of Exchequer in England as late as the nineteenth century, to certify payment of taxes and keep a record of income and expenditures, with different kinds of notches representing pounds, shillings, and pence, as well as multiples of them (fig. 4–3). In a speech on administrative reform, Charles Dickens commented on the practice not long after it was discontinued:

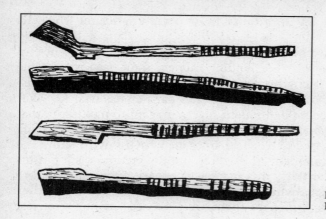

Fig. 4–3. Thirteenth-century English tally sticks.

"Ages ago a savage mode of keeping accounts on notched sticks was introduced into the Court of Exchequer and the accounts were kept much as Robinson Crusoe kept his calendar on the desert island. A multitude of accountants, bookkeepers, and actuaries were born and died. . . . Still official routine inclined to those notched sticks as if they were pillars of the Constitution, and still the Exchequer accounts continued to be kept on certain splints of elm-wood called *tallies*. In the reign of George III an inquiry was made by some revolutionary spirit whether, pens, ink and paper, slates and pencils being in existence, this obstinate adherence to an obsolete custom ought to be continued, and whether a change ought not to be effected. All the red tape in the country grew redder at the bare mention of this bold and original conception, and it took until 1826 to get these sticks abolished. In 1834 it was found that there was a considerable accumulation of them; and the question then arose, what was to be done with such worn-out, worm-eaten, rotten old bits of wood? The sticks were housed in Westminister, and it would naturally occur to any intelligent person that nothing could be easier than to allow them to be carried away for firewood by the miserable people who lived in that neighbourhood. However, they never had been useful, and official routine required that they should never be, and so the order went out that they were to be privately and confidentially burned. It came to pass that they were burned in a stove in the House of Lords. The stove, over-gorged with these preposterous sticks, set fire to the panelling; the panelling set fire to the House of Commons; the two houses were reduced to ashes; architects were called in to build others; and we are now in the second million of the cost thereof."

The practice of notching was also used for contracts. In Arabic, the root of the verb *farada* can mean either "to make notches (in wood)" or "to give someone his share (according to a contract or an inheritance)". Notched sticks were apparently the original form of contracts in China, and were displaced when writing was developed. The Chinese word for "contract" is represented by a combination of characters, one of which means "notched stick," another "knife."

In France, tallies were once used to certify that merchandise had been delivered. The body of law known as the *Code Civil*, promulgated by Napoleon in 1804, states: "Matching tallies are legally binding between persons who customarily use this means to give specific acknowledgment of the delivery or receipt of goods."

Another use of notching for numerical purposes was found among Alpine and Celtic shepherds, who cut signs into flat pieces of wood to keep account of the various categories into which they divided their animals.

"On a Walachian tally dating from 1832, the shepherd charged with keeping account of the sheep under his guard distinguished the milch ewes by a special notation: on the one hand, barren animals and, on the other, those that gave only half of normal production. In some parts of the Swiss Alps, herdsmen used carefully formed and decorated wooden tallies for recording various kinds of information, particularly the number of animals in their keeping, but they kept separate counts for cows, barren animals, ewes, goats. . . .

"We may probably assume that herdsmen all over the world must keep these same kinds of records and that only their notations differ. Depending on where they live, they will use the multiple knotted strings of the *quipu,* or a primitive form of the tally, or a small board that bears, for example, the words *Küo* (cows), *Gallier* (barren animals) and *Geis* (goats) engraved beside true tally figures. One thing remains the same: the herdsman, responsible for animals that need care and food, must know their number. But he must also keep account of them according to various categories, separating those that give milk from those that do not, and distinguishing them by age and sex. This means that he must keep not a single count, but three or four 'parallel counts' (and sometimes more) juxtaposed on the recording device" (Gerschel 4).

A large number of special-purpose tallies were once used in Switzerland. Gerschel gives us an idea of their variety:

1. *The milk tally.* "At Ulrichen, a single large tally bore the ownership marks* of all those who delivered milk, with an indication of the

---

*As Gerschel points out, "The ownership mark was necessary to users of tallies, since it connected a *name* with a *count*. . . . After first creating rights, it also proved to be capable of engendering obligations; in this sense it was the ancestor of the signature and we can see why it has remained the traditional signature of the illiterate."

quantity delivered opposite each mark. At Tavetsch, according to Gmür, everyone had his own tally, showing each quantity of milk for which he owed, under the ownership mark of the person to whom he owed for it, and what was shown under his own mark on others' tallies indicated how much was owed to him; the amount of compensation was established by comparing the tallies."

2. *The mole tally.* "In some regions, the authorities kept tallies from all burghers, with their ownership marks on them. Whenever a burgher brought in a mole, or at least its tail, it was recorded on the tally that bore his ownership mark. At the end of the year, the totals were added up and everyone was paid what was coming to him."

3. *The pasture tally.* "Pasture rights (a right, a half-right, a quarter-right . . .) were recorded next to each ownership mark. Gmür mentions a tally of this kind, dating from 1624, in the Basel Museum."

4. *The water tally.* "It entitled its holder to use water for irrigating grassland during a specified period; besides the ownership mark, it bore strokes or other signs corresponding to an hour, half an hour, twenty minutes, etc." (fig. 4–4).

Fig. 4–4. Water tally from Wallis, Switzerland. Museum für Völkerkunde, Basel.

5. *The loan tally.* "The commune or the ecclesiastical institutions had money that was lent in small sums to burghers. The debtor gave his creditor a tally bearing his ownership mark and the amount, in tally figures, of the sum borrowed.

"Whereas ordinary tallies had only numerical indications, special-purpose tallies had ownership marks as well as signs for number or quantity" (Gerschel 3).

There were also "tallies that had only ordinal numbers and ownership marks. They served to indicate the order in which everyone had to satisfy certain obligations. There were such *Kehrtesseln* . . . for the duties of night watchman, municipal worker, sexton, banner bearer, country policeman, etc." (Gerschel 3).

Calendar tallies for computing the dates of religious holidays were used in Scandinavia, Austria, Bavaria, and England from the Middle Ages till the seventeenth century.

Another use of notching is reported by Jules Harmand, with regard to the Khas Bolovens of Indochina in the nineteenth century: "Along the way I saw, a few paces from the beginning of a little path, a big gate made of felled trees and bamboo, adorned with hexagons and tufts of grass. Above the path hung a little board with regular rows of large and small notches on both edges. On the right edge, a row of twelve small notches, a row of four large ones, and another row of twelve small ones. Translation: 'Twelve days from now, any man who dares to come through our fence will either become our prisoner or have to pay us a ransom of four water buffaloes and/or twelve ticals.' On the left edge, eight large notches, eleven medium ones and nine small ones, which meant, 'Our village has eight men, eleven women, and nine children' " (Harmand 2).

In altered forms, the use of notched sticks still lingers on, for example, in the practice of French café owners who record a customer's unpaid bill by making a chalk mark on a slate each time he orders a drink. And its influence was seen during both world wars, when pilots showed the number of their aerial victories or bombing missions by painting airplanes or bombs on their fuselages. The old counting techniques are surprisingly long-lived.

5

From Pebbles to Calculation

T he word "calculate" comes from the Latin *calculus*, "small stone." (The unchanged Latin word has its original meaning in medicine, where "calculus" means "stone," as in "gallstone" or "kidney stone.") The Arabic word *iḥsā'*, meaning "enumeration" or "statistics," has a similar derivation, since it has the same root as *ḥaswa*, "pebble."

The use of pebbles or other objects as numerical devices is on the same elementary level as the use of notches in wood or bone. It is an accounting method that requires no words or memory and is based only on the principle of one-to-one correspondence, so there is no need to be intellectually prepared for the concept of counting in order to use it. As we have seen, a herdsman can use one-to-one correspondence to keep account of his animals without understanding the abstract concept of number.

A certain amount of abstraction is, however, involved in this method: twenty pebbles can just as well represent twenty men as twenty sheep or twenty measures of grain. And when the human mind became capable of conceiving the artifice of succession, it was able to consider pebbles from the standpoint of abstract counting. By means of the "pebble principle," it was then possible to count even absent persons or things. Armies could evaluate the number of their casualties or prisoners by considering only piles of stones, as was once done in Ethiopia. Before going off on an expedition, each warrior placed a stone on a pile; afterward, each survivor took a stone off the pile, and losses were judged by the number of stones remaining.

People gradually realized that this method did not go far enough to satisfy their increasing needs. To count up to 1000, for example, they would have had to gather a thousand pebbles.

That is why, once the principle of the base had been grasped, the usual pebbles were replaced with stones of various sizes to which different orders of units were assigned. If a decimal system was used, the number 1 could be represented by a small stone, 10 by a larger one, 100 by a still larger one, and so on. Other numbers could be represented by combina-

		CLAY ENVELOPES	CYLINDERS	DISKS	SPHERES	CONES	
Syria	HABUBA KABIRA	★		★	★	★	4th millennium B.C.
Iraq	TEPE GAWRA		★	★	★	★	4th mill.
Iraq	URUK	★	★	★	★	★	4th mill.
Iran	SUSA	★	★	★	★	★	4th mill.
Iran	CHOGHA MISH	★	★	★	★	★	4th mill.
Iran	TALL-I MALYAN	★	★		★	★	4th–3rd mill.
Iran	SHAHDAD	★					4th–3rd mill.
Iran	TEPE YAHYA	★		★	★	★	4th–3rd mill.
Iraq	JEMDET NASR			★		★	4th–3rd mill.
Iraq	UR			★	★		4th–3rd mill.
Iraq	TELLO		★	★	★	★	4th–3rd mill.
Iraq	FARA			★	★	★	3rd–2nd mill.
Iraq	KISH			★	★	★	2nd mill.
Iraq	NUZI	★		★	★	★	2nd mill.

TABLE 5–1. Near Eastern archaeological sites where clay envelopes and tokens of various shapes and sizes have been found.

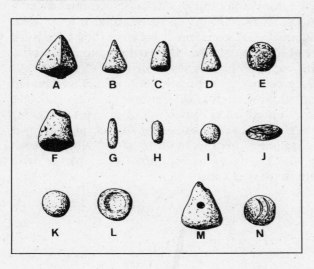

Fig. 5–1. Clay tokens dating from about 3300 B.C., found at Susa: A, B, C, E, F, J, and L, Salle Iranienne, Louvre; D, G, H, I, K, and M, discovered by the Délégation Archéologique Française en Iran, under the direction of A. Lebrun, during excavations in 1977–78.

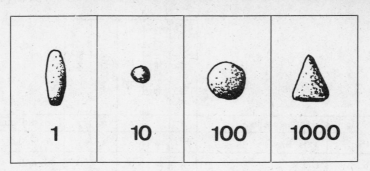

Fig. 5–2.

tions of those standard-sized stones: 30, for example, by three "10 stones," and 54 by five "10 stones" and four "1 stones."

This was a practical method, but it needed improvement to eliminate the difficulty of finding enough stones of the same size and shape for each order of units. The problem was solved, in one part of the world, at least, by making clay objects in various shapes: cones, spheres, disks, cylinders, tetrahedrons, etc. (table 5–1, fig. 5–1). Objects of this kind, made from the sixth to the second millennia B.C., have been discovered at archaeological sites in the Near East, from Asia Minor to the Indus Valley. We now know that they were used in a concrete system of numeration. Each of them was a symbol for a specific numerical value; assuming the use of a decimal system, a cylinder, for example, could be used for 1, a small sphere for 10, a larger sphere for 100, a cone for 1000, and so on (fig. 5–2).

This system was the remote ancestor of our present monetary conventions and it was also a precursor of written accounting, at least as far as the Sumerians and Elamites are concerned, because they used such numerical objects for the needs created by their intense economic activity in the fourth millennium B.C., before they had writing.

The use of pebbles for counting eventually led to the development of devices to facilitate calculation, such as the counting board, or abacus. The counting board was invented when someone had the idea of placing pebbles or other objects in columns marked on a flat surface, or in grooves dug into the ground, and assigning an order of units to the objects in each column or groove. Later, loose objects in columns were replaced with beads that could slide along parallel rods.

6

Numbers on Strings

The archives of the Incas

Early in the sixteenth century, when the Spanish conquistadores led by Francisco Pizarro arrived in South America, they found a vast empire extending nearly 2500 miles north and south in territory that is now Bolivia, Ecuador, and Peru, with an area of nearly 400,000 square miles (larger than France, Belgium, Luxembourg, the Netherlands, Switzerland, and Italy combined). At that time the civilization of the Incas, whose origins went back to the beginning of the twelfth century, was at its peak. Its prosperity and high degree of culture seemed all the more surprising at first sight because the Incas did not have the wheel, or draft animals, or even writing in the strict sense of the term.

Their success can be partly explained by their ingenuity in keeping precise records with an elaborate system of knotted strings. A *quipu* (from an Incan word meaning "knot") consisted of a main strand, about two feet long, with thinner, variously colored strings hanging from it. These hanging strings had different kinds of knots in them (fig. 6–1).

Sometimes wrongly called abacuses, *quipus* were actually memory aids answering the needs of the very efficient Incan administration. They could be used for religious, chronological, or statistical matters, and on occasion could serve as a calendar or a means of transmitting messages, with certain colors of the strings representing objects or abstract ideas (white for silver or peace, yellow for gold, red for blood or war, etc.). In particular, *quipus* were used as accounting devices, or rather as a form of concrete numeration: the colors of the strings, the number and relative positions of the knots, and the size and spacing of groups of knots had precise numerical meanings (figs. 6–2, 6–3, 6–4). They could show the results of various counts (for which a decimal system was used) and were therefore a valuable statistical tool (fig. 6–5) for dealing with such things as military operations, tribute, harvests, evaluating the number of animals killed in the immense annual drives, deliveries, censuses, births and deaths,

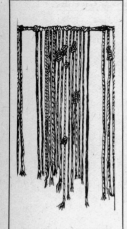

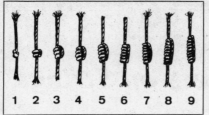

| 1 | 2 | 3 | 4 | 5 | 6 | 7 | 8 | 9 |

(*left*) Fig. 6–1. A Peruvian *quipu*.

(*above*) Fig. 6–2. Incan *quipu* method of indicating the first nine numbers on string.

(*right*) Fig. 6–3. Peruvian *quipu* showing the number 3643.

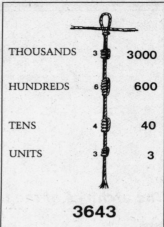

THOUSANDS	3	**3000**
HUNDREDS	6	**600**
TENS	4	**40**
UNITS	3	**3**

3643

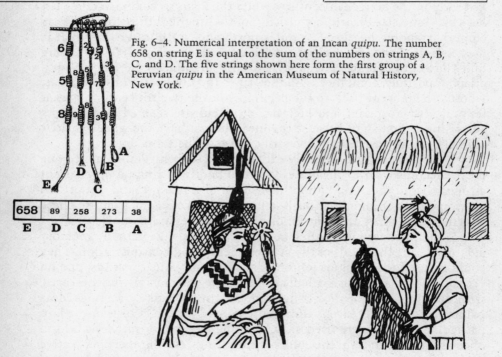

Fig. 6–4. Numerical interpretation of an Incan *quipu*. The number 658 on string E is equal to the sum of the numbers on strings A, B, C, and D. The five strings shown here form the first group of a Peruvian *quipu* in the American Museum of Natural History, New York.

| 658 | 89 | 258 | 273 | 38 |
| E | D | C | B | A |

Fig. 6–5. An Incan *quipucamayoc* reporting to a royal functionary and using a *quipu* to describe the result of a statistical enumeration.

determining the tax base of an administrative district, budgets, and inventories of resources.

"This concern for statistics has been cited as proof of *the socialistic nature of the Incan empire*. But we must not let ourselves be misled by words. Censuses of the population, divided into age groups, and evaluation of the wealth produced by forced labor answered very simple needs. The Incas could not have undertaken their conquest or built their numerous palaces and fortresses if they had not had information on available manpower and the resources necessary for maintenance" (Métraux).

Fig. 6–6. An Incan *quipucamayoc* manipulating a *quipu*. From a drawing by the seventeenth-century Indian chronicler G. Poma De Ayala.

Each district of the empire had royal officials called *quipucamayocs* ("knot keepers," fig. 6–6) who "were required to furnish the government with information on various important matters. One had charge of the revenues, reported the quantity of raw material distributed among the laborers, the quality and quantity of the fabrics made from it, and the amount of stores, of various kinds, paid into the royal magazines. Another exhibited the register of births and deaths, the marriages, the numbers of those qualified to bear arms, and the like details in reference to the population of the kingdom. These returns were annually forwarded to the capital, where they were submitted to the inspection of officers acquainted with the art of deciphering these mystic records. The government was thus provided with a valuable mass of statistical information, and the skeins of many-colored threads, collected and carefully preserved, constituted what might be called the national archives" (Prescott).

Use of the Incan *quipu* persisted a long time in Bolivia and Ecuador. In the nineteenth century, on the high Peruvian plateaus and on some farms and ranches, herdsmen still counted by means of *quipus*. In a group of white strings they noted the sheep and goats, usually putting rams on the first string, lambs on the second, goats on the third, kids on the fourth, ewes on the fifth, and so on. They used a group of green strings for cattle, with bulls on the first, milch cows on the second, barren cows on the third, then calves according to age and sex (fig. 6–7).

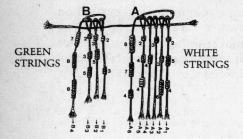

GREEN
STRINGS

WHITE
STRINGS

Fig. 6–7. Use of the *quipu* by herdsmen on the high Peruvian plateaus in the nineteenth century to keep account of their animals: Group A (white strings): $A_1 = 254$ rams, $A_2 = 36$ lambs, $A_3 = 300$ goats, $A_4 = 40$ kids, $A_5 = 244$ ewes, $A_6 = $ total, 874 sheep and goats; Group B (green strings): $B_1 = 203$ bulls, $B_2 = 350$ milch cows, $B_3 = 235$ barren cows, $B_4 = $ total, 788 cows and bulls.

The Indians of Bolivia and Peru still use a similar device: the *chimpu*, a direct descendant of the Incan *quipu* (fig. 6–8). On the *chimpu*, numbers below 10 are formed by making as many knots in a string as necessary. Two strings knotted together are used for tens, three strings for hundreds, four for thousands, and so on. Six knots, for example, represent 6, 60, 600, or 6000, depending on whether they are tied in one, two, three, or four strings at once.

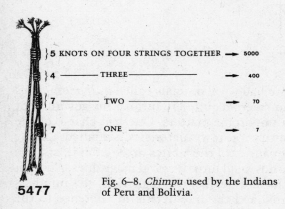

5 KNOTS ON FOUR STRINGS TOGETHER → 5000

4 ———— THREE ———— → 400

7 ———— TWO ———— → 70

7 ———— ONE ———— → 7

5477

Fig. 6–8. *Chimpu* used by the Indians of Peru and Bolivia.

Use of knotted strings in other parts of the world

Knotted strings were used as a numerical device in Greece and Persia in the first millennium B.C. The Greek historian Herodotus (485–425 B.C.) writes that when King Darius I of Persia (reigned 522–486 B.C.) led an expedition against a group of Scythian horsemen he ordered some Greek soldiers allied with him to guard a strategically important bridge. Before leaving them, he gave them a thong with sixty knots tied in it, telling them to untie one knot each day, and that if he had not come back by the time the last knot was untied they were to board their ships and go home.

When Palestine was under Roman domination in the second century A.D., tax collectors used a large rope (probably formed by combining a certain number of strings) as a register. The receipt given to each taxpayer was a string knotted in a particular way.

Such record-keeping systems were most likely used in China for a long time before writing was fully developed. According to Chinese tradition, the semilegendary Shen Nong, one of the three emperors supposed to have established the foundations of Chinese civilization, took a hand in developing a system of knotted strings and taught it as a means of keeping accounts and recording events. There is a reference to it in the *I Ching* ("Book of Changes"), a classic work thought to date from the first half of the first millennium B.C.: "In the most ancient times, men were governed by means of the system of knotted strings." There is also a mention of it in the *Tao Te Ching*, traditionally attributed to Lao-tse (c. 604–531 B.C.).

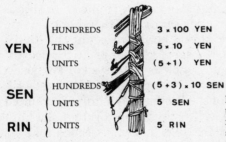

YEN	HUNDREDS	3 × 100 YEN
	TENS	5 × 10 YEN
	UNITS	(5 + 1) YEN
SEN	HUNDREDS	(5 + 3) × 10 SEN
	UNITS	5 SEN
RIN	UNITS	5 RIN

356 YEN , 85 SEN , 5 RIN

Fig. 6–9. System for noting sums of money used in the Ryukyu Islands, notably by workers in Okinawa and tax collectors in Yaeyama. The sum shown here is 356 yen, 85 sen, and 5 rin (1 yen = 100 sen; 1 sen = 10 rin).

In the Far East, the use of knotted strings for numerical purposes has not yet totally disappeared (fig. 6–9). It is still found, for example, in the Ryukyus, an island chain that includes Okinawa, between Taiwan and Kyushu, Japan. "It is with such a system of knots in straw ropes that workers in certain mountainous areas of Okinawa keep account of their days of work, noting the sums owed to them, etc. . . . In the town of Shuri, pawnbrokers record a transaction by means of a long string made of reeds or bark, divided into two halves by another string tied to its middle. The knots in the upper half indicate the month when the loan was made, those in the lower half indicate the amount. On the island of Yaeyama, the yields of harvests were once calculated and recorded by similar procedures, and each taxpayer received notification in the form of a knotted string indicating the amount due" (Février).

Similar practices are found in the Caroline Islands (near Tahiti), the Hawaiian Islands, western Africa (particularly among the Yebu of Nigeria), and, on the other side of the world, in certain tribes of North American Indians, including the Yakima of eastern Washington, the Walapai and Havasupai of Arizona, the Miwok, Northern Maidu, and Southern Maidu of California, and the Apaches and Zuñis of New Mexico.

Till the end of the nineteenth century German millers used knotted strings in their transactions with bakers (fig. 6–10).

Fig. 6–10. Knotted strings used by German millers till the late nineteenth century. (Shown here is the method used in Baden.)

Several religions use rosaries composed of knots to indicate the number and nature of prayers. In their religious services, Tibetan monks use a bundle of 108 knots (or a necklace of 108 beads) to count the sacred number 108. Only a few decades ago, the same kind of practice was still common among certain Siberian peoples, such as the Ostiaks, Tunguses, and Yakuts. And there is a Moslem tradition, transmitted by Ibn Sa'ad, according to which Fatima, daughter of the Prophet Mohammed, counted the ninety-nine attributes of Allah, as well as the supererogatory blessings said after obligatory prayer, on knotted strings (not on a beaded rosary).

Fig. 6–11. Phylacteries and prayer shawl worn by a praying Jewish man.

Finally, we can mention a practice of the Jewish religion. In observance of the Law (Exodus 13:16, Deuteronomy 6:8 and 11:18), during morning prayer every adult Jewish male is required to wear phylacteries (small leather boxes containing strips of parchment on which passages from the Hebrew Scriptures are written) strapped to his forehead and his left arm, and a fringed prayer shawl over his shoulders (fig. 6–11). The four outer strands of the fringes on the prayer shawl are always knotted to represent a certain number: 26 in the Sephardic (eastern) tradition, 39 in the Ashkenazic (western) tradition. According to a system of numerically evaluating the letters of the Hebrew alphabet, 26 is the numerical value of YHWH (Yahweh, the ineffable name of God), and 39 is that of *YHWH eḥad* ("Yahweh is one"; fig. 6–12). Some rabbis point out that 39 is also the numerical value of *tal* ("dew"), the Hebrew word from which *tallith* ("prayer shawl") is derived, and that when a Jew wears his prayer shawl with 39 knots he expresses the oneness of God and is able to hear all of God's words, which "shall distil like dew, like fine rain upon the grass" (Deuteronomy 32:2).*

Fig. 6–12.

   Thus knotted strings not only have served as a means of concrete numeration but have also been used in such things as business transactions, administrative archives, contracts, receipts, and calendars. And although they do not constitute writing in the strict sense of the term as it is used by linguists, "they are comparable to writing in that they share its function: to preserve memory of the past and assure the permanence of contractual bonds between members of society" (Alleton).

---

*All biblical quotations are from *The New English Bible*, Oxford University Press and Cambridge University Press, 1970 (translator's note).

7

Number, Value, Money

When people lived in small communities that took everything they needed from nature, there must have been very little communication between different societies. But because of the uneven distribution of various natural products, commercial exchange gradually became necessary.

The first type of commercial exchange was barter, in which things were traded directly, without the use of money. Sometimes, between groups on unfriendly terms with each other, these direct exchanges were carried out in the form of "silent barter." In Siberia, for example, which had a barter economy until relatively recent times, "the foreign merchant laid out the merchandise he wanted to exchange and abandoned it; the next day he found beside it, or in place of it, the products of the region—mainly furs—that were offered in exchange. If he was satisfied with their value, he took them; if not, the next day he found a larger amount of them and he either took it or left it, depending on his estimate of its worth. The operation might go on for days and sometimes it ended without an exchange, if the parties could not agree" (Hambis).

As communication between groups intensified and the size of transactions increased, barter—in which goods are often traded according to an individual's whim, or by virtue of custom, or after endless discussion—became cumbersome. A need was felt for a relatively stable system of establishing equivalences, based on the definitions of a few fixed units or standards in terms of which it would always be possible to estimate a certain value, not only for purely economic transactions but also—and perhaps especially—for settling important juridical problems, such as the "bride price," the "blood price" (to be paid for violence resulting in death), and the "theft price."

The first trade unit accepted in pre-Hellenic Greece, and among the Romans before the fourth century B.C., seems to have been the cow or bull. In Homer's *Iliad* (seventh century B.C.), a woman "skilled in countless tasks" is evaluated at four head of cattle, Glaucus's brass armor at nine, and Diomed's gold armor at 100. In a list of rewards, an engraved silver cup, a cow, and half a talent of gold are given in decreasing order of value. The Latin word *pecunia*, "wealth," from which our word "pe-cuniary" is derived, came from *pecus*, "cattle." And the original meaning of *pecunia* was "wealth in cattle."

In India, at the time when the Vedic texts were written, fees and ritual sacrificial offerings were evaluated in cows. During the Shang period in China (sixteenth to eleventh centuries B.C.), values were estimated in stone weapons and tools, tortoise shells, seashells, leather, furs, animal horns, and grain. Among the Aztecs in pre-Columbian Mexico, "certain commodities normally served as criteria of value and mediums of ex-change: the *quachtili*, a roll of cloth, with its multiple the 'load' (twenty rolls); the cacao bean, a form of 'small change,' with its multiple the *xiquipilli*, a bag containing, or supposed to contain, 8000 beans; quills filled with gold dust" (Soustelle 2). The Mayas used cotton, cacao, bi-tumen, jade, stone beads, pottery, jewels, and gold.

Until recently, the Dogons of Mali still used cowry shells as their main standard for evaluating merchandise. "A chicken," said a Dogon named Ogotemmeli, quoted by M. Griaule, "was worth three times eighty cowries; a goat or a sheep, three times 800; a donkey, forty times 800; a horse, eighty times 800; an ox, 120 times 800." But, Ogotemmeli pointed out, "cowries were not used that way in very old times. People began by trading pieces of cloth for animals or things. The cloth was money. It was measured in hand's-breadths, twice eighty threads wide. A sheep was worth eight cubits of three hand's-breadths. . . . Later, the values of things were set in cowries by Nommo the Seventh, the master of words."

In some Pacific islands, goods were evaluated with the aid of pearl and seashell necklaces. For some Indian tribes of northeastern America, notably the Iroquois and Algonquins, the standard was wampum, strung shells. In some parts of Siberia, till the beginning of this century, "pur-chases were made by trading furs, which were regarded as monetary units. Till 1917, this system was used by the Russian government for collecting taxes from the people of those regions" (Hambis).

Such methods had serious practical drawbacks, however. As trade developed, different metals—either as ingots or in the form of tools, ornaments, or weapons—played an increasingly large part in commercial transactions and gradually became the medium of exchange preferred by buyers and sellers. Evaluation was in terms of standard units of weight for various metals. Negev points out that the shekel of silver (1 shekel = 11.4 grams) repeatedly appears in the Old Testament as a standard of value. Abraham bought a plot of land from Ephron the Hittite for 400 shekels of silver (Genesis 23:16). A servant of Saul's father offered to give a quarter-shekel of silver to a "man of God" in exchange for his help (1 Samuel 9:8). Fines were established in shekels of silver (Exodus 21:32).

In ancient Egypt, merchandise was often evaluated and paid for in metal (copper, bronze, and sometimes gold or silver), which was weighed in the form of nuggets, grains, ingots, or rings. The main unit of weight was the *deben* (ninety-one grams). To facilitate certain evaluations, fractions of the *deben* were also used as units: the *shat*, a twelfth of a *deben*, during the Old Kingdom (2780–2280 B.C.), and the *quite*, a tenth of a *deben*, during the New Kingdom (1552–1070 B.C.).

An Old Kingdom contract, fixing terms of payment for the temporary hiring of a servant, gives values by the standard of the *shat* of bronze:

        8 bags of grain: value 5 *shats*
              6 goats: value 3 *shats*
               silver: value 5 *shats*
                total value 13 *shats*

Another example is an account dating from the New Kingdom in which values are given in *debens* of copper:

Sold to Hay by Nebsmen:
1 ox, value 120 *debens* of copper

Received in return:
2 jars of grease, value 60 *debens*
5 fine loincloths, value 25 *debens*
1 garment of southern linen, value 20 *debens*
1 tanned hide, value 15 *debens*

The ox was sold for the equivalent of a certain amount of copper, measured in grease, loincloths, etc. This was not barter, in the sense of an exchange based only on values assigned by the two parties, or on customary and often arbitrary practices; instead, it was a transaction within a genuine economic system that had established "fair prices" for merchandise.

In a letter dating rom 1800 B.C. and preserved in the royal archives of the Mesopotamian city of Mari, Ishki Addu, King of Qatna, reproaches his "brother" Ishme Dagan, King of Ekallatim, for having sent him inadequate payment in tin for two valuable horses: "Thus speaks Ishki Addu, your brother. This should not be said, but I must say it to relieve my heart. You asked me for two horses which you desired, and I had them taken to you. Now you have sent me only twenty minas of tin. Did I not grant your desire completely and without discussion? Yet you have sent me that small amount of tin. The value of those horses, here in Qatna, is 600 shekels of silver, and you have sent me twenty minas of tin. If someone learned of this, what would he say?"

Since a shekel of silver was worth three to four minas of tin at that time, Ishki Addu's indignation is understandable.

Such exchanges did not involve the use of money, in the sense of metal coins specifically intended for commercial transactions, with a certain weight and value guaranteed by a stamp that only the government is allowed to use. Money in that sense was invented in the first millennium B.C. (probably by the Lydians). Till then, there was only a standard weight taken as a unit of value in terms of which payment for goods and services was made. Various metals (in ingots or some other form), weighed in reference to that unit, could be used for such things as wages and fines, and serve as a medium of exchange.

To give an idea of what commerce was like in the days when goods were bought and sold this way, here is a description of a market in ancient Egypt, written by Gaston Maspero on the basis of documents that have come down to us:

> Early in the morning, peasants came in from the surrounding countryside in endless lines and took up their positions in a public square reserved for their use since time immemorial. Sheep, geese, goats, and long-horned cattle were grouped in the center, to await buyers. Market gardeners, fishermen, potters, craftsmen, and hunters of birds and gazelles squatted at the edges of the square, along the houses, and displayed their wares in reed baskets or on low tables: fruit and vegetables, bread and cakes baked the night before, meat either raw or cooked in various ways, cloth, perfumes, jewelry, all the necessities and superfluities of everyday life. It was an opportunity for workers as well as more affluent people to buy more cheaply than they could do in permanent shops, and they took advantage of it according to their means.
>
> The buyers brought with them products of their work—a new tool, a pair of shoes, a mat, jars of ointment or liqueur—or strands of cowries or little boxes filled with rings of copper, silver, or even gold, weighing one *tabnu* [since Maspero's time, this term has been replaced with the more accurate reading *deben*], which they proposed to trade for whatever they needed.
>
> When a buyer was interested in a large animal or an object of considerable value, the bargaining was long, ardent, and tumultuous: he and the seller had to agree not only on the price but also on its composition, and draw up an "invoice" in which beds, canes, honey, oil, pickaxes, and clothing served as equivalents of a bull or a donkey.
>
> Small retail commerce did not require such complicated calculations. Let us imagine two men who have stopped in front of a peasant displaying onions and wheat in baskets. [Some details of the scenes that follow are taken from the Egyptian tomb painting shown in figure 7–1.] The first man's trading capital seems to consist only of two necklaces of multicolored glass or enameled clay beads; the second man holds two fans: a rounded one with a wooden handle and a triangular one of the kind used by cooks to make the fire burn hotter.

Fig. 7–1. Market scenes in an Egyptian tomb painting of the Old Kingdom, Fifth or Sixth Dynasty (c. twenty-fifth century B.C.), Tomb of Teteka, at the north end of the Saqqara necropolis (between Abusir and Saqqara).

"Here's a beautiful necklace that will suit you perfectly. It's just what you need," the first customer says to the peasant.

"Here are two fans," says the second.

The peasant is not disconcerted by this double assault. Proceeding methodically, he takes one of the necklaces to examine it at leisure.

"Let me have a look at it to see what it's worth."

One asks too much, the other offers too little; by way of concessions on both sides, they will eventually reach agreement on the number of onions or the amount of grain that exactly corresponds to the value of the necklace or the fan.

Farther on, a customer wants to exchange a pair of sandals for perfume, and praises his property in good faith.

"These are very well-made sandals."

But, having no need for sandals at this time, the merchant demands a string of cowries for his little jars.

"It takes only a few drops of this perfume to make a delightful smell," he says persuasively.

A woman with two jars, probably containing ointment that she has made, holds them under the nose of a squatting man.

"Here's a fragrance you can't resist."

Behind this group, two men debate the relative merits of a bracelet and a packet of fishhooks; a woman holding a box bargains with a necklace merchant; another woman tries to bring down the price of a fish that is being cleaned in front of her.

Exchanging merchandise for metal requires two or three more operations than ordinary trading. The rings or folded sheets that represent the *tabnu* and its multiples do not always contain the standard amount of gold or silver, often being too light. They must be weighed in each transaction to determine their real value, and the interested parties do not miss such a good chance for a heated argument. When they have spent a quarter of an hour shouting that the scales are not working right, that the weighing was done carelessly and must be done over, they finally come to terms and separate, more or less satisfied with each other. Now and then a shrewd, unscrupulous trader adulterates the precious metal in the rings by mixing it with as much base metal as it can take without showing his fraud. An honest merchant who thinks he is trading an object for a certain amount of gold, eight *tabnus*, let us say, and actually receives only eight *tabnus* of an alloy that looks like pure gold but is one-third silver, loses nearly a third of his merchandise without realizing it.

For a long time, fear of fraud was a factor in limiting the use of *tabnus* among the common people and maintaining the system of buying and selling in public markets by trading natural products or handmade articles.

Money in the modern sense of the term made its appearance when metal was cast in coins of uniform weight, small enough to be handled

Fig. 7–2. Greek coins: (*left*) Silver tetradrachma from Agrigento, c. 415 B.C.; (*right*) Tetradrachma from Syracuse, c. 310 B.C. (Agrigento Museum.)

easily, and stamped with the official mark of a public authority that certified their weight and composition.

This invention was made in Greece and Anatolia in the seventh century B.C., and evidently in China at about the same time, in either the seventh or sixth century B.C., during the Chou period. Who first had the idea for it? Some believe that Phidon, King of Argos, in the Peloponnesus, introduced the new system in his own city and Aegina in about 650 B.C. But most specialists agree that credit for the invention should go to Asian Greece (or Asia Minor), and probably to Lydia (fig. 7–2). Be that as it may, because of its many advantages the use of money spread rapidly in Greece, Phoenicia, Rome, and numerous other places.

8

The Abacus

"Abacus" is an unchanged Latin word, from the Greek *abax*, "board," "table," or "tablet," and perhaps also from the Semitic *abq*, "sand" or "dust." The Romans applied the word *abacus* to various objects, all with the common characteristic of having a flat surface: tables used in different kinds of games, sideboards, dressers, and the calculating device still known as the abacus.

There have been different forms of the abacus: the counter abacus used by the Greeks (fig. 8–7), the Etruscans (fig. 8–1), the Romans (fig. 8–3), and the Christian peoples of Europe from the Middle Ages to the end of the eighteenth century (figs. 8–19, 8–22, and 8–23); the dust abacus, used by the Greeks and Romans (fig. 8–2) as well as the Arabs and Persians, and perhaps also by Indian mathematicians; the abacus with counting rods, once used in China and Japan (fig. 8–25); the hand abacus of the Romans (figs. 8–12, 8–14).

Greek and Roman abacuses

The most common were tables or boards on which divisions into several parallel lines or columns separated the decimal orders (units, tens, hundreds, etc.). Counters (called *calculi* in Latin and *psephoi* in Greek) were used to represent numbers and perform calculations.

In the Roman counter abacus (figs. 8–3, 8–4), each column or row usually symbolized one of the successive powers of 10. Going from right to left, the first column was associated with units, the second with tens, the third with hundreds, and so on. To represent a given number, the reckoner placed in each column as many counters as there were units in the corresponding decimal order (fig. 8–5A). Sometimes each column was divided into two parts: in the lower part a counter designated a unit corresponding to the decimal order of the column, and in the upper part a counter designated half of a unit of the next column on the left: 5 for the upper part of the first column on the right, 50 for the second, 500 for the third, and so on (fig. 8–5B).

Fig. 8–1. Etruscan cameo, carved
stone, showing a man reckoning
on an abacus and holding a tablet
with Etruscan numerals on it.
National Library, Paris.

Fig. 8–2. Another type of abacus used by the Greeks and
Romans, mentioned by Plutarch and Apuleius, among
others: a tablet in a frame with raised edges, filled with
fine sand. The reckoner wrote on it with his finger or a
pointed instrument. Mosaic showing Archimedes (287?–
212 B.C.) trying to protect an abacus of this type from a
Roman soldier who is about to kill him. Städtisches
Kunstinstitut, Frankfurt am Main.

With these divisions, arithmetical operations could be performed by
moving counters: adding or taking them away, shifting them from one
column to another. To add one number to another on the original Roman
abacus, both numbers were indicated with counters and the result, obtained concretely, was read. If the number of counters in a column was
ten or more, ten of them were replaced with one in the next column to
the left (fig. 8–5A). (With the simplified Roman abacus, as shown in fig.
8–5B, this procedure was modified: five counters in the lower part of a
column were replaced with one in its upper part, and two counters in the
upper part of a column were replaced with one in the lower part of the
next column to the left.) Subtraction was done by an analogous procedure,
and multiplication by taking the sum of several partial products.

Let us take the example of multiplying 720 by 62 on the simplified
Roman abacus. We begin by forming the multiplicand, 720, and the multiplier, 62, as shown in figure 8–6A. Next, by the 7 of the multiplicand
(which has the value of 700), we multiply the 6 of the multiplier (value
of 60); we find 42,000 and we now place two counters in the fourth column
and four in the fifth (fig. 8–6B). By the 7 of the multiplicand (which still
has the value of 700), we multiply the 2 of the multiplier and indicate

Fig. 8–4. Roman abacus counters. (From originals in the Städtisches Museum, Wels, Austria.)

Fig. 8–3. Roman counter abacus. (Reconstruction.)

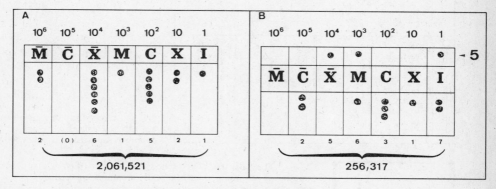

Fig. 8–5. The principle of the Roman counter abacus. A: original abacus; B: simplified abacus.

the result, 1400, on the abacus by placing four counters in the third column and one in the fourth (fig. 8–6C), then we eliminate the 7 of the multiplicand. Next we multiply the 6 of the multiplier (value of 60) by the 2 of the multiplicand (value of 20) and indicate the result, 1200, on the abacus by placing two counters in the third column and one in the fourth (fig. 8–6D). We multiply the 2 of the multiplier by the 2 of the multiplicand (value of 20) and place four counters in the second column (fig. 8–6E). Finally, by reducing the counters we obtain the result of multiplying 720 by 62: 44,640 (fig. 8–6F).

The Salamis abacus (fig. 8–7), found in 1846 on the island of Salamis by the Greek archaeologist Alexandre Rangabé (1810–1892), will give us some idea of the counter abacuses used by the ancient Greeks. It is a large slab of white marble, about five feet long and thirty inches wide, with five parallel lines chiseled into it and, below them, eleven parallel lines divided in half by a perpendicular line. The third, sixth, and ninth of these eleven lines are marked by crosses at their points of intersection with the perpendicular line.

Three nearly identical series of Greek characters are on three sides

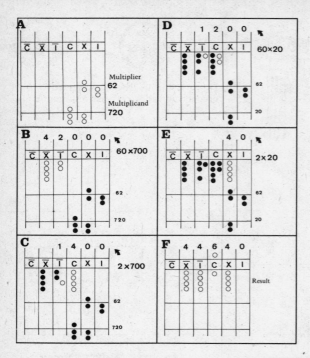

Fig. 8–6. Procedure for multiplying 720 by 62 on the Roman counter abacus (simplified type).

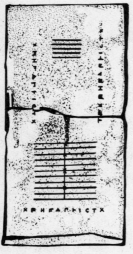

Fig. 8–7. The Salamis abacus. Once wrongly identified as a game board of some sort, it is actually a computing device analogous to the Roman counter abacus. Date uncertain: fifth or sixth century B.C. (Epigraphical Museum, Athens.)

of the slab. The most complete of the three series contains the following thirteen symbols:

Τ 𝐗̅ Χ 𝖥 Η 𝖥̅ Δ Γ Ⱶ Ι C Τ Χ

These are numerical signs for noting sums of money in talents, drachmas, obols, and khalkoses (1 talent = 6000 drachmas; 1 obol = 1/6 drachma; 1 khalkos = 1/8 obol). The values of the signs are shown in table 8–1.

Sign	Value	Description
T	1 talent	Initial of Τάλαντον , "talent"
Ⴖ	5000 drachmas	
X	1000 drachmas	Initial of Χίλιοι "one thousand" (drachmas)
Ⴖ	500 drachmas	
H	100 drachmas	Initial of Ηέκατον, "one hundred" (drachmas)
Ⴖ	50 drachmas	
Δ	10 drachmas	Initial of Δέκα , "ten" (drachmas)
Γ	5 drachmas	Initial of Πέντε , "five" (drachmas)
Ⱶ	1 drachma	
I	1 obol	Mark for 1 (1 obol)
C	1/2 obol	Half of the letter O, initial of Ὀβόλιον
T	1/4 obol	Initial of Τεταρτημόριον
X	1 khalkos	Initial of Χαλκοῦς
	1 talent = 6000 drachmas	
	1 drachma = 6 obols	
	1 obol = 8 khalkoses	

TABLE 8–1.

On the Salamis abacus, the talent and the khalkos, the largest and smallest units of the ancient Greek monetary system, are at the two ends of the scale. The Greek historian Polybius (210–128 B.C.) was no doubt referring to such an abacus when he put these words into the mouth of Solon (c. 638–559 B.C.): "Those who live at the courts of kings are like counters on a counting board: the will of the reckoner makes them worth either a khalkos or a talent."

A reckoner using the Salamis abacus must have stood facing one of its long sides and placed counters in the columns marked off by its parallel lines. The counters changed value according to their positions. The principle was analogous to that of the Roman counter abacus, in which each column was associated with a decimal order.

The four columns on the right, as shown in figure 8–8, were reserved for fractions of the drachma, the first one on the right corresponding to khalkoses (1/48 of a drachma), the second to quarter-obols (1/24 of a drachma), the third to half-obols (1/12 of a drachma), and the fourth to obols (1/6 of a drachma). The next five columns (to the right of the center cross in fig. 8–8) were associated with multiples of the drachma: the first on the right corresponded to units, the second to tens, the third to hundreds, and so on. In the lower part of each of these columns, a counter stood for a unit corresponding to the order of that column, and in the upper part it stood

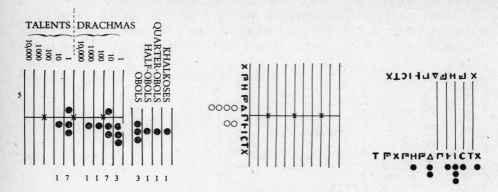

(left) Fig. 8–8. Principle of the Salamis abacus. The sum shown here is 17 talents, 1173 drachmas, 3 obols, 1 half-obol, 1 quarter-obol, and 1 khalkos.

(right) Fig. 8–9. On the Salamis abacus, to multiply, for example, 121 drachmas, 4 obols, 1 half-obol, and 1 khalkos by 42, the reckoner first indicated the multiplier, 42, by placing the corresponding counters (white here) under the appropriate numerical signs in the series on the left. Then he indicated the multiplicand in the same way, placing the corresponding counters (black here) under the numerical signs of one of the two series on the right. Finally, proceeding in a way analogous to the Roman technique (fig. 8–6), he obtained the result.

for five units of the column. The last five columns (to the left of the center cross in fig. 8–8) were respectively associated with talents, tens of talents, hundreds, and so on. Since a talent was equivalent to 6000 drachmas, counters representing 6000 drachmas were replaced with one counter in the talents column. By means of these different divisions, the reckoner could perform addition, subtraction, and multiplication (figs. 8–9, 8–10).

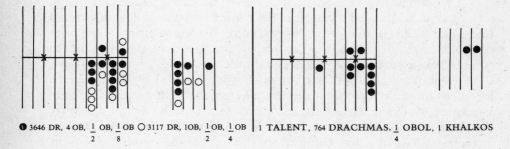

Fig. 8–10. Addition on the Salamis abacus. In this example, the sum of 3646 drachmas, 4 obols, 1 half-obol, and 1 khalkos (black counters) is added to the sum of 3117 drachmas, 1 obol, and 1 quarter-obol (white counters). By reducing the counters in accordance with established monetary conventions, the result was obtained: 1 talent, 764 drachmas, 1 quarter-obol, and 1 khalkos.

The first pocket calculator

Roman *calculatores* (the word *calculator* could designate either a calculating teacher or, in a large partrician household, a man who kept the accounts) also used another kind of abacus, better designed than those we have just seen: the hand abacus, which we might call the "pocket abacus." It consisted of a small metal tablet with parallel grooves, usually nine of them, each corresponding to an order of units, in which spherical counters slid (fig. 8–11). A bas-relief on a Roman gravestone of the first century A.D. shows a young *calculator* reckoning on an abacus of this type, at his master's dictation.

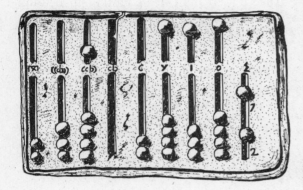

Fig. 8–11. Roman hand abacus. National Library, Paris.

To see the principle of the hand abacus at work let us take the example of the one shown in figure 8–11. For the moment, we will disregard the two grooves on the right. Each of the seven others is divided into a lower part containing four sliding counters and a shorter upper part containing only one. Between these two rows of grooves is a series of signs, each associated with a groove; they are numerals representing successive powers of 10 in the classic Roman numeration, and by means of them bankers and tax collectors reckoned in asses, sesterces, and denarii.*

X̄	(((ɔ)))	((ɔ))	(ɔ)	C	X	I
10^6	10^5	10^4	10^3	10^2	10	1

*The Roman monetary unit was the bronze as, whose weight diminished from the beginning of Roman money, in about the sixth century B.C., till the end of the empire. Its successive weights were, in grams: 273, 109, 27, 9, and 2.3. Its multiples were the sesterce (silver in the third century B.C., bronze under Caesar, brass during the empire), the denarius (silver), and, from the time of Caesar onward, the aureus (gold). In the second century B.C.: 1 denarius = 2.5 asses; 1 sesterce = 4 denarii = 10 asses. After the monetary reform in that same century: 1 sesterce = 4 asses; 1 denarius = 4 sesterces = 16 asses; 1 aureus = 25 denarii = 400 asses.

Each of these seven grooves was thus associated with a power of 10. Starting from the right, the third of the nine grooves was associated with units, the fourth with tens, the fifth with hundreds, and so on (fig. 8–12). When the units of a certain order did not go beyond four, the reckoner indicated them in the corresponding lower groove by pushing up as many counters as necessary. When they reached or went beyond five, he first pushed down the counter in the upper groove (representing five units in the order of that groove), then pushed up as many counters as necessary in the lower groove. Assuming that he was counting in denarii, the number indicated on the abacus in figure 8–12 corresponds (disregarding the first two grooves on the right) to the sum of 284 denarii: four counters pushed up in lower groove III represent four units, or four denarii; the counter in upper groove IV pushed down, and three counters pushed up in lower groove IV, represent (5 + 3) tens, or 80 denarii; and two counters pushed up in lower groove V represent two hundreds, or 200 denarii.

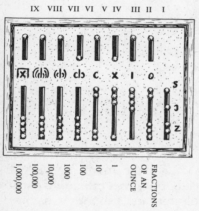

Fig. 8–12. The principle of the Roman hand abacus. The one represented here once belonged to the German Jesuit Athanasius Kircher (1601–1680) and is now in the Museo delle Terme, Rome.

$\frac{1}{2}$ OUNCE

$\frac{1}{4}$ OUNCE

$\frac{1}{3}$ OUNCE

Fig. 8–13. European reckoning tables of the Renaissance.

Now for the first two grooves on the right. They were used for divisions of the as, or of the units it represented.* The second groove, marked with the sign O, had an upper part containing one counter and a lower part containing five; it served to indicate multiples of an ounce or twelfths of an as, each lower counter representing one ounce and the upper counter six ounces (a convention that made it possible to count up to $^{11}/_{12}$ of an as). The first groove, marked off into three parts and containing four counters, served to indicate a half-ounce, a quarter-ounce, and a duella, or third of an ounce. If the upper counter was raised to the highest of the three signs next to the groove, it indicated a half-ounce or $^{1}/_{24}$ of an as. If the second counter from the top was placed beside the middle sign, it indicated a quarter-ounce or $^{1}/_{48}$ of an as. Finally, each of the last two counters indicated a third of an ounce or $^{2}/_{72}$ of an as when it was placed next to the bottom sign.

The four counters in the first groove were probably distinguished by three different colors (one for a half-ounce, another for a quarter-ounce, and another for a third of an ounce) if they were together in the same groove (as in the abacus shown in fig. 8–12). In some abacuses, however, these counters were in three separate small grooves.

S OR **Ƨ** OR **Ƹ** $\left[\text{SIGN FOR } \textbf{(As) Semuncia: } \frac{1}{24}\text{(OF AN AS)} \right]$

Ɔ OR **❱** OR **7** $\left[\text{SIGN FOR } \textbf{(As) Sicilicus:} \frac{1}{48}\text{(OF AN AS)} \right]$

Z OR **2** OR **2** $\left[\text{SIGN FOR} \left\{ \begin{array}{l} \textbf{(As) Duæ sextulæ} : \frac{2}{72} \text{ (OF AN AS)} \\ \text{AND \quad FOR} \\ \textbf{Duella} : \text{DUELLA (THIRD OF AN OUNCE)} \end{array} \right. \right]$

Abacists versus algorists: the Renaissance

Use of the abacus was widespread in Europe from the Middle Ages till relatively recent times. The kind of abacus still in use during the Renaissance was a table on which lines marked off the different decimal

---

*In Roman commercial arithmetic, fractions of a monetary unit or a unit of weight were always related to the as—the basic arithmetical unit, subdivided into twelve equal parts called ounces (*unciae* in Latin)—which gave its name to the corresponding whole number, whatever its nature. Each multiple or submultiple of the as, or of the unit it represented, was given its own name. The submultiples, for example, were:

$^{1}/_{2}$: *as semis*	$^{1}/_{7}$: *as septunx*	$^{1}/_{12}$: *as uncia*
$^{1}/_{3}$: *as triens*	$^{1}/_{8}$: *as octans*	$^{1}/_{24}$: *as semuncia*
$^{1}/_{4}$: *as quadrans*	$^{1}/_{9}$: *as dodrans*	$^{1}/_{48}$: *as sicilicus*
$^{1}/_{5}$: *as quincunx*	$^{1}/_{10}$: *as dextans*	$^{1}/_{72}$: *as sextula*
$^{1}/_{6}$: *as sextans*	$^{1}/_{11}$: *as deunx*	

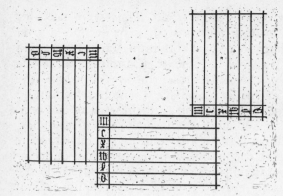

Fig. 8–14. Reckoning tables with three divisions, sixteenth or seventeenth century, used in Switzerland and Germany for computing taxes. The letters on it are, from bottom to top: *d* for pence (denarii), *s* for sols or shillings, *lb* for pounds (*libras*), then X, C, and M for 10, 100, and 1000 pounds. Basel Historical Museum.

Fig. 8–15. Metal counter bearing the coat of arms of Michel de Montaigne (and surrounded by the collar of the Order of Saint-Michel de Montaigne). This counter was found about twenty years ago in the ruins of the Château de Montaigne, but the die with which it was made had been found a century earlier.

orders (figs. 8–13, 8–14). The counters were made of various materials. All administrations, merchants, bankers, lords, and rulers had reckoning tables with counters bearing their own marks, made of base metal, gold, or silver, depending on their rank or social position. As an illustration, figure 8–15 shows a counter with the coat of arms of Michel de Montaigne (1533–1592).

The value of a counter depended on its position. On consecutive lines, going from bottom to top, it was worth 1, 10, 100, 1000, and so on. Between two consecutive lines, it was worth five units of the line immediately below it (fig. 8–16).

Addition was done in the left (or forward) part of the frame by placing counters corresponding to the numbers involved, then reducing them and taking account of the value assigned to each location. To multiply two numbers, the reckoner began by representing the first one in the left part of the frame, then he eliminated one by one the counters representing this number as he replaced them in the right (or rear) part of the frame with their products by the second number.

Computation on the reckoning table was taught till the eighteenth century. The fact that it was discussed in many European arithmetic books in the sixteenth, seventeenth, and eighteenth centuries gives an idea of how common it was. It was so firmly anchored in European tradition that even when written computation with Hindu-Arabic numerals was becoming general, results obtained in that way were always checked

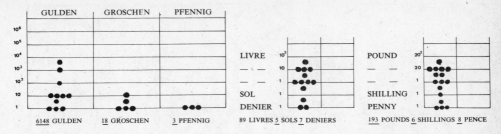

Fig. 8–16. Some of the many uses of the European reckoning table, sixteenth to eighteenth centuries: (*left*) Germany, (*middle*) France, (*right*) England.

on a reckoning table, just to be sure. In England, till the end of the eighteenth century, taxes were computed on reckoning tables called exchequers because of the divisions on them ("exchequer" is an old word for "chessboard"), and the British minister of finance is still known as the Chancellor of the Exchequer.

It is therefore not surprising that many authors during the Renaissance, and into the seventeenth century, referred to the art of computing on the reckoning table. Montaigne, for example, in his *Essays:* "I do not know how to calculate with either counters or a pen." And: "We judge him not according to his worth, but as we judge counters: according to the prerogative of his rank."

The reckoning table made addition and subtraction easy, but for more complex operations it was slow and required long training. This drawback must have been what prompted the fierce controversy that began early in the sixteenth century between the abacists, who clung to the reckoning table and the archaic Greek and Roman numerations, and the algorists, who advocated a form of written computation that was the ancestor of our modern methods (figs. 8–17, 8–18).

Here, for example, is what Simon Jacob (died in Frankfurt in 1564) wrote about the reckoning table: "It does seem to have some advantage in domestic calculations, where addition and subtraction are often required, but in more complicated calculations it is often cumbersome. I do not say that such calculations cannot be done on the lines [of the reckoning table], but calculating with numerals has the same advantage over calculating on the lines as walking free and unburdened has over walking while carrying a heavy load."

Written computation was quickly adopted by mathematicians and astronomers, while the reckoning table continued to be used in business and finance.

The shortcomings of the reckoning table gave rise to the idea of the calculating machine, which developed into our modern electronic computers. In 1639, at the age of sixteen, Blaise Pascal (1623–1662) imagined

Fig. 8–17. The quarrel between the abacists and the algorists. Illustration in one of the editions of *The Ground of Artes*, by the English mathematician Robert Recorde (1510–1558).

Fig. 8–18. Woodcut from *Margarita Philosophica*, by Gregorius Reich (Freiburg, 1503). Arithmetic, symbolized by the woman standing in the center, seems to be deciding the debate between the abacists and the algorists: she looks toward the reckoner using Hindu-Arabic numerals (with which her dress is adorned). Museum of the History of Science, Oxford.

the first French machine for performing addition and subtraction, using a system of gears, and it was put into operation in 1642. He got the idea for it when he had to do long, laborious series of additions and multiplications on a reckoning table for his father, who was then a treasury official in Rouen.

Pascal, however, was not the inventor of the calculating machine. In 1623 (the year of Pascal's birth) Wilhelm Schickard (1592–1635), a German mathematician, astronomer, and Orientalist, made a machine that could *perform all four arithmetical operations and extract square roots.* He described it in a letter to Johannes Kepler (1571–1630) dated February 25, 1624, and enclosed a drawing of it. Unfortunately this machine, far ahead of its time, was destroyed by fire on February 22, 1624.

Rods on the Chinese counting board

In China (and later in Japan), rods made of ivory or bamboo were placed on a counting board to represent numbers and perform arithmetical operations (figs. 8–19, 8–20, 8–21). Our information on their use goes back only to the second century B.C., but the system was probably invented much earlier.

On this kind of abacus, each column represented a decimal order: going from right to left, the first column was for units, the second for tens, the third for hundreds, and so on. To indicate a given number, the reckoner could place in each column, along a single horizontal line, as many rods as there were units in the corresponding decimal order. For example, the number 2645: five rods in the first column on the right, four in the second, six in the third, and two in the fourth.

In practice, however, a simplifying convention was used: "For the number 5 and those below it," wrote the Chinese mathematician Mei Wen-ting (1633–1721), "the reckoner lined up vertically as many rods as necessary. For 6 and above [up to 9], he placed a rod horizontally to represent 5 and completed the representation with vertical rods below the horizontal one [fig. 8–22]. Tens, hundreds, thousands, and ten thou-

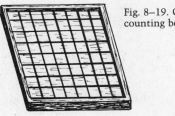

Fig. 8–19. Chinese counting board.

(*left*) Fig. 8–20. Reckoner using rods on a counting board. From an illustration in the 1795 edition of the Japanese work *Shojutsu Sangaku Zuye*, by Miyake Kenryu.

(*right*) Fig. 8–21. Chinese teacher instructing two pupils in the art of reckoning on the counting board. From an illustration in *Suan Fa Thung Tsung*, published in 1593.

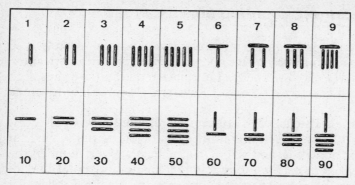

Fig. 8–22. Representation of units and tens with rods on a counting board.

sands were formed by going from left to right, as in the columns of the abacus or the relative positions of numerals in the written computation of central Asia and Europe."

To avoid errors, the Chinese had the idea of orienting the rods vertically in odd-numbered columns, starting with the first column on the right, and horizontally in even-numbered columns (figs. 8–22, 8–23, 8–24). The rods were therefore vertical for units, hundreds, ten thousands, etc., and horizontal for tens, thousands, etc. (Because of this convention, as we will see later, the system gave rise to a highly interesting place-value numeration.) A third-century Chinese work states the rule as follows: "The units are vertical and the tens horizontal, the hundreds stand while the thousands lie down; thousands and tens therefore look the same, as also the ten thousands and the hundreds" (Needham).

In ancient times, this system enabled the Chinese to perform various kinds of arithmetical and algebraic operations (multiplication, division, extraction of square and cube roots, solution of algebraic equations and systems of equations, etc.). And it is interesting to note that modern characters for the Chinese word *suan*, "calculation," bears traces of the ancient method (fig. 8–25).

Addition and subtraction were done simply: the numbers to be added (or subtracted) were indicated on the counting board, then the corresponding rods were joined (or removed) column by column. For multiplication, the multiplicand was indicated at the bottom of the board and the multiplier at the top; partial products were then indicated on the middle line and automatically added as they were obtained. Division was done in an analogous way, with the divisor indicated at the bottom and the dividend on the middle line; the quotient was indicated at the top, and rods representing partial products were taken away from the dividend.

Algebraic equations with one or more unknowns could be solved on the counting board. *K'iu-ch'ang Suan-shu* ("Calculation in Nine Chap-

	UNITS IN ODD-NUMBERED COLUMNS (columns for even powers of 10)	UNITS IN EVEN-NUMBERED COLUMNS (columns for odd powers of 10)
1		
2		
3		
4		
5		
6		
7		
8		
9		

ten thousands ▼	thousands ▼	hundreds ▼	tens ▼	units ▼	
8	1	2	2	1	◄ - - 81,221
	1	1	1	1	◄ - - 1111
3	(o)	1	(o)		◄ - 3010
6	(o)	(o)	(o)		◄ - - 6000

(*left*) Fig. 8–23. Representation of units of different orders in odd-numbered and even-numbered columns of the counting board, according to the orientation of the rods: vertical for units, hundreds, ten thousands, etc. (odd-numbered columns); horizontal for tens, thousands, hundred thousands, etc. (even-numbered columns).

(*right*) Fig. 8–24. Representation of four numbers on the counting board.

MODERN FORMS		
A	B	C
筭	算	祘
ANCIENT FORMS		
A'	B'	C'

Fig. 8–25. Ancient and modern characters for the Chinese word *suan*, "calculation." They bear witness to the ancient method of calculating with bamboo rods on a counting board or ruled table. Ideas expressed by character A or B: two hands, a ruled table, bamboo. Idea expressed by character C: representation of numbers on a counting board with rods, according to the orientation of the rods. The ancient forms A', B', and C' are highly evocative.

ters"), an anonymous work compiled during the Han period (206 B.C.–A.D. 221), gives details of algebraic operations on the counting board, including the solution of a system of *n* equations with *n* unknowns.

One column of the board was assigned to each equation of the system, and one horizontal line to the coefficients of each unknown in the equations. The ordinary rods used for positive (*cheng*, "correct") numbers were replaced with black rods each time a negative (*fu*, "deceptive") number appeared. Thus the system:

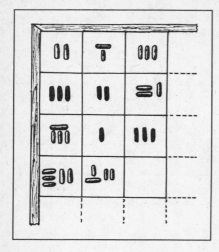

Fig. 8–26. Representation of a system of equations with three unknowns by means of rods on a counting board. (From a mathematical treatise of the Han period, 206 B.C.–A.D. 220). First column on the left: equation $2x - 3y + 8z = 32$. Second column: equation $6x - 2y - z = 62$. Third column: equation $3x + 21y - 3z = 0$.

x	2	6	3
y	-3	-2	21
z	8	-1	-3
	32	62	0

Fig. 8–27.

$$2x - 3y + 8z = 32$$
$$6x - 2y - z = 62$$
$$3x + 21y - 3z = 0$$

was noted as shown in figures 8–26, 8–27. Such a system could be solved by subtle manipulations of the rods.

The following anecdote, dating from the middle of the ninth century, shows algebra being used in selecting minor functionaries:

"Yang Sun (a high official) was famous for selecting and promoting members of the civil service not through private influence or personal preference but by taking general opinions on their merits and weighing all criticisms which might be brought forward. He applied this principle even to petty clerks and menial officers. . . .

"Once there were two clerks who held the same rank and had equal lengths of government service. They had even acquired the same commendations, and the criticisms in their personal dossiers were identical. The responsible official of middle rank was quite baffled by the problem of their promotion, and appealed to Yang Sun, who was his superior. Yang Sun thought the matter over, and then said, 'One of the best merits of minor clerks is to be quick at computations. Let both the candidates now listen to my question. Whoever first gets the right answer will obtain the advancement. The problem is this: Someone, when walking in the woods, overheard a number of robbers discussing how best to share the rolls of cloth which had been stolen. They said that if each had six rolls there would be five rolls left, but if each had seven rolls they would be eight rolls short. How many robbers were there, and what was the total number of the rolls of cloth?' This problem was taken down by another minor clerk, and Yang Sun then asked the two candidates to reckon it out with counting-rods on the stone steps of the hall. After a short time one of them in fact got the right answer. He was duly given the better position, and the officials dispersed having nothing to complain of or criticise in the decision" (Needham).

The bead abacus

Today, when electronic calculators are becoming more and more common in the West, a centuries-old little "calculating machine" still has a very important place in the Far East and some countries of eastern Europe. In China the bead abacus (*suan pan*) is used by nearly everyone: peddlers who do not know how to read or write, merchants, accountants, bankers, hotelkeepers, mathematicians, and astronomers. It is so deeply rooted in Far Eastern tradition that even the "Westernized" Chinese and Vietnamese of Bangkok, Singapore, Taiwan, Polynesia, Europe, and America— who have easy access to modern calculators—usually continue to do all sorts of computation on an abacus (figs. 8–28, 8–29). Even more striking is the example of Japan, one of the world's foremost producers of pocket calculators, where the abacus (*soroban* in Japanese) is still the main calculating instrument in everyday life, and every child learns how to use it in school (fig. 8–30).

In the Soviet Union the abacus (*s'choty* in Russian) appears beside modern cash registers and is often used for computing what customers must pay in shops, supermarkets, hotels, and department stores. A friend who spent a month in the Soviet Union told me that one day when he went to a foreign exchange office to convert some French francs into Soviet rubles, he was surprised to see the clerk first use a modern calculator for the conversion, then check the result on an abacus.

Of all the ancient calculating devices that have come down to us, the bead abacus is the only one on which all four arithmetical operations can be done rapidly and in a relatively simple way. For someone who knows how to use it, it is an efficient means of adding, subtracting, multiplying, and dividing, and even extracting square and cube roots. Westerners are usually amazed to see how quickly complex calculations can be done on it by an expert.

This was vividly illustrated by a calculating contest held in Tokyo on November 12, 1945, between Kiyoshi Matsuzaki, a clerk in the Japanese Communications Ministry with seven years of special abacus training, and Pvt. Thomas Woods, a finance clerk in the U.S. Army with four years' experience in the use of electric calculating machines. The contest consisted of five events: addition (numbers with three to six digits), subtraction (six to eight digits), multiplication (five to twelve digits), division

Fig. 8–28. Chinese merchant reckoning on a bead abacus. From an illustration exhibited at the Palais de la Découverte, Paris.

(five to twelve digits), and a composite event of thirty additions, three subtractions, three multiplications, and three divisions of numbers with six to twelve digits. Matsuzaki, using an abacus, defeated Woods, using an electric calculating machine, in four out of the five events; Woods won only in multiplication.

J. Tricot, writing in the November 1978 issue of the French magazine *Science et Vie*, has commented on this contest: "For someone who has seen a competent Japanese *soroban* operator in action, there can be no doubt that if such a contest were held today, with an electronic calculator rather than an electric one, the result would be the same, at least as far as addition and subtraction are concerned. Most of us cannot punch the keys of a calculator fast enough to compete with the dexterity of a *soroban* operator."

An article by Clyde Haberman in the August 6, 1983, issue of the

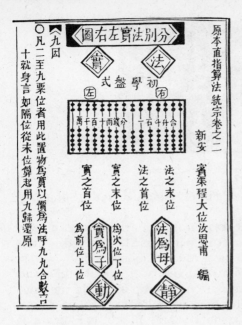

Fig. 8–29. Explanations concerning the *suan pan* in *Suan Fa Thung Tsung*, a Chinese work published in 1593.

Fig. 8–30. Japanese accountant reckoning on a *soroban*. Illustration in an eighteenth-century Japanese work.

New York Times, titled "In the Land of Sony, the Abacus Is Still King," shows that the *soroban* continues to hold its own:

"Japan produced 58 million electronic calculators last year, but that does not necessarily mean that people count on them for anything.

"Despite all the state-or-the-art gimcracks that this country cranks out, daily life is still governed by a rectangular gadget that never blinks or blips, never runs down and has never been known to go wrong just because a 6-year-old tried to test its mettle underwater.

"The simple, elegant abacus is not only surviving, but by some measures is even thriving. . . . At railway stations clerks punch out tickets on gleaming computers of heavy plastic and metal, but when it is time to tote up the fare they flick fingers across tiers of beads. Salespeople routinely ignore cash registers in favor of the abacus, or soroban, as it is known here. . . .

"The resilience of the soroban has taken even some of its strongest advocates by surprise. When inexpensive, easy-to-carry calculators flooded the market a decade ago, abacus use dropped noticeably and sales dipped by 10 percent or more. Suddenly, four or five centuries after its impor-

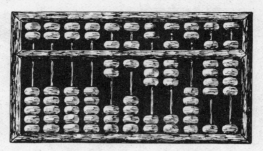

Fig. 8–31. The Chinese bead abacus (*suan pan*).

tation from China, the soroban seemed a likely candidate for the obituary page.

"Now production has stabilized at 2.1 million of the devices a year, most of them about a foot long and costing between $12 and $20. At the same time popular interest seems to have revived considerably.

"Nationwide abacus contests draw hundreds of competitors, and some companies give their own examinations to employees. In the mid-1970's an average of 2.4 million junior high school students a year took proficiency tests sponsored by the National League for Soroban Education. Last year's figure was 3 million, and the league expects to do as well in 1983."

The Chinese *suan pan* has a rectangular wooden frame with thin rods across it. On each rod are seven slightly flattened glass or wooden beads: five below the strip of wood that divides the frame into two unequal parts, and five above it (fig. 8–31).

The rods correspond to numerical values increasing tenfold from right to left. (Base 10 is not obligatory, of course; another base, such as 12 or 20, could be used if an adequate number of beads were placed on the rods.) If the first rod on the right corresponds to units, the second one corresponds to tens, the third to hundreds, and so on. Users of the *suan pan*, however, do not always begin with the first rod on the right to represent whole numbers: they sometimes begin with the third from the right, reserving the first two for centesimal and decimal fractions.

Most *suan pans* have ten or twelve rods, but their number can be increased according to the needs of the user. A *suan pan* with fifteen rods, for example, has a numerical capacity of $10^{15} - 1$—that is, a quadrillion minus one.

Each of the five beads on the lower part of a rod has the value of one unit of the order corresponding to the rod, and each of the two beads on the upper part has the value of five units. Numbers are formed by moving beads toward the crossbar separating the upper and lower parts of the rods. To form the number 3, for example, the user simply raises three beads on the lower part of the first rod on the right. For 9, he lowers one upper bead on that rod and raises four lower ones (fig. 8–32).

Fig. 8–32.

9

UNITS 1st rod on the right			9
TENS 2nd rod			3
HUNDREDS 3rd rod			7
THOUSANDS 4th rod			5
result			5739

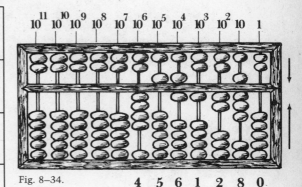

Fig. 8–34.

4 5 6 1 2 8 0

Fig. 8–33.

For 5739, he raises four lower beads on the first rod and lowers an upper one; then he raises three lower beads on the second rod; then, on the third rod, he raises two lower beads and lowers an upper one; and finally he lowers an upper bead on the fourth rod (fig. 8–33).

For 4,561,280, he moves no beads on the first rod (representation of zero, or absence of units); on the second rod, he raises three lower beads and lowers an upper one; and so on (fig. 8–34).

As we have seen, nine units of the order corresponding to a given rod are represented by one upper bead with the value of five units and four lower beads with the value of one unit each. Five beads are thus enough to represent nine units. This raises the question of why each rod has seven beads with the total value of 15. "The reason is that in doing division on an abacus it is often helpful to indicate, temporarily, a number greater than 9 on a single rod. For the three other operations, five beads on each rod are enough. In the case of division, however, calculation may be simplified if a partial result greater than 9 is temporarily indicated on one rod" (Vissière).

Beginning in the middle of the last century, the Japanese *soroban* (of Chinese origin) gradually lost one of the two upper beads on each of

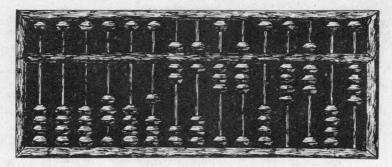

Fig. 8–35. The Japanese *soroban* in the form it has had since World War II. The number represented here is 763,804,804 (with the first rod on the right corresponding to units).

Fig. 8–36. The Russian *s'choty*. A similar abacus is used in Armenia and some parts of Iran (where it is called the *choreb*) and in Turkey (*coulba*).

its rods, and after World War II it lost one of the five lower beads (fig. 8–35). This new form of the *soroban* required longer training for its users because they had to acquire a more elaborate and precise "fingering technique" than users of the Chinese *suan pan*.

The Russian *s'choty* (fig. 8–36) has ten beads on each rod. Two of these beads (the fifth and sixth) are of a different color. To represent a given number, as many beads as necessary are moved up on each rod. In figure 8–37 the Russian abacus is shown representing 5,123,012.

Finally, there is the French *boulier-compteur* (probably derived from the Russian abacus), which was used in nineteenth-century schools for teaching arithmetic. It has ten beads on each rod and they move horizontally (fig. 8–38).

Since I have undertaken to give only a general idea of the different kinds of abacuses, detailed explanation of their use is beyond my purpose. I will simply point out that although one has to know only the addition and multiplication tables from 1 to 9 in order to do arithmetical operations on the bead abacus, this ingenious calculating device does have some serious drawbacks: it requires intensive training, a very precise touch (though this becomes instinctive with practice), and a perfectly steady support. Furthermore the slightest error—when it is detected—makes it necessary to do the whole calculation over, since intermediate results (partial products in multiplication, remainders in division, etc.) disappear in the course of the operations.

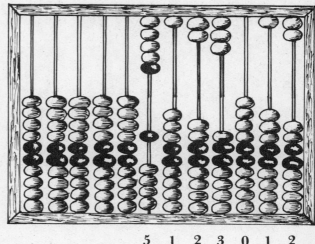

Fig. 8–37.

5   1   2   3   0   1   2

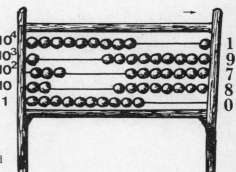

Fig. 8–38. French *boulier-compteur* used in nineteenth-century schools.

PART III

The Invention of Numerals

9

Roman Numerals: A Vestige of Primitive Origins?

Many things are so familiar to us that we never think about them. That is true of the Roman numerals we sometimes still use today, even though they now seem old-fashioned and quaint. But interesting questions can be raised about their origin and their history, which runs through a considerable period in the history of Western civilization.

The numerals of Roman civilization

As we all know, these numerical signs are:

I	V	X	L	C	D	M
1	5	10	50	100	500	1000

They are obviously letters of the Roman alphabet, but this does not mean that they have an alphabetic origin. The signs L, C, D, and M are not the original forms of the numerals for 50, 100, 500, and 1000; they are altered forms of much older numerals. Known instances of the use of L, D, and M as numerals do not go back much farther than the first century A.D. To the best of my knowledge, the oldest Roman inscription using L for 50 dates from 44 B.C. The oldest known use of M and D as numerals is in a Latin inscription dated 89 B.C., in which the number 1500 appears as M D.

Before M was adopted for 1000, the Romans used one or another of the signs in figure 9–1, mainly during the republican period. As we will see later, these signs are simply variants of a single original sign, ɸ, which appears on some Roman hand abacuses (fig. 8–11). Various signs for 1000 are found in texts from the imperial period, and some of them even survived long after the collapse of the empire. They are used in numerous books published in the sixteenth and seventeenth centuries.

For 500 the Romans first used one of the three following signs:

500

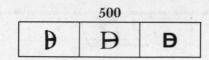

They resemble the letter D (with which they were later replaced), but they approximately correspond to half of one of the signs formerly used for 1000. They too are sometimes found in texts from the imperial period (fig. 9–3).

1000

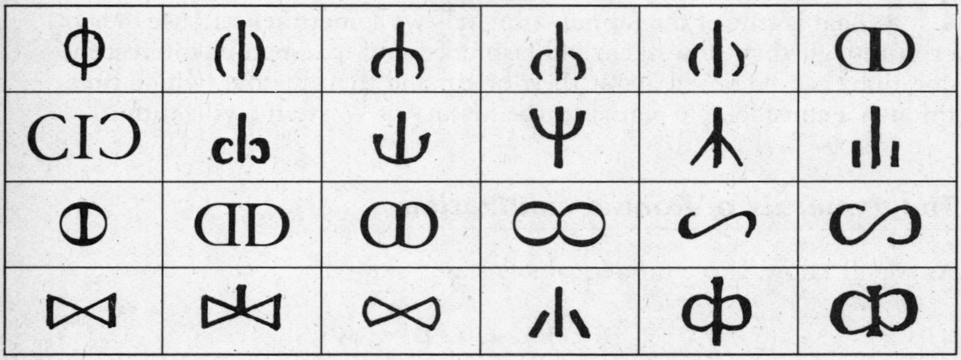

Fig. 9–1.

28	XXIIX	140	CXL
45	XⅬV	268	CCⅬXIIX
69	LXIX	286	CCXXCVI
74	LXXIV	340	CCCXⅬ
78	LXXIIX	345	CCCXⅬV
79	LXXIX	1290	∞CCXC

Fig. 9–2. Latin inscriptions from the republican period, illustrating the use of the subtractive principle, which probably resulted from influence of the common system on the monumental system and was rather rare in carefully made inscriptions.

As for the numeral letter L, we will later see that its origin was the following sign which the Romans used for 50 during the republican period: ↓ or Ѱ. This sign was sometimes still used during the reign (27 B.C.– A.D. 14) of Augustus (figs. 9–2, 9–4, 9–5).

1	I	26	XXΛI		
2	II	40	XXXX	837	ÐCCCXXXVII
3	III	50	↓ or ↧ or ⊥ or .L or L	1000	Φ or Ψ or ∞ or ⋈ or ⋈ or M or M
4	IIII				
5	Λ or V				
6	VI				
7	VII	51	↓I	1200	∞CC
8	VIII	74	↓XXIIII	1500	MD
9	VIIII	95	LXXXXV	2000	∞∞
10	X	100	C	2320	∞∞CCCXX
14	XIIII	100	Ɔ	3700	ΦΦÐCC
15	XV	300	CCC	5000	Þ or I) or ... or ʰ
19	XVIIII	400	CCCC		
20	XX	500	Đ or D		
24	XXIIII				

5000	IƆƆ
7000	ʰ ƆƆ
8670	ʰ) ƆƆƆDC.LXX
10,000	CCIƆƆ
12,000	... ƆƆ
21,072	... LXXII
30,000	CCIƆƆ CCIƆƆ CCIƆƆ
30,000	...
50,000	IƆƆƆ
100,000	... or CCCIƆƆƆ

Fig. 9–3. Numerical notations in Latin monument inscriptions, republican period to the beginning of the imperial period.

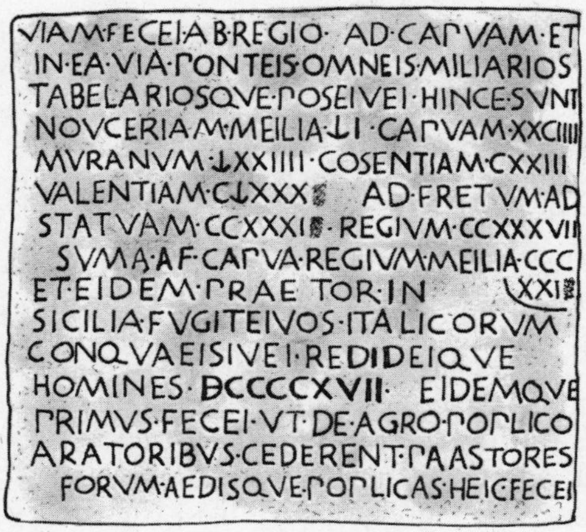

line 4	↓I	51
line 4	XXCIIII	84
line 5	↓XXIIII	74
line 5	CXXIII	123
line 6	C↓XXX	180
line 7	CCXXXI	231
line 7	CCXXXVII	237
line 8	CCCXXI	321
line 12	ÐCCCCXVII	917

(*left*) Fig. 9–4. Military inscription found at Forum Popilii in Lucania (province of Salerno), now in the Museo della Civiltà Romana, Rome. Made by order of C. Popilius Laenas, consul in 172 B.C. and 158 B.C.

(*right*) Fig. 9–5. Numerical notations in the inscription shown in figure 9–4.

Questionable ancestors

According to a hypothesis that is now commonly accepted, all these numerals are of Greek origin. It is useful to recall that Latin writing came from Etruscan writing. The Etruscans, whose origin and non-Indo-European language are still not well known, dominated Italy from the Po to Campania between the seventh and fourth centuries B.C. During the Roman imperial period they were gradually assimilated by their conquerors and disappeared.

The Etruscan alphabet is related to archaic Greek alphabets of the western type. (Archaic Greek alphabets are usually divided into two categories: those of the western type, which—like the alphabet of Chalcidice, for example—give the sound *kh* to the letter Ɣ or ↓ or Ψ , and those of the eastern type, which—like the alphabets of Miletus and Corinth, for example—give this letter the sound *ps* and express the sound *kh* by the letter + or ×.) Some specialists have therefore supposed—and tradition seems to support this hypothesis—that the Etruscan alphabet "may have been borrowed from a Greek alphabet of the western type within Italy itself, since Cumae, the oldest Greek colony having such an alphabet, goes back to about 750 B.C., half a century before the blossoming of Etruscan civilization" (Bloch).

On the basis of comparisons between graphic forms, many specialists have favored the opinion that the ancient Roman signs for 50, 100, and 1000 were derived from the following letters, which belong to the Chalcidic alphabet:

CHI   : Ɣ or ↓ or Ψ

THETA : ⊞ or ⊕ or ⊖ or ⊂

PHI   : Φ or ⊕ or ⅭⅮ

These are letters whose sounds were not used in Etruscan or Latin. According to the theory, they developed into the Roman signs in question; thus the Greek letter theta (originally ⊞ or ⊕ , then ⊖ or ⊖) gradually became C, under the influence of the initial of the Latin word *centum* ("hundred").

This theory of the Greek origin of the Roman numerals L, C, and M is still presented as an established fact by some Latinists, Hellenists, epigraphers, and historians of science. Attractive though it may be at first sight, it should be approached with serious reservations. For what reason would three foreign characters, and only three, have been introduced into the Roman system? And why letters? Probably, supporters of the theory would answer, because the Greeks often used letters as numerical signs.

It is true that the Greeks used two written numerations whose signs were letters of their alphabet. The first one employed initials of number words: the letter delta, δ, which has the value of 10 in this system, is the initial of the word δεκα, *deka*, "ten"; and the letter chi, χ, with the value of 1000, is the initial of χιλιοι, *khilioi*, "thousand." The second system used all the letters of the alphabet:

A	alpha	1	I	iota	10	P	rho	100
B	beta	2	K	kappa	20	Σ	sigma	200
Γ	gamma	3	Λ	lambda	30	T	tau	300
Δ	delta	4	M	mu	40	Y	upsilon	400
E	epsilon	5	N	nu	50	Φ	phi	500
. . . .			Ξ	xi	60	X	chi	600
H	eta	8	O	omicron	70	Ψ	psi	700
Θ	theta	9	. . . .			Ω	omega	800

But in the Greek world these systems scarcely seem to be attested before the second half of the first millennium B.C. Furthermore the letter chi, supposedly borrowed for 50 in Latin, has the value of 1000 in the first of the two systems and 600 in the second; theta, supposedly borrowed for 100 in Latin, has the value of 9 in the second system; and phi, supposedly borrowed for 1000, has the value of 500 in the second system. Why these divergences? No one knows.

It is true that the archaic Roman numerals for 50 and 1000 bear a resemblance to the Chalcidic letters chi and phi, but may that not be simply a coincidence?

Moreover, if to represent 50 and 100 the Romans had really borrowed the signs:

    chi: Ψ or ν or ↓

    theta: ⊕ or ⊖ or ⊂

the Etruscans would also have borrowed them, for the reason mentioned above. But in fact the Etruscans used totally different signs for 50 and 100 (fig. 9–6).

    Λ or ↑ for 50

    ✳ or ✗ for 100

The traditional theories, still supported by several "serious" but dogmatic scientific authors, are proving to be less and less solidly based. Thanks to the work of Lucien Gerschel (which is unfortunately not well known) and the interesting suggestions he has made in this area, it is now possible to develop another theory that seems to me much more plausible, insofar as it derives less from abstract or doctrinal notions than from concrete cases.

Value	Sign	Value	Sign
1	I	38	XII XXX
2	II	42	IIX XXX
3	III	44	IIIIXXXX
4	IIII	50	↑ or Λ or ⇑
5	Λ or ∩	52	II↑
6	IΛ	55	Λ⇑
7	IIΛ	60	X↑
8	III∩	75	ΛX·X↑
9	IIIIΛ	82	II+++↑
10	X or X or +	86	IIIIIIXXX↑
19	XIX	100	✳ or ✶ or ЖC
36	IΛXXX	106	IΛЖ
38	IIIΛXXX		

Fig. 9–6.
Numerical notations
in Etruscan
inscriptions.

The origin of Roman numerals

The question remained obscure for a long time, but there can now be no doubt that the signs I, V, and X are by far the oldest of the series. Earlier than any form of writing (and therefore earlier than any alphabet), these signs and their corresponding values naturally occur to the human mind under certain conditions.

To understand this, we will begin by making a "world tour" through different times and cultures to examine some notations for numbers 1 to 9.

Figure 9–7 shows numerical notations that begin by indicating a whole number with a row of as many identical signs as necessary, but make a break at 4 because the people who used those notations were unable to read a row of more than four identical signs at a glance. Beyond that number they either arranged the signs in more than one group or used a special sign for 5.

The Egyptians, Cretans, Hittites, and "Proto-Indians," for example, used the following notation, which has one group of vertical strokes for each number up to 4, then two groups for each number from 5 to 8 (or 9), and in some cases three groups for 9:

I	II	III	IIII	III II	III III	IIII III	IIII IIII	IIIII IIII	OR	III III III
1	2	3	4	5	6	7	8		9	
				3+2	3+3	4+3	4+4	5+4		3+3+3

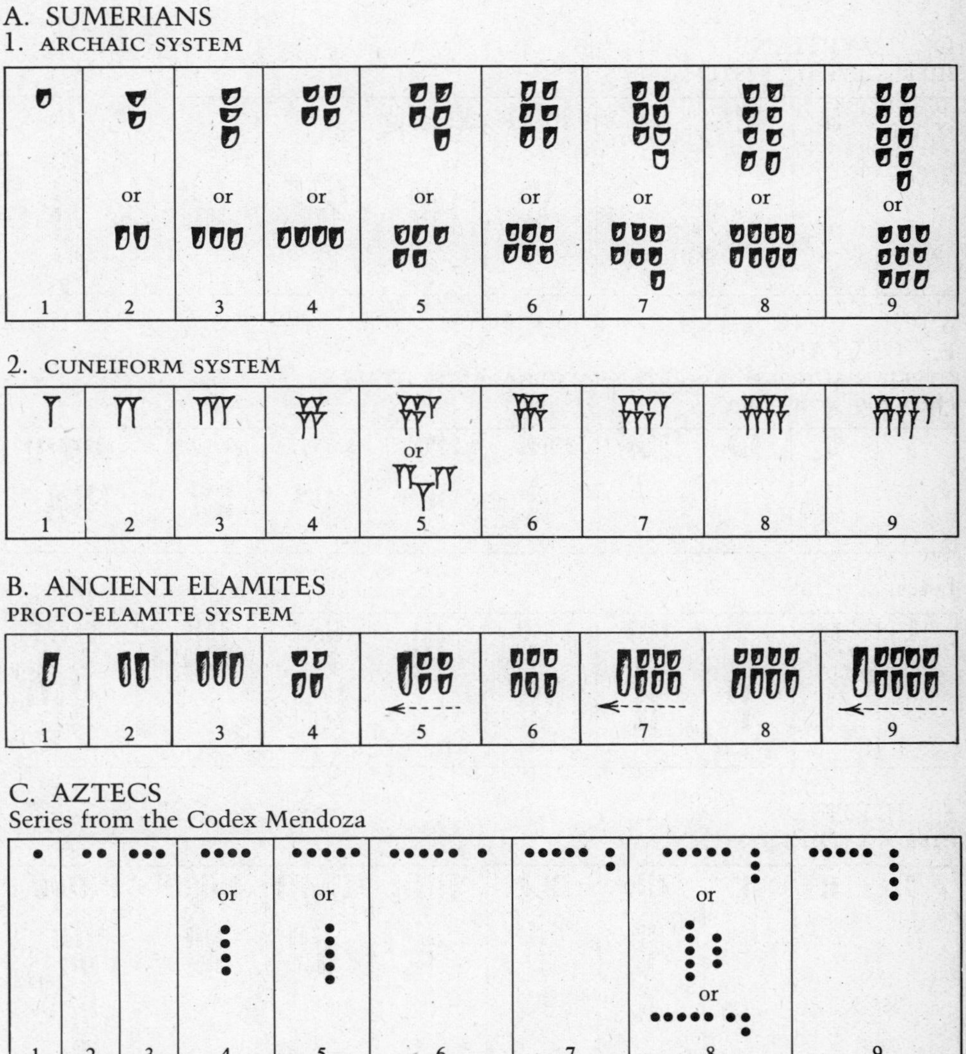
A. SUMERIANS
1. ARCHAIC SYSTEM
or
1 2 3 4 5 6 7 8 9
2. CUNEIFORM SYSTEM
or
1 2 3 4 5 6 7 8 9
B. ANCIENT ELAMITES
PROTO-ELAMITE SYSTEM
1 2 3 4 5 6 7 8 9
C. AZTECS
Series from the Codex Mendoza
or
1 2 3 4 5 6 7 8 9

Fig. 9–7.

AZTECS (continued)
Series from the Codex Telleriano Remensis

Fig. 9–7 continued.

D. EGYPTIANS
HIEROGLYPHIC SYSTEM

E. CRETANS
CYPRIAN-MINOAN & CRETAN-MYCENAEAN SYSTEMS
Hieroglyphic system

Linear systems

F. HITTITES
HIEROGLYPHIC SYSTEM

G. INDUS CIVILIZATION
PROTO-INDIAN SYSTEM

Fig. 9–7 continued.

1	2	3	4	5	6	7	8	9

H. URARTIANS
HIEROGLYPHIC SYSTEM

1	2	3	4	5	6	7	8	9
								?

I. ANCIENT GREEK WORLD
1. INSCRIPTIONS FROM EPIDAURUS, ARGOS, AND NEMEA

1	2	3	4	5	6	7	8	9

2. INSCRIPTIONS FROM TROEZEN, CHALCIDICE, AND TAURIC CHERSONESE

1	2	3	4	5	6	7	8	9

3. INSCRIPTIONS FROM ATTICA, THEBES, ORCHOMENUS, AND CARYSTUS

1	2	3	4	5	6	7	8	9

*Letter Π (pi), initial of ΠΕΝΤΕ (*pente,* "five").

J. MINAEAN EMPIRE & SABAEAN KINGDOM
(ANCIENT SOUTHERN ARABIA)

Fig. 9–7 continued.

I	II	III	IIII	ʊ*	ʊI	ʊII	ʊIII	ʊIIII
1	2	3	4	5	6	7	8	9

*Southern Arabic letter ʊ (kha), initial of ʊ〱ᴎX (khamsat, "five").

K. LYCIANS
(ASIA MINOR)

I	II	III	IIII	∠	∠I	∠II	∠III	∠IIII
1	2	3	4	5	6	7	8	9

L. MAYAS

•	••	•••	••••	—	•/—	••/—	•••/—	••••/—
1	2	3	4	5	6	7	8	9

M. ARAMAEANS OF PALMYRA

'	''	'''	''''	𐑫	'𐑫	''𐑫	'''𐑫	''''𐑫
1	2	3	4	5	6	7	8	9

In the Proto-Elamite and Sumerian cuneiform notations, division of the signs into two groups began at 4 and continued to 9. Other peoples who used similar notations, and encountered the same difficulty from 4 onward, used three groups of signs from 7 to 9 (fig. 9–8).

Still other peoples, such as the Greeks, Sabaeans, Lycians, Mayas, Palmyrans, Etruscans, and Romans, had a special sign for 5 and used the quinary principle from 6 to 9, an idea that probably arose from elementary finger counting.

This illustrates the basic rule that the human capacity for immediately recognizing concrete quantities very seldom goes beyond the number 4. From 5 on, we must nearly always use the artifice of abstract counting.

This digression will help us to understand the origin of the Roman numerals I, V, and X.

Let us now turn back to the simple numerical technique of notching, and take the example of a shepherd who wants to verify the number of sheep in his flock. Using only his natural ability to recognize concrete

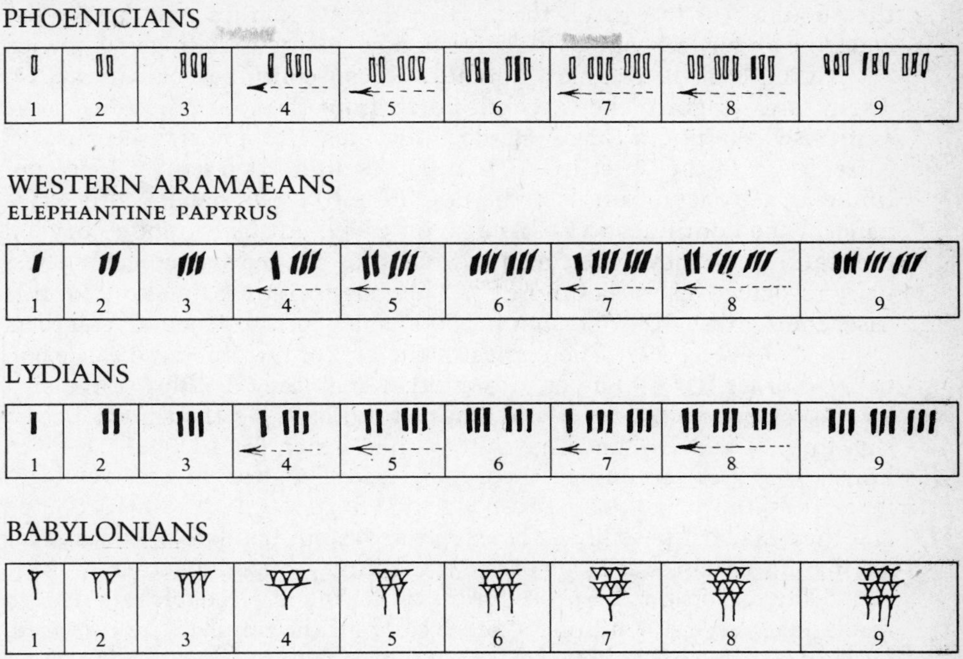
PHOENICIANS
1 2 3 4 5 6 7 8 9
WESTERN ARAMAEANS
ELEPHANTINE PAPYRUS
1 2 3 4 5 6 7 8 9
LYDIANS
1 2 3 4 5 6 7 8 9
BABYLONIANS
1 2 3 4 5 6 7 8 9

Fig. 9–8.

quantities, he cannot go beyond 4, because no one, however "civilized" he may be, can "read" at a glance a series of five (IIIII), six (IIIIII), seven (IIIIIII), or more.

If our shepherd uses the technique of making notches on a stick, how will he know, when he has made the last notch, whether the number of notches corresponds exactly to the number of sheep, if that number is greater than 4? The answer is simple: once he has made four consecutive similar notches on his stick, he will change the form or orientation of the fifth notch, so that the series will still be "legible." If the fifth one has a different form or orientation and those following it are the same as the first four, the fifth one will mean "five" by itself. The shepherd will thus have created a new counting unit, which will be all the more familiar to him because it corresponds to the number of fingers on one hand. He can now go on counting up to 9 with numbers that do not go beyond 4: 5 + 1, 5 + 2, 5 + 3, and 5 + 4.

But when he reaches 10 (or the fifth number after 5), he must use a notch different from the first four. He can, of course, use the notch he has already adopted for 5, but since he is dealing with the total number of fingers on both hands he will naturally think of using something like the "double" of the notch for 5. He will thus mark a new break, and counting on his tally stick will correspond to elementary finger counting.

Going back to ordinary notches, he can continue counting to 14. Then, to distinguish the fifteenth notch from the four preceding it, he again uses the notch for 5. (He does not create a new sign because 15 corresponds to the "first hand" that comes after "two hands.") He continues to 19, then again uses the notch for 10 ("two hands" after "two hands") and continues to 24. And so on to 49. But he cannot go beyond that number without creating a new sign for 50, because he cannot visually recognize a series of more than four signs of the same kind (in this case, the sign for 10). The sign for 50 enables him to reach all numbers up to 50 + 49 = 99. (If he comes to the end of one stick, he can begin using another.) At 100 he creates another sign (the "double" of the sign for 50, for example) which takes him up to 499. He then introduces a sign for 500 which takes him to 999, then a sign for 1000 which takes him to 4999, and so on—in theory, at least. Thus the successive stages marked on the tally stick correspond to 1, 5, 10, 50, 100, 500, 1000, etc., and they make it possible to develop the system regularly without ever having more than four signs of the same kind grouped together.

"The reason for the procedure is clear: the tally enables its user to reach relatively great numbers, practically all the numbers he can need, without ever having to count a series of more than four similar elements. And so, without knowing how to 'count beyond four,' with the help of his tally he can count a collection of 50 or 100 elements. The tally increases his mental capacity as a lever increases his physical capacity by enabling him to lift a weight which his unaided muscles could not move" (Gerschel 4).

But what signs are to be made on the tally? Cutting notches into wood or bone presents the same difficulties everywhere and therefore leads to the same solutions, whether it is done in Oceania, Europe, Africa, America, or Asia. We therefore find the same signs everywhere (fig. 9–9).

In nearly all parts of the world, we find notches with these shapes: I, V or Λ, X or +. Used as numerical signs in the way described above, would they not lead to something identical with, or at least similar to, the Roman and Etruscan notations? In other words, were not the Roman numerals I, V, and X, and the Etruscan numerals I, Λ, and X or +, derived from the use of tally sticks? The hypothesis seems so natural that one would be inclined to accept it even if there were no evidence for it, but such evidence exists.

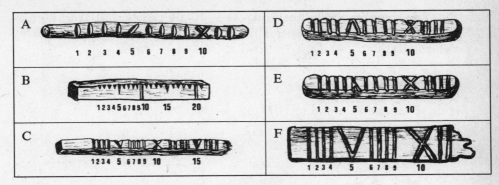

Fig. 9–9. Any user of a tally will be led to indicate the numbers 1, 5, 10, 15, etc. in one of these ways.

Anthropological and historical evidence

Franjo Škarpa has published a detailed study of different types of tallies used from time immemorial by Dalmatian shepherds in Yugoslavia. In one of them, the number 1 is represented by a small notch, 5 by a slightly larger one, and 10 by a much larger one (which recalls the markings on our thermometers). In another type, 1 is represented by a vertical notch, 5 by a slanting one, and 10 by a cross. In another one, numbers 1, 5, and 10 are as follows:

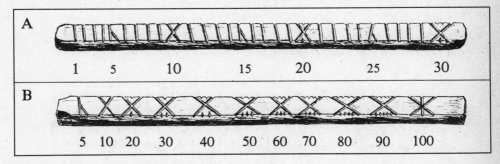

Fig. 9–10. Shepherds' tallies found in Dalmatia.

The signs on the tallies in figure 9–10 are rather similar to Roman and Etruscan numerals, especially since the same type of tally has this sign for 100:

is the Etruscan numeral for 100 (fig. 9–12).

Fig. 9–11. Etruscan coins from the fifth century B.C., bearing the numerals Λ(5) and X(10). (Landesmuseum, Darmstadt.)

Fig. 9–12. Fragment of an Etruscan inscription with the notations X↑X, 213 III∩XX, 15 ∩X.

One might wonder why this type of tally used such a sign for 100 but not a "half-sign" for 50, as the Etruscans did (fig. 9–6). Examination of the tally gives us the answer: the tens, from 20 to 90, are distinguished by small notches in the upper and lower edges (fig. 9–13). This system can do without a special sign for 50. If a shepherd using it wants to record the numbers 83 and 77, he can notch a piece of wood as shown in figure 9–14.

Fig. 9–13.

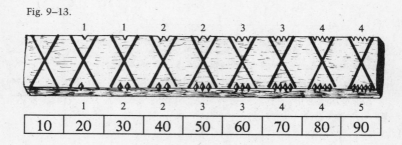

| 10 | 20 | 30 | 40 | 50 | 60 | 70 | 80 | 90 |

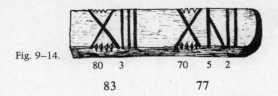

Fig. 9–14.

80   3   70   5   2

83   77

Other types of Dalmatian tallies (fig. 9–15) have the following signs:

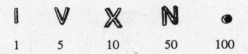

1   5   10   50   100

Use of the sign N for 50 seems natural, since it is simply the sign for 5 with a vertical line added, just as the sign for 100 shown in figure 9–10 is the sign for 10 with a vertical line added.

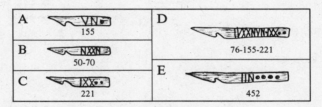

Fig. 9–15. Dalmatian shepherds' tallies.

Series closely resembling those of the Dalmatian tallies are also found in western Austria and parts of Switzerland: at Saanen, in tallies used for "parallel counts" (fig. 9–16); at Ulrichen, in sticks that were used for measuring milk; and at Visperterminen, in the loan tallies that used these signs to record loans (fig. 9–17).

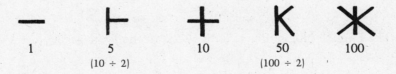

1   5   10   50   100
          (10 ÷ 2)   (100 ÷ 2)

Further evidence comes from "calendar figures," strange numerical signs on the wooden calendar boards and sticks used from the late Middle Ages till the seventeenth century in England and northern Europe from Austria to Scandinavia (figs. 9–18, 9–19).

38

39

Fig. 9–16. Swiss shepherds' tallies, eighteenth century, found at Saanen. Museum für Völkerkunde, Basel.

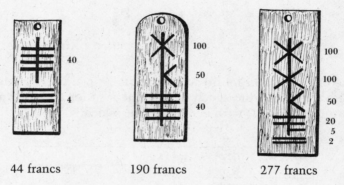

44 francs   190 francs   277 francs

Fig. 9–17.

Allowing for graphic variations, these calendar boards give the following series of signs to indicate the "golden number" of the nineteen-year Metonic cycle:

1 2 3 4	5	6 7 8 9	10	11 12 13 14	15	16 17 18 19

The English clog almanacs (fig. 9–20) of the Renaissance used these signs:

1 2 3 4	5	6 7 8 9	10	11 12 13 14	15	16 17 18 19	20	21 22

And the following signs appear on Scandinavian runic calendars:

·	··	::	::	>	>>>>	+	++++	+	++++	+	++
1	2	3	4	5	6 7 8 9	10	11 12 13 14	15	16 17 18 19	20	21 22

These notations may seem dissimilar at first sight, but examination of them shows that their origin lies in the notched tally stick. For numbers 1, 5, and 10 they all have signs that bear resemblances to the Roman numerals I, V, and X and the Etruscan numerals I, Λ, and + or X.

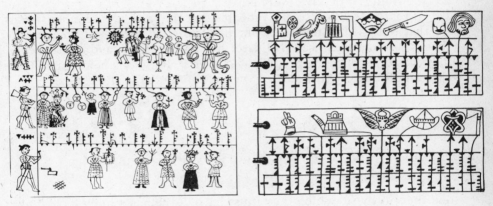

(*left*) Fig. 9–18. "Page" of a wooden calendar of 1526. Figdorschen Collection, Vienna.

(*right*) Fig. 9–19. Two "pages" of a fifteenth-century Tyrolean wooden calendar. Figdorschen Collection, Vienna.

Fig. 9–20. English clog almanac (calendar stick). Ashmolean Museum of Oxford.

The last piece of evidence is particularly convincing because it comes from a tradition that is unquestionably free of any Roman influence. In the nineteenth century, for their irrigation tally sticks, the Zuñi Indians of North America still used a series of signs closely resembling Roman numerals: a vertical notch for 1, a slightly deeper V-shaped notch (or a slanting notch) for 5, and an X-shaped notch for 10 (fig. 9–21).

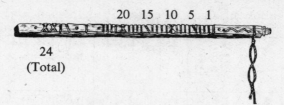

20 15 10 5 1

24
(Total)

Fig. 9–21. Zuñi irrigation tally stick. The count made from right to left totals 24, which is marked at the left end in the form XXI\, which recalls the Roman notation XXIV, using the subtractive principle.

There can be no doubt that Roman and Etruscan numerals, whose shapes for numbers 1, 5, and 10 are found all over the world, are directly derived from the notched tally stick.

It therefore seems highly probable that the same technique of notching that was used by the Zuñis and by Dalmatian shepherds (to cite only those two examples) was also used in early antiquity, and perhaps even in prehistoric times, by peoples who lived in Italy long before the Etruscans and Romans. At first they were naturally led to these signs:

1      $\mathbf{I}$

5   / or \ or /\ or ⟩ or V or ⟨ or ⋋ or ⋌

10   X or X̸ or X̶ or ✛ or X̄

and then to these (which they must have developed simply by adding a vertical line to the signs V and X).

50   N̈ or V̈ or A̅ or V̅I

100   Ẍ or ✳ or X̸I

Heirs to this old tradition, the Etruscans and Romans kept only the following signs:

	ETRUSCANS				ROMANS	
1	I				I	1
5	∧				V	5
10	X	or X̸	or ✛		X	10
50	⋀				⋁	50
100	✳				(✳)	100

before developing these signs for the number 1000:

⊗   or   ⊗   or   ⊕

This hypothesis is supported by the fact that a similar form appears in the Etruscan cameo shown in figure 8–1. Furthermore, as we have seen (figs. 9–2, 9–4), the archaic Roman numeral for 500 is half of one of those signs.

We can now reconstruct the origin of the Roman numerals for 50, 100, 500, and 1000. The sign previously used (on tallies) for 50 went through a series of changes (fig. 9–2) and finally took the form of the letter L:

Ѱ → ↓ → ⬇ → ⊥ → ⊥ → L

The sign for 1000 (from which the Roman numeral D, for 500, was derived) must have evolved into the form Φ, obviously the origin of all the graphic stylizations once used by the Romans for that value, and then it was gradually replaced by the letter M, under the influence of the initial of the word *mille* (fig. 9–22).

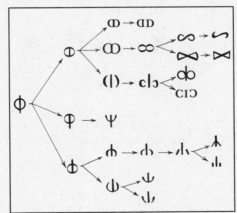

Fig. 9–22.

The archaic sign for 100 must have evolved into this form (found in Etruscan inscriptions): ⋈

and then into the following forms, which were used by the Romans and other peoples of Italy (the Oscans, for example):

Ɔ   or   C

The Romans identified the second of these signs with the letter C, under the influence of the initial of the word *centum*.

This is the most plausible explanation of the origin of Roman and Etruscan numerals. And it is certainly not contradicted by what A. P. Ninni reported in the nineteenth century: that Tuscan peasants and shepherds still used the signs they called *cifre chioggiotti* (fig. 9–23) much more often than Hindu-Arabic numerals. These have been interpreted as survivals of Etruscan or Roman numerals, but it seems more reasonable to regard them as a continuation of the tradition of notching, which is older than any form of writing and has been found in all parts of the world.

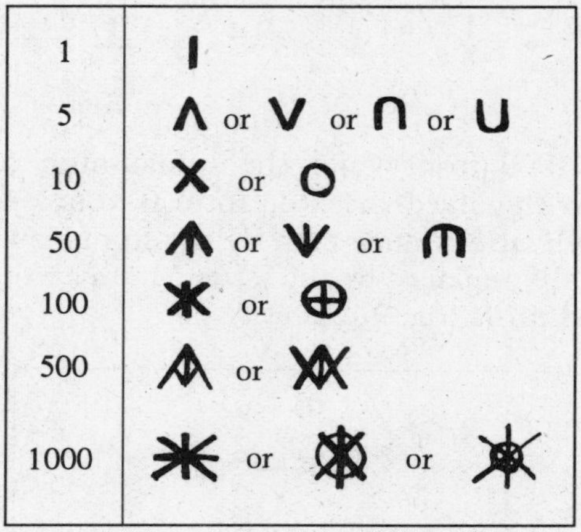

Fig. 9–23.

Perhaps some day an archaeologist excavating a prehistoric site will discover notched bones bearing numerical signs similar to Roman numerals. . . .

10

Was Writing Invented by Accountants?

Five thousand years ago, the Sumerians and Elamites learned to write

Disregarding Egyptian hieroglyphs for the moment, the oldest known forms of writing were invented shortly before the end of the fourth millennium B.C., not far from the Persian Gulf, in Lower Mesopotamia and Elam.

It was in Lower Mesopotamia, or Sumer, that the civilization known as Mesopotamian began developing in the fourth millennium, with the appearance of the Sumerians, a non-Semitic people whose origin is still unknown. Elam was a region next to Mesopotamia, occupying the western part of the plateau of Iran and the eastern part of the plain on which Sumer was situated.

Sumerian writing originated about 3200–3100 B.C., and the writing known as Proto-Elamite about 3000. Comparisons have been made between a few Proto-Elamite signs and a certain number of Sumerian pictographs, but most of them are so different that the hypothesis of a common origin is not tenable. This would seem to indicate that the Elamites invented their writing independently of the Sumerians. If they borrowed anything from the Sumerians in that respect, it can only have been the general idea of writing, and perhaps a few signs, but certainly not their whole system.

In both Mesopotamia and Elam, writing seems to have been invented for strictly utilitarian reasons. In the second half of the fourth millennium, economic needs evidently led the people of the two countries to become gradually aware that their respective civilizations were encountering difficulties in continuing their growth, and that work had to be organized in a different way.

"In the expanding societies of Sumer and Elam, writing was invented by accountants who had to deal with economic operations much too complex and varied to be entrusted to a single memory. Writing bears witness to a radical transformation of the traditional way of life within the new social and political framework heralded by the great construction projects of the preceding period" (Amiet). These projects presupposed great wealth and, in particular, enough surplus food to feed the many workers they employed.

The royal Sumerian city of Uruk, originally situated on the right bank of a branch of the Euphrates (the water gradually subsided and the location of the city is now about twelve miles north of the river, in Iraq), is the best known and earliest excavated of the Sumerian sites, and it serves as a kind of "chronological yardstick." Deep probes in certain parts of it have revealed strata to which archaeologists refer in dating discoveries, with the order of the different levels corresponding to stages in the development of Sumerian civilization.

It was here, at the level designated Uruk IVa, that the oldest known examples of Sumerian writing were discovered. They are on small clay tablets that were made according to a "standard model" (fig. 10–1).

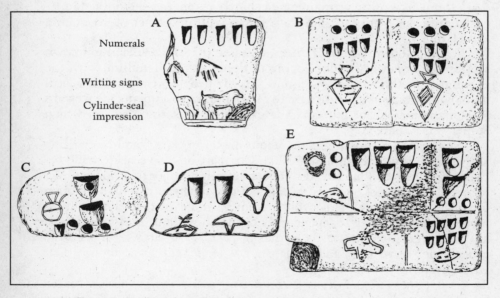

Fig. 10–1. Sumerian tablets from 3200–3000 B.C., found at Uruk. They are the earliest examples of the oldest known forms of writing.

The oldest known examples of Proto-Elamite writing, also on clay tablets, were found at several Iranian sites, chiefly Susa, at the archaeological level called Susa XVI (fig. 10–2). Susa, about 180 miles east of Sumer, was the capital of Elam, and under Cyrus the Great (reigned 550–529 B.C.) it became the administrative capital of the Persian Empire.

FRONT

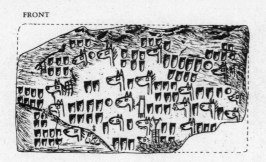

BACK

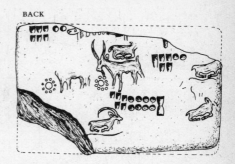

Fig. 10–2. Prote-Elamite tablet found at Susa, dating from c. 3000 B.C. According to V. Scheil, it is an inventory of horses in three categories—stallions (manes slanting forward), mares (manes slanting rearward), colts (no manes)—with the corresponding numbers indicated by marks of different shapes and sizes. On the back of the tablet is the impression of a cylinder seal representing standing and lying goats. Louvre Museum, Paris.

Each of these tablets has on one side, or sometimes on both sides, a certain number of marks of different shapes and sizes, made by pressing a tool into the clay while it was still soft (fig. 10–3). The same marks appear in many later Sumerian and Proto-Elamite documents. They have been identified as numerical symbols and are thus the oldest known "figures" in history. In the left-to-right order shown in figure 10–3, they correspond to the consecutive orders of units in a system of numeration.

Although they shared certain signs, the Sumerian and Proto-Elamite systems were quite different, as we will later see in detail. The large wedge, for example, indicated 60 in Sumer but 300 in Elam; the large circular sign, 3600 in Sumer but 100 in Elam; and so on.

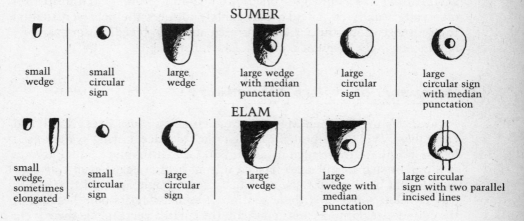

Fig. 10–3. The main numerical symbols in Sumerian and Proto-Elamite writing.

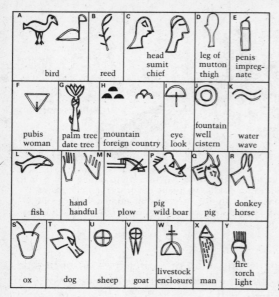

Fig. 10–4. Pictographs in archaic Sumerian writing.

Beside the numerical symbols are small drawings, made with a pointed tool, representing various kinds of objects and animals: these are writing signs (fig. 10–4). Some of the tablets also bear the impressions of cylinder seals that were rolled across them.

In both Elam and Sumer, these tablets seem to have been used to note quantities of various commodities, and they probably served as records of deliveries, inventories, and exchanges.

And so, by the beginning of the third millennium B.C., the Sumerians and Elamites had adopted the practice of recording numerical information on small, usually rectangular clay tablets to meet the needs created by their intense economic activity (farming and stockbreeding, for example, and increasingly frequent commercial exchanges).

Precursors of written accounting

Excavations in Elam and Mesopotamia, chiefly at Susa (level XVIII) and Uruk (level IVb), have brought to light an older accounting system used by the people of those regions about 3300 B.C. It involved small objects of different shapes and sizes, placed in spherical or egg-shaped clay containers that have come to be called envelopes (table 5–1). At first, these clay envelopes attracted the attention of archaeologists only because of the cylinder-seal impressions on them; the small objects inside them were considered to be talismans or symbols of the commodities specified in a contract. Actually, as we will see, they were used to indicate the amount of a transaction.

In this accounting system, directly derived from the widespread practice of pebble counting, the small clay tokens symbolized numerical values by their shapes and sizes as well as by their number. Each of them corresponded to an order of units in a numerical system: a cylinder (or a small cone) for a unit of the first order, a sphere for a unit of the second order, a disk (or a large cone) for a unit of the third order, and so on.

When tokens had been placed inside a soft clay envelope, its opening was closed and one or two cylinder seals were rolled across it to guarantee its origin and authenticity.

From clay envelopes to account tablets

This transition has been elucidated by archaeological discoveries recently made in Iran by the Délégation Archéologique Française en Iran during excavations at the site of the Susa Acropolis, under the direction of Alain Lebrun.*

Direct evidence of the transition has so far been found only in Elam, but there are good reasons for believing that it also took place in Sumer. Though each had its own distinctive elements, the Elamite and Sumerian civilizations were equivalent. They followed similar lines of development, under the same conditions, in the second half of the fourth millennium B.C. From the beginning, they both used clay for symbolic expressions of thought, and they later used it for representing language. As we have seen, there are obvious similarities between Sumerian tablets from about 3200–3100 B.C. and the earliest known tablets bearing Proto-Elamite writing, from about 3000–2900. And finally, the system of envelopes and tokens was used in both Elam and Sumer about 3500–3300. It is therefore reasonable to hope that new archaeological discoveries in Sumerian territory will someday settle the question.

Let us now follow, in chronological order and in the light of recent archaeological discoveries at Susa, the evolution of the Elamites' accounting system, comparing it with the evidence we have from Sumer.

FIRST STAGE
Archaeological level: Susa XVIII.
Period: c. 3500 B.C.

At Susa, the system of tokens and envelopes was used for recording numbers (the amount of a commercial transaction, for example). When an

---

*I will take this opportunity to express my deep gratitude to Alain Lebrun and François Vallat, who allowed me to examine those important discoveries and authorized me to report on them here, before their official publication.

envelope had been closed, with the tokens inside it (fig. 10–5), it was stored in the "archives." If verification was later needed, as in the case of a dispute between the parties concerned, the envelope was broken open and the tokens were counted. (Lebrun and Vallat report finding five different kinds of tokens at Susa: cylinders, spheres, disks, small cones, and large punched cones, as shown in fig. 10–6.) A closely similar system was used in Sumer during the same period. Evidence of it has been found at the archaeological level of Uruk IVb.

While the clay was still soft, the envelope was marked with one or two cylinder seals to identify it and prove that it was genuine (fig. 10–7). In both Sumer and Elam, each man of a certain social rank had his own seal, a small stone cylinder bearing an engraved symbolic image, often of a religious nature. The seal represented the man himself and was therefore associated with all economic or juridical activities in which he was involved. Its impression functioned as either a signature or an ownership mark, and was produced by rolling the cylinder on soft clay.

Let us imagine ourselves at Susa in 3500 B.C. A shepherd is about to take a flock of 299 sheep to pasture for several months. The flock has been entrusted to him by a rich stockbreeder of the region. Before leaving, the shepherd goes with his employer to have the number of sheep verified by an accountant who administers the breeder's property.

Fig. 10–5. Intact clay envelope, as seen in an X-ray photograph.

Fig. 10–6. Tokens found at Susa, level XVIII, by the Délégation Archéologique Française en Iran, during the excavations of 1977–78.

or

cm
0   1   2   3

Fig. 10–7. Spherical clay envelope marked by two cylinder seals, c. 3500–3300 B.C. Found at Susa. Louvre Museum, Paris.

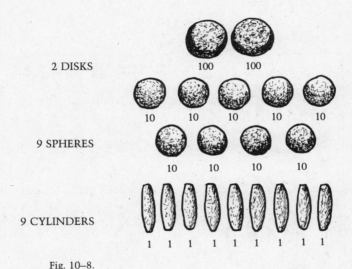

2 DISKS

9 SPHERES

9 CYLINDERS

Fig. 10–8.

After counting the sheep, the accountant fashions a round clay envelope about the size of a modern tennis ball, making it hollow by pressing it with his thumb from the inside. When he withdraws his thumb, it leaves an opening in the envelope. Through this opening he drops two clay disks each symbolizing 100 sheep, nine spheres each symbolizing ten sheep, and nine cylinders each symbolizing one sheep, making a total of 299 (fig. 10–8). He then closes the opening. To authenticate the record he has just made, he rolls the stockbreeder's seal over the surface of the envelope. There is now no possibility of falsification.

When the clay has dried, the accountant stores the envelope with others of its kind. It and the tokens inside it guarantee the count that has been made, for the shepherd as well as for the owner of the flock. When the shepherd returns, the accountant will break open the envelope and the tokens will enable him to determine whether or not the whole flock has been brought back.

SECOND STAGE

Archaeological level: Susa XVIII.
Period: 3300 B.C.

The system just described had one drawback: the envelope had to be broken whenever someone wanted to verify the record it contained. To avoid that drawback, a procedure related to the already ancient practice of notching was adopted: the tokens inside an envelope were symbolized by marks of different shapes and sizes on its outer surface, which was also marked by a cylinder seal as before (fig. 10–9).

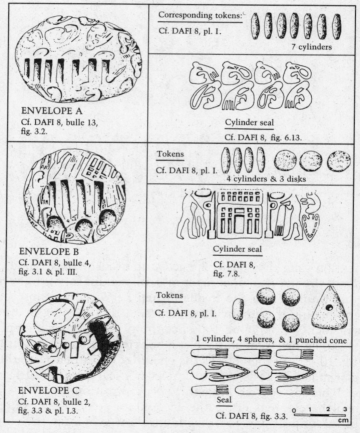

Fig. 10–9. Clay envelopes, each containing tokens symbolized by marks on the surface, which is also marked by a cylinder seal. They date from c. 3000 B.C. and were found at Susa, level XVIII, by the Délégation Archéologique Française en Iran during excavations in 1977–78. "DAFI" and the indications following it refer to the article by Lebrun and Vallat listed in the bibliography.

These marks were as follows: a small wedge-shaped sign (produced by pressing against the clay with a small stylus held at an angle) for a cylinder; a small circular sign (small stylus held perpendicular to the surface) for a sphere; a large circular sign (large stylus held perpendicular, or sometimes a fingertip) for a disk; a large wedge (large stylus held at an angle) for a cone; a large wedge with a small circular mark inside it for a punched cone (fig. 10–10).

An envelope containing three disks and four cylinders, for example, had three large circular signs and four small wedges on its outer surface (fig. 10–9B). One containing seven cylinders was marked with seven small wedges (fig. 10–9A).

It was no longer necessary to break an envelope in order to learn or verify its contents: "reading" the signs on it was sufficient.

To the best of my knowledge, no envelope of this type has yet been found in Sumerian territory.

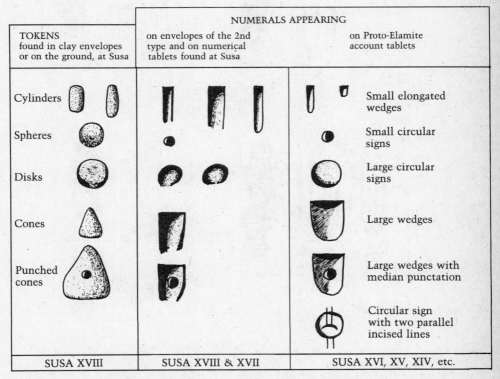

Fig. 10–10. The different marks on clay envelopes corresponded to the tokens they contained. These marks are similar not only to those on the numerical tablets found at Susa but also to the numerals on Proto-Elamite tablets of later periods.

THIRD STAGE
Archaeological level: Susa XVIII.
Period: c. 3250 B.C.

Since the marks on the outside of an envelope made the tokens inside it superfluous, both envelopes and tokens fell into disuse and were replaced with clay tablets bearing, on one surface, the same marks that had formerly been made on envelopes. At first, these tablets were roughly circular or oval (fig. 10–11).

A cylinder-seal impression still guaranteed their authenticity and the numerical symbols on them conveyed the same information as the tokens they had once represented. It was therefore in this period that the first "account tablets" were created in Elam.

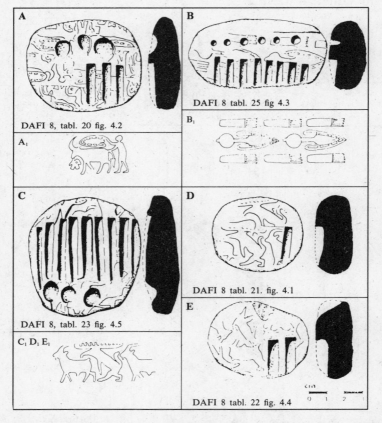

Fig. 10–11. Oblong or approximately round tablets bearing numerical signs of the same kind as those on clay envelopes and marked by one or two cylinder seals. They date from c. 3250 B.C. and were found at Susa, level XVIII, by the D.A.F.I. during excavations in 1977–78. "DAFI" refers to the article by Lebrun and Vallat listed in the bibliography.

The three stages we have just examined took place in a relatively short time, since the objects that attest them were all found at the same archaeological level: Susa XVIII. Furthermore, the same seal was used to mark one envelope and two tablets (see the seal shown to the right of envelope C in fig. 10–9 and below tablet B in fig. 10–11).

FOURTH STAGE
Archaeological level: Susa XVII.
Period: c. 3200–3000 B.C.

Like those of the preceding stage, the tablets of this one bore only numerical symbols and cylinder-seal impressions, with no writing in the

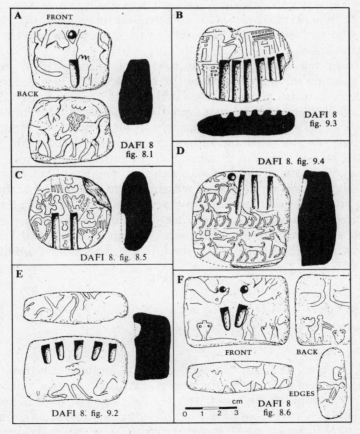

Fig. 10–12. Numerical tablets, c. 3200–3000 B.C., found at Susa, level XVII, by the D.A.F.I. during excavations in 1972. "DAFI" refers to the article by Lebrun and Vallat listed in the bibliography.

strict sense of the term (fig. 10–12). These tablets, however, show certain changes in relation to those found at level XVIII and indicate increased use of the new accounting system. Their shapes are less crude, their numerical symbols are more regular and engraved less deeply, and they are marked with seals not on one side only but on both sides and also on edges.

ANALYSIS OF THE NOTATION

The system of notation used on envelopes and tablets up to this point does not constitute true writing. The seals and numerical symbols convey information, but the things directly involved in a transaction are not indicated: they are designated only by their quantities, with no specific signs showing their nature. And the transaction itself is not specified; we have no way of knowing whether it was a sale, an exchange, a distribution, or something else. Nor will we ever know anything about the people who took part in the transaction: their names, where they lived, their occupations, etc.

It seems reasonable to assume, however, that the nature of the transaction was implicitly indicated by the seal or seals. Since the economy was still in its infancy, people who had business dealings with each other must have been personally acquainted and must also have known each other's seals. They were therefore able to identify a given transaction when they saw the seals and numerical symbols related to it. This explains the lack of detail in those clay documents.

They were one of the last stages in the "prehistory" of writing and they disappeared when the use of writing in the true sense became common. It should be pointed out, however, that their numerical symbols constituted a genuine written numeration, since they were part of an elaborate notational system and each of them was associated with an order of units. When writing appeared, it kept this same numerical notation.

FIFTH STAGE

Archaeological level: Susa XVI.
Period: c. 3000–2900 B.C.

The tablets used in Elam during this period are generally thinner and more nearly rectangular, so that they practically have a standardized format. Most importantly, they bear the first signs of what is known as Proto-Elamite writing, along with numerical symbols and sometimes cylinder-seal impressions. The writing signs were evidently used to indicate the nature of the objects involved in the economic operation associated with the tablet.

Several of the tablets found at Susa level XVI have no seal impressions, which seems to support the hypothesis that the nature of the things referred to by the envelopes and tablets of levels XVII and XVIII must have been implicitly indicated by the seal impressions on them (fig. 10–13A, B, C).

SIXTH STAGE
Archaeological levels: Susa XV and Susa XIV.
Period: c. 2900–2800 B.C.

The writing signs on the Proto-Elamite tablets of this period are beginning to cover a larger area than the numerical symbols (fig. 10–13D, E, F, G). Does this mean that Proto-Elamite writing was able to indicate the sounds and grammatical structure of the corresponding language? It seems likely, but we cannot know because the meaning of that writing still escapes us. Its signs probably represented all sorts of objects and living creatures. With a few exceptions, they are highly simplified drawings that are no longer directly pictorial (fig. 10–14). Since we do not know the language that corresponded to them, their phonetic values are also unknown to us.

ARCHAIC SUMERIAN TABLETS

About 3200–3100 B.C. (that is, during the fourth stage in the development of the Susa tablets), the oldest known Sumerian tablets appeared at Uruk, level IVa. As we have seen (fig. 10–1), they were small, generally rectangular clay tablets bulging on both sides. Some of them bear engraved marks as well as seal impressions and small drawings that functioned as pictographs or ideograms. Others have the same drawings and numerical symbols but no seal impressions. The purpose of those more or less realistic drawings was to specify the nature of the things or animals involved in a given economic operation, their number being indicated by the corresponding numerical signs.

Although these tablets all seem to have had a purely economic function, some of them reveal that Sumerian writing (the word "writing" here refers to a systematic effort to note spoken language) was now based not on vague thought translated into "pictures" but on more precise thought, analyzed, ordered, and decomposed into elements, as it may appear in language. This is shown by the fact that some of the tablets (fig. 10–1E) have horizontal and vertical lines that form "boxes" containing writing signs and numerical symbols together.

Sumerian tablets used about 3200–3100 B.C. therefore show a marked advance over tablets used in Susa during the same period. The former already had writing signs that specified the nature of the commodities

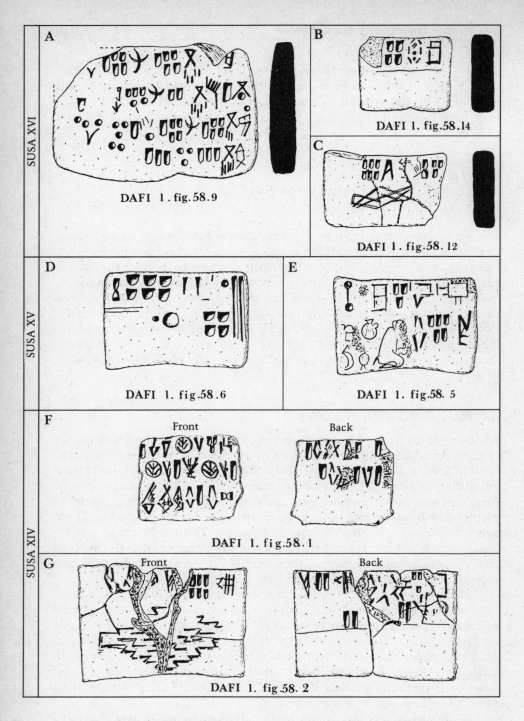

Fig. 10–13. The oldest known Proto-Elamite tablets, c. 3000–2800 B.C., with writing signs beside the corresponding numerals. They were found at Susa by the D.A.F.I. during excavations in 1967–71. "DAFI" refers to the article by Lebrun and Vallat listed in the bibliography.

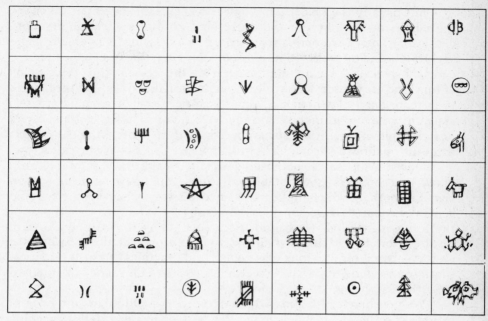

Fig. 10–14. Proto-Elamite writing signs.

involved in various transactions, while the latter were still limited to symbolism.

The mnemonic function of Sumerian writing*

To try to understand the way in which these civilizations invented and developed their respective writings, we will examine more closely the signs that appear on archaic Sumerian tablets and attempt to discern the primary character of the corresponding writing system.

The signs are drawings that represent the appearance of the things with which they are associated. Some are quite realistic (fig. 10–4). Others are highly simplified but still have the outlines of the things they represent. An ox, a donkey, a pig, and a dog, for example, are each depicted by an evocative though simplified drawing that shows the head of the animal for the animal itself.

---

*I owe the material in this section to the Assyriologist Jean Bottéro, as the result of many conversations with him. I thank him for having been kind enough to read the section in preliminary form, and I have taken full advantage of his criticisms and observations.

But usually the object is no longer recognizable; a part is taken for the whole, or an effect for the cause, in a kind of stylization and condensation. Thus a woman is represented by a drawing of a pubic triangle (fig. 10–4F) and the verb "to impregnate" by a drawing of a penis. In many cases, however, we see no relation between the sign and the object. Signs of this kind are simple, more or less geometrical drawings, and the objects they represent (known to us from semantic and paleographic studies) apparently have nothing in common with them. The sign for a sheep is one example of this: it consists of a cross within a circle (fig. 10–4U). Is it meant to show a sheepfold? A stockbreeder's mark? We do not know.

One striking feature of these signs is their uniformity: they are firmly drawn and show very few notable variations from one example of a given sign to another, which indicates that their forms had been fixed and generally adopted. If we consider the number of variations that appear in the following period, we cannot help concluding that this uniformity resulted from the origin of writing as an invention, a deliberately devised system—based on previous discoveries and practices but also on an essential new element—that was either accepted voluntarily or imposed by authority.

We are therefore dealing with a system of graphic signs intended to express specific ideas. But it is not yet writing in the strictest meaning of the term. If we limit ourselves to that meaning, a simple "visual representation of human thought by material signs" cannot be considered true writing, which is more directly related to speech than to thought itself. If there is to be true writing, there must be *a systematic effort to render speech*, because writing, like language, is a system and not a series of occasions or accidents. "Writing is a system of human communication by means of conventional signs which have definite meanings and represent a language, these signs being such that they can be transmitted, received, understood by both interlocutors, and associated with the words of spoken language" (Février).

With the Sumerian signs of this period, we are still in the "prehistory," or rather the "protohistory," of writing (or at the pictographic stage in the history of writing). All these signs, whether we know their meanings or not, are visual representations of material objects.

But from this we should not conclude that they could note only material objects, because each such object could be employed not only for the acts and activities it implied but also for related concepts. A leg, for example, could be used for "to walk," "to go," or "to stand"; a hand for "to take," "to give," or "to receive" (fig. 10–4M); the rising sun for "day," "light," or "clarity"; a plow for "to plow," "to sow," or "to till the soil," "plowman," or "farmer."

The content of each ideogram could also be enriched by the already ancient use of symbolism, which arose from social conventions. Two parallel lines expressed the idea of "friend" or "friendship," two crossed lines that of "enemy" or "hostility." And meanings could be extended by combining two or more signs to express new ideas or realities that did not lend themselves easily to direct visual representation. The combination "mouth + bread" was used for "to eat" or "to devour"; "mouth + water" for "to drink"; "mouth + hand" for "to pray" (in accordance with Sumerian ritual); "eye + water" for "to weep"; and in that country situated on a plain, where mountains suggested a foreign or enemy country (fig. 10–4H), "woman + mountain" indicated "foreign woman" (literally "woman from the mountains") and, by extension, "female slave," since women were brought from foreign countries, after being either bought or captured in war, to serve as slaves in Sumer. There was the same association of ideas for "male slave," designated by "man + mountain" (fig. 10–15).

The pictographic system was a methodical attempt to express everything expressed by language, but it was far from being able to do so precisely and unambiguously. Still too closely bound to material objects that could be given direct visual representation, it required a very large number of signs. It has been estimated that about 2000 were in use during the first stage of writing in Mesopotamia.

The system was not only difficult to use, it was also ambiguous. If a drawing of a plow, for example, could mean "plow," "to plow," or

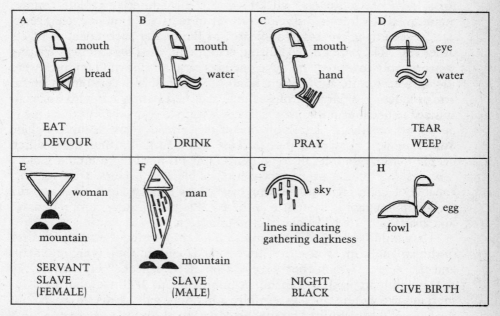

Fig. 10–15. Examples of combined signs (or "logical aggregates") used in archaic Sumerian writing.

"plowman," how could one know which of those meanings was intended? How could a sign be made to indicate important distinctions and shades of meaning, carefully expressed by language and necessary for perfect comprehension of thought, such as the ideas of plurality or singularity, sex or quality, or the countless relations between objects in time and space, or the multiple variations of acts in time?

Although this writing rendered spoken language to some extent, it was still limited to what could be expressed by images—that is, by designations of objects and acts capable of being directly represented or suggested. That is why Sumerian writing in its first stage is not completely intelligible to us and probably never will be.

Consider, for example, the drawing of an ox head on tablet D, figure 10–1. Was it really meant to represent an ox head? Or (and this is more likely) did it indicate an ox, or a head of cattle in general, or one of the products of cattle, such as leather, milk, horn, or meat? Or did it designate a person who had a name analogous to Oxford or Oxenham? And what exactly was the operation indicated here: sale, purchase, exchange, distribution? Probably only the people involved in that operation could fully understand the sign without having its context explained to them.

So at this stage Sumerian writing was much less suited to conveying new information than to reminding people of things they already knew in general but might have forgotten in detail.

This did not prevent it from answering the needs of the time. Except for a few "lists of signs," all known archaic Sumerian tablets contain summaries of administrative operations of redistribution or exchange, as can be seen from the totals indicated at the end of each tablet or on its other side. They are all accounting documents used by temples. Writing was probably invented and developed because at that time temples were the only economic authorities for all of Sumer, where constant, systematic production of large surpluses required increasingly complex and centralized redistribution. Accounting consists in keeping records that enable people to recall the details of operations already carried out. Archaic Sumerian writing served that purpose quite well. Its first use—which deeply marked its later development—was primarily as a memory aid.

Before it could be perfectly intelligible and become true writing, capable of unambiguously noting everything expressed by language, the archaic pictographic system had to be made much more comprehensive and greatly improved in clarity and precision.

It would be beyond my purpose here to describe the main features of the language for which the Sumerians developed their graphic system and the way in which they gave it a phonetic aspect. I will simply say that notable progress was made, beginning about 2850 B.C., but that Sumerian writing remained essentially a memory aid; it was improved but not radically transformed by the introduction of a phonetic aspect, which did little more than extend the meanings of pictographs and make them more precise. After this development, the Sumerians kept a great number of their archaic signs, each of them still indicating a word that designated a person, animal, or object, or several words connected by relations of meaning with varying degrees of subtlety, such as symbolism and causality, for example.

In Mesopotamia and Elam, the idea of writing seems to have arisen in response to strictly utilitarian needs within an essentially economic context. In any case it is clear that in Elam, and probably also in Sumer, written accounting must have preceded the transcription of language.

11

Sumerian Numerals

To represent whole numbers, the Sumerians used a system that had a special sign for each of these units: 1, 10, 60, 600 (= 60 × 10), 3600 (= 60^2), 36,000 (= 60^2 × 10)—that is, for each of the following numbers:

$$1$$
$$10$$
$$10 \times 6$$
$$(10 \times 6) \times 10$$
$$(10 \times 6 \times 10) \times 6$$
$$(10 \times 6 \times 10 \times 6) \times 10$$

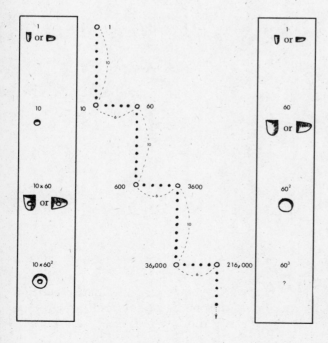

Fig. 11–1. Structure of the Sumerian numeration: a sexagesimal system built on a compromise between the alternating bases 10 and 6.

It was thus a sexagesimal written numeration built on the *alternating bases* 6 and 10 (see Chapter 2).

In early periods, the number 1 was represented by a small wedge (sometimes elongated), 10 by a small circular sign, 60 by a large wedge, 600 (= 60 × 10) by a combination of the latter two signs, 3600 (= 60 × 60) by a large circular sign, 36,000 (= 3600 × 10) by a large circular sign with median punctuation (fig. 11–1).

At first, these numerals were inscribed on clay tablets with this orientation:

| 1 | 10 | 60 | 600 | 3600 | 36,000 |

But in the twenty-seventh or twenty-sixth century B.C. the noncircular signs underwent a 90° counterclockwise rotation:

| 1 | 10 | 60 | 600 | 3600 | 36,000 |

Numbers 1 to 9 were represented by repeating the sign for 1 as often as necessary; numbers 10, 20, 30, 40, and 50 by repeating the sign for 10; numbers 60, 120, 180, 240, etc. by repeating the sign for 60; and so on. In general, to represent a given number the Sumerians were content to repeat as often as necessary the corresponding numerals within each order of units.

An account tablet from Uruk, late fourth millennium B.C. (fig. 10–1C), represents the number 691 in this way:

On a tablet found at Fara (Shuruppak), dating from about 2650 (fig. 11–2), we find the notation for the number 164,571 as shown in figure 11–3.

Thus the Sumerian system sometimes required inordinate repetition of signs, since it was essentially based on the principle of juxtaposing signs and adding their values. To note the number 3599, for example,

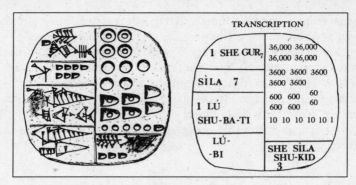

	TRANSCRIPTION
1 SHE GUR₇	36,000 36,000 36,000 36,000
SÌLA 7	3600 3600 3600 3600 3600
1 LÚ SHU-BA-TI	600 600 60 600 600 60 10 10 10 10 10 1
LÚ- -BI	SHE SÌLA SHU-KID 3

Fig. 11–2. Tablet from Shuruppak (Fara), dating from about 2650 B.C. It was probably an administrative document in the archives of the ancient city of Shuruppak, or perhaps a calculation exercise assigned to students. It represents a problem of division, since it mentions a dividend, a divisor, a quotient, and a remainder. In the left-hand column, the first compartment contains a numerical sign and a cuneiform group signifying "barley granary." In the second compartment the notation for the number 7, in archaic Sumerian figures, is associated with a writing sign designating the *silà*, a measure of volume equivalent to 0.842 liter. In the third compartment are two cuneiform groups, one meaning "a man," the other something like "receives in hand." Finally, the inscription in the bottom compartment means "these men." The numerical signs in the first compartment of the right-hand column represent the number 164,571. The inscription in the bottom compartment of this column means "Three *silà* of barley are left."

Since the "barley granary" was a unit of volume equal to 1,152,000 *silà* of barley (969,984 liters), and since the number 164,571 is the quotient of the division of 1,152,000 by 7, with a remainder of 3, the tablet indicates that a distribution has been carried out. It can be translated as: "We have distributed 1 'barley granary' [i.e., 1,152,000 *silà* of barley] among a certain number of men, giving 7 *silà* to each of them. These men numbered 164,571, and at the end of the distribution we had 3 *silà* of barley left."

⊙ ⊙ ⊙ ⊙	36,000 + 36,000 + 36,000 + 36,000	144,000
◯ ◯ ◯ ◯ ◯	3600 + 3600 + 3600 + 3600 + 3600	18,000
	600 + 600 + 600 + 600 + 60 + 60	2400 120
	10 + 10 + 10 + 10 + 10 + 1	50 1
		164,571

Fig. 11–3.

twenty-eight signs were needed! So for the sake of simplification Sumerian scribes often used the subtractive method, noting 9, 18, 38, and 57, for example, in the forms:

10 – 1	20 – 2	40 – 2	60 – 3
9	18	38	57

(The sign ⊢ or ⌐ , pronounced *lá* in the Sumerian language, was the equivalent of our minus sign.) And these forms were used for 2360 and 3110:

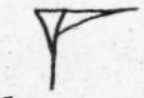

2400 – 40

3120 – 10

The way in which the Sumerians arranged their numerals within each order of units made it easy for them to discern the values represented by different groups of identical signs. To take the example of numbers 1 to 9, there were generally two types of arrangements (fig. 11–4), one based on visual recognition of odd or even, the other on a special status for the number 3.

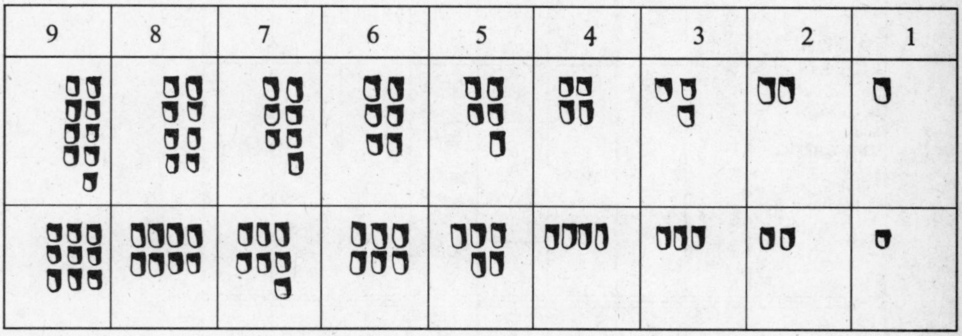

9	8	7	6	5	4	3	2	1
▯▯▯	▯▯	▯▯	▯▯	▯▯	▯▯	▯▯	▯▯	▯
▯▯▯	▯▯▯	▯▯▯	▯▯▯	▯▯				
▯▯▯	▯▯▯	▯▯▯	▯▯					
▯▯▯	▯▯▯▯	▯▯▯	▯▯▯	▯▯▯	▯▯▯▯	▯▯▯	▯▯	▯
▯▯▯	▯▯▯▯	▯▯▯	▯▯▯	▯▯				
▯▯▯		▯						

Fig. 11–4.

The development of cuneiform numerals

With the advent of cuneiform ("wedge-shaped") writing, Sumerian numerals naturally took on an entirely different appearance.

This change resulted from replacement of the old round stylus, blunt on one end and pointed on the other, with one whose tip had the same cross section as a modern ruler. This new type of instrument led the Sumerians to replace curves with as many vertical wedges or crescents as necessary (fig. 11–5). The number 1 was now represented by a vertical wedge (instead of the small wedge-shaped sign that stood for a cylinder), 10 by a crescent (instead of a small circular sign), 60 by a slightly larger vertical wedge (instead of the old large wedge-shaped sign), 600 by a larger vertical wedge and a crescent (instead of a large wedge with median punctation), 3600 by a polygon formed by combining four vertical wedges (instead of a large circular sign), 36,000 by this type of polygon with a crescent inside it (instead of a large circular sign with median punctation).

In this new system, the Sumerians even introduced a sign for 216,000 (60^3) by combining the signs for 3600 and 60.

It goes without saying that although Sumerian numerals took on different forms, the mathematical structure of the corresponding system was not altered. It was still a sexagesimal system built on a compromise between the two alternating bases 10 and 6, with a special sign for each of these numbers:

$$1, 10, 60, 10 \times 60, 60^2, 10 \times 60^2, 60^3.$$

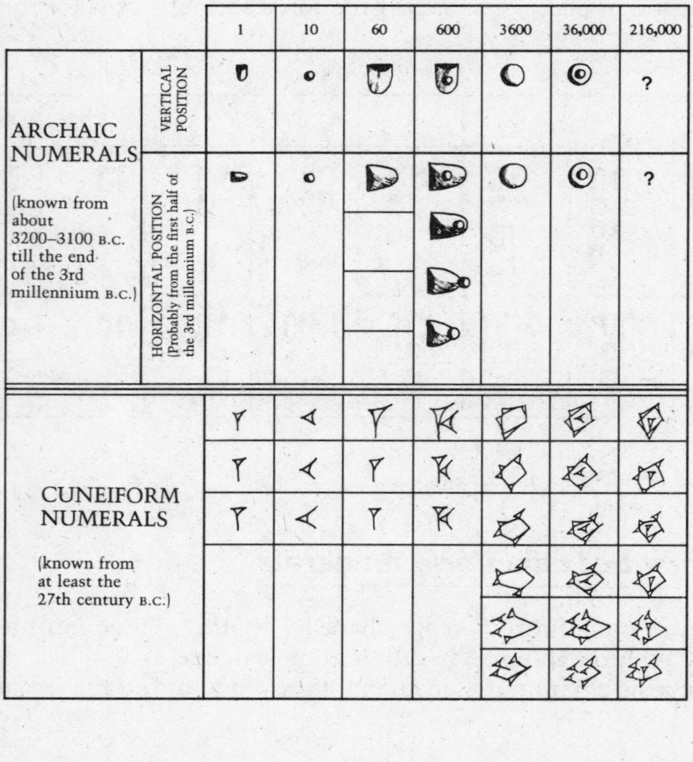

Fig. 11–5.
Changes
in Sumerian
numerals.

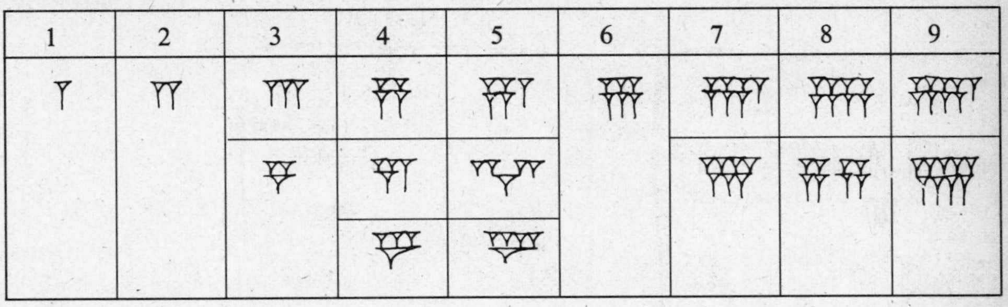

1	2	3	4	5	6	7	8	9
Y	YY	YYY	YYYY	YYYYY	YYYYYY	YYYYYYY	YYYYYYYY	YYYYYYYYY
		YYY	YYYY	YYYYY		YYYYYYY	YYYY YYYY	YYYYYYYYY
			YYYY	YYYYY				

Fig. 11–6.

YY	30 8	60 50 7	180 40 1	240 40 1	120 10 9
4	38	117	221	281	139

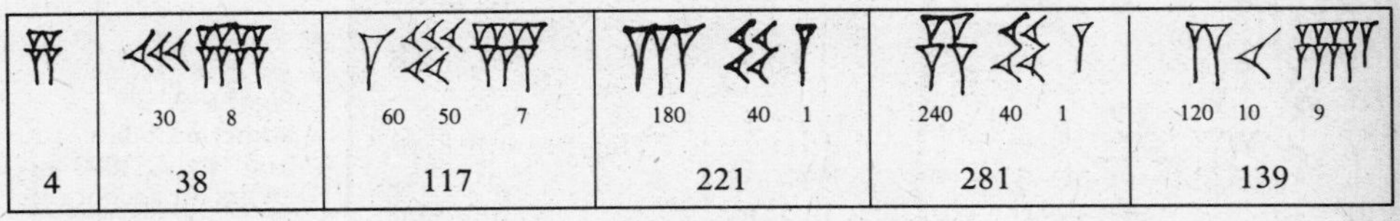

Fig. 11–7.

To represent a given number with cuneiform numerals, the Sumerians still repeated signs within each order of units as many times as necessary. And they continued to represent the successive units of each order by means of arrangements generally based on visual recognition of odd and even. For example, the numbers 1 to 9 were indicated as shown in figure 11–6.

On a Sumerian tablet from about 2000 B.C. (fig. 11–9, see next page) we find numerical notations for different kinds of livestock as shown in figure 11–7 (above).

On a tablet contemporary with figure 11–7, from a clandestine excavation at Tello, numbers 54,492 and 199,539 are expressed as shown in figure 11–8.

Cuneiform signs, which were used at least as early as the twenty-seventh century B.C., coexisted for several centuries with the archaic

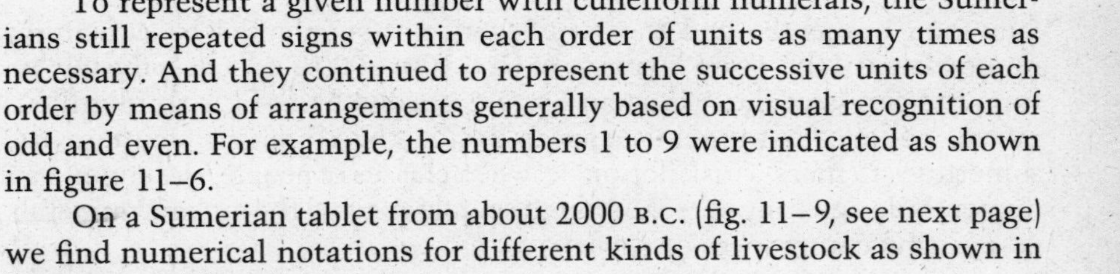

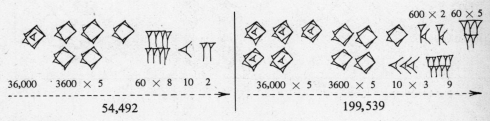

| 36,000 | 3600 × 5 | 60 × 8 | 10 | 2 | | 36,000 × 5 | 3600 × 5 | 600 × 2 | 60 × 5 | 10 × 3 | 9 |

54,492   199,539

Fig. 11–8.

TRANSLATION

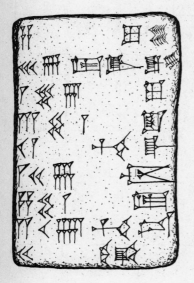

4	Fattened sheep
38	Lambkins
117	Rams
221	Ewes
11	He-goats
88	She-goats
281	Lambs
139	Nearly adult kids
20	Young kids

Fig. 11–9.
Sumerian tablet
from about 2000 B.C.,
giving an account of
livestock in
cuneiform signs.

Sumerian numerals, and the latter were not supplanted by cuneiform numerals until the Third Dynasty of Ur (2100–2000 B.C.). On some tablets from the period of the Akkadian Empire (second half of the third millennium), cuneiform numerals appear beside their archaic counterparts, sometimes to mark a distinction between classes of people, the cuneiform numerals designating persons of high social rank and the archaic numerals designating commoners or slaves.*

Beginning about 2600 B.C., certain irregularities appeared in the cuneiform numerical notation. Besides the use of the subtractive principle that has already been mentioned, there was an irregular notation for the consecutive multiples of 36,000. Instead of noting numbers 72,000, 108,000, 144,000, etc. in the usual way—that is, by repeating the sign for 36,000

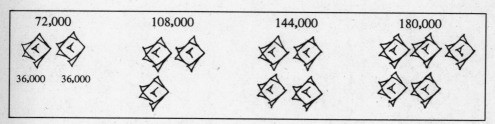

Fig. 11–10.

*Personal communication from Jean-Marie Durand.

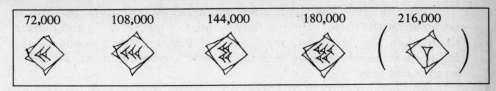

| 72,000 | 108,000 | 144,000 | 180,000 | 216,000 |

Fig. 11–11.

(fig. 11–10)—the Sumerian scribes had the idea of sometimes noting them as shown in figure 11–11.

The forms above obviously correspond to:

72,000 = 3600 × 20 (instead of 36,000 + 36,000)
108,000 = 3600 × 30 (instead of 36,000 + 36,000 + 36,000)
144,000 = 3600 × 40 (instead of 36,000 + 36,000 + 36,000 + 36,000)
180,000 = 3600 × 50 (instead of 36,000 + 36,000 + 36,000 + 36,000 + 36,000)

Putting it in modern terms, this procedure involved finding a common factor. Having noticed that the sign for 36,000 was composed of the signs for 3600 and 10, the Sumerian scribes used 3600 as a common factor. To represent the number 108,000, for example, they went from the form

$$(3600 \times 10) + (3600 \times 10) + (3600 \times 10)$$

to the simpler form

$$3600 \times (10 + 10 + 10).$$

Difficulties of the cuneiform numeration system

One difficulty lies in the fact that numbers 70 and 600 are both represented by a combination of the signs for 60 (large vertical wedge) and 10 (crescent):

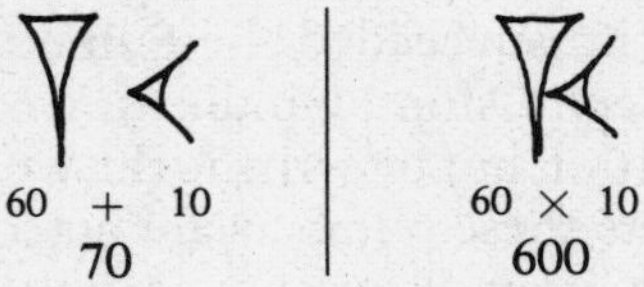

$$\begin{array}{c|c} 60 + 10 & 60 \times 10 \\ 70 & 600 \end{array}$$

This ambiguity arises because the combination of signs is based on the additive principle for 70 and on the multiplicative principle for 600. In the earlier notation, there was almost no risk of such ambiguity:

60 + 10   OR   OR
70   60 × 10
          600

The Sumerians usually avoided confusion by separating the vertical wedge from the crescent when they wanted to represent 70 and joining them when they wanted to represent 600. But this rule was not always followed; some tablets have been found on which both numbers are noted in the same way.

Another difficulty of this notation concerns numbers 61, 62, 63, etc., since 60 and 1 both have the same sign (a vertical wedge), so that there is no way to distinguish, for example:

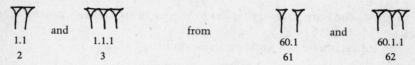

| 1.1 | | 1.1.1 | from | 60.1 | and | 60.1.1 |
| 2 | | 3 | | 61 | | 62 |

At an early stage, however, the Sumerians began indicating 1 by a small vertical wedge and 60 by a larger one:

60 1	60 2	60 3	60 4	60 5	60 6	60 7	60 8	60 9
61	**62**	**63**	**64**	**65**	**66**	**67**	**68**	**69**

And later, when the signs for 1 and 60 had the same size, a space was left between them:

60 1	60 2	60 3	60 4	60 5	60 6	60 7	60 8	60 9
61	**62**	**63**	**64**	**65**	**66**	**67**	**68**	**69**

The Akkadians, who succeeded the Sumerians and inherited their culture, eliminated all confusion by indicating 61, 62, 63, etc., not with a purely numerical notation but by writing the words for them. On several Assyro-Babylonian tablets these numbers are noted in the following forms, in which the cuneiform group "*shu-shi*" corresponds to the word for 60:

| 1   SHU-SHI   1 | 1   SHU-SHI   2 | 1   SHU-SHI   6 |
| **61** | **62** | **66** |

This method was extended to other numbers:

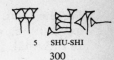

2 SHU-SHI	3 SHU-SHI	5 SHU-SHI
120	180	300

Continuance of the Sumerian system

Just as some French people still think in terms of the old francs that became obsolete in 1960, the inhabitants of Mesopotamia remained strongly attached to the tradition of counting by sixties long after the eclipse of the Sumerians. The Sumerian sexagesimal cuneiform numeration was still in common use at least as late as the end of the first Babylonian dynasty, in the fifteenth century B.C. Although the decimal system of the Assyrians and Babylonians was well established in Mesopotamia by the eighteenth century B.C., a tablet dating from that period, found at the site of Larsa, near Uruk, contains a count of sheep with numbers written in the Sumerian cuneiform system (fig. 11–12).

The scribe of another tablet from northern Babylonia, seventeenth century B.C., used the old system in drawing up a list of calves and cows for a landowner. For 277 cows and 209 calves, making a total of 486, he used the notations:

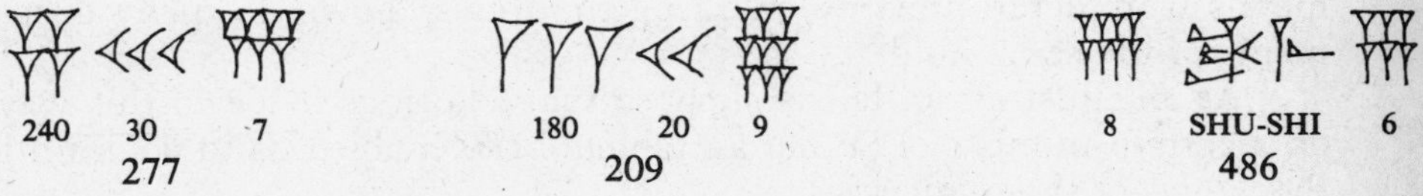

240 30 7	180 20 9	8 SHU-SHI 6
277	209	486

61 (ewes)	60 1	96 (ewes)	60 30 6
84 (rams)	60 20 4	105 (rams)	60 40 5
145 (sheep)	120 20 5	201 (sheep)	180 20 1

Fig. 11–12.

For the total, the number 486 is written in a form that can be translated as "8 sixties, 6 [units]." The scribe did this to avoid any misinterpretation of the total, just as we write the amount of a check in both figures and words.

How modern scholars deciphered Sumerian numerals

To understand the method used for this purpose, we must first note that among the many Sumerian account tablets that have been found, several are what can be called bills of lading. On one side, each of them has a list of various goods or commodities whose nature is indicated by Sumerian writing signs and whose number is indicated by groups of wedges or circular signs (or, in the case of tablets from later periods, by cuneiform numerals). On the other side is a summary of the information on the first side, giving totals of the items listed.

The front of a tablet discovered at Uruk, dating from about 2850 B.C. (fig. 11–13), lists fowls and bags of various commodities, and its back gives the total number of fowls and bags.

Sumerian scribes were in the habit of writing totals on the backs of their account tablets. By taking advantage of this fact, modern scholars were able to decipher the Sumerian numeration: the values of the numerals were determined by checking the totals shown on many documents of the same kind.

As an illustration, let us suppose that additions made in this way on a certain number of Sumerian tablets have enabled us to determine the values of the numerals:

$$\rhd \ =1 \qquad\qquad \mathsf{O} \ =10$$

We will also assume that in trying to find the still unknown value (which we will designate by the letter x) of the large wedge, we happen to come upon the tablet reproduced in figure 11–13, which bears the two numerals above, and also the unknown one:

 $=x?$

We begin by disregarding the notation for the fifteen fowls (a small circular sign, five small wedges and a pictograph), since the same notation appears unchanged on the back of the tablet. We consider only details related to the bags because they form a list of items associated with a single kind of writing sign. By adding up the numbers on the front of the tablet, using the values we already know, we obtain this total:

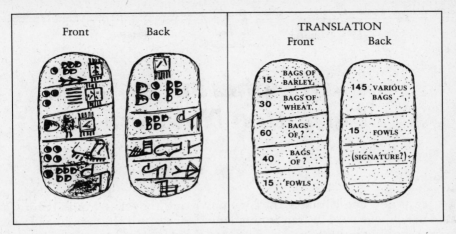

Fig. 11–13. Sumerian "bill of lading" from about 2850 B.C., discovered at Uruk.

$$10 + 5 + 30 + x + 40 = x + 85$$

And, on the back:

$$2x + 20 + 5 = 2x + 25$$

By forming an equation with these two results:

$$x + 85 = 2x + 25$$

we obtain the result we want:

$$x = 60$$

This does not allow us to conclude that the large wedge really had the value of 60, however, unless that value gives perfectly consistent results with other tablets of the same kind. But, in fact, this is known to be the case.

Such is the method that has enabled scholars to establish the values of the Sumerian numerals.

12

Deciphering a Forgotten Notation

L et us now come back to the ancient writing known as Proto-Elamite, which was used in Iran from about 3000 B.C. till the second half of the third millennium B.C.

It has not yet been deciphered, since the language associated with it is still unknown to us and most of its signs are highly simplified drawings with no direct pictorial value, though it is likely that they originally represented all sorts of objects, persons, and animals (fig. 10–14).

And the problem of the numerical signs in Proto-Elamite writing still awaits a convincing solution. The table drawn up by W. C. Brice, and recently examined by A. Lebrun and F. Vallat, shows that most of these signs (fig. 12–1) have been given contradictory interpretations (fig. 12–2).

In spite of the serious difficulties raised by this problem, I will try to present its main points as methodically as possible, distinguishing things that my own research has led me to regard as more or less certain from those that seem to me only probable or simply hypothetical.

I will examine in detail the Proto-Elamite account tablets that, since the end of the last century, have been discovered at Susa by the Mission Archéologique Française en Iran and published in Volumes VI, XVII, XXVI, and XXXI of the MDP.*

The method used in deciphering the Sumerian written numeration can also be applied here because, like their Sumerian counterparts, Elamite scribes often gave totals on the backs of their account tablets. This can be seen in some of the tablets shown in figure 12–3, even though the values of Proto-Elamite numerals are still unknown. Consider tablet B, for example. Items are indicated on it with writing signs (whose meanings,

---

*The letters MDP designate the following consecutive publications of the Mission Archéologique Française en Iran, which have changed names several times but have kept the same numbers:
   Volumes I to V: *Mémoires de la Mission Archéologique en Susiane.*
   Volumes VI to XIII: *Mémoires de la Délégation en Perse.*
   Volumes XIV to XXX: *Mémoires de la Mission Archéologique en Perse.*
   Volumes XXXI et seq.: *Mémoires de la Mission Archéologique en Iran.*

(*left*) Fig. 12–1. Table of Proto-Elamite numerals.

(*right*) Fig. 12–2. Contradictory proposals, put forward in this century, for the values of Proto-Elamite numerals. (See text for explanation of MDP.)

	F	G	H	M	N
1923 V. Scheil MDP XVII	1	10	100	60	600
1923 V. Scheil MDP XVII	1	10	100	600	6000
1905 V. Scheil MDP VI	1	10	100	1000	10,000
1935 V. Scheil MDP XXVI	1	10	100	1000	10,000
1925 S. Langdon	1		100	1000	10,000
1949 R.De Mecquenem MDP XXXI	1	10	100	300	1000

let me repeat, still escape us in most cases). The numbers associated with these items are indicated by groups of numerical signs, as in the Sumerian documents shown in figures 11–7 and 11–13.

The table in figure 12–4 gives what I will call a "rearranged transcription" of tablet B. On the front of the tablet, the large wedge is reproduced twice, the large circular sign twice, the small circular sign nine times, the small elongated wedge once, the arc twice, and the "double circle" (fig. 12–1D) once.

That is exactly what we find on the back. The number indicated on the back therefore corresponds to the total of the numbers indicated on the front.

In figure 12–5 is a rearranged transcription of tablet C in figure 12–3. Considering only what can be seen in the present state of the tablet, there are eighteen small wedges on the front and nine on the back, and three small circular signs on the front and four on the back. If we assume that these signs have the same values as Sumerian signs of the same form—that is, 1 for the small wedge and 10 for the small circular sign—the total of the numbers on the front, $18 + 3 \times 10 = 48$, is one less than the total of those on the back, $9 + 4 \times 10 = 49$. The break at the left edge of the tablet offers a possible explanation of the difference: the missing part may have contained one small wedge on the front. Since other tablets using the same two numerical signs (tablet D in fig. 12–3, for example) give identical results for both sides, this possibility can be

SOURCES
A Teheran Museum (MDP. XXVI, tabl. 437)
B Louvre Museum (MDP. VI, tabl. 358)
C Teheran Museum (MDP. XXVI, tabl. 297)
D Louvre Museum (MDP. XVII, tabl. 3)
E Teheran Museum (MDP. XXXI, tabl. 3)
F Louvre Museum (MDP. XVII, tabl. 45)
G Teheran Museum (MDP. XXVI, tabl. 118)
H Louvre Museum (MDP. VI, tabl. 220)
I Teheran Museum (MDP. XXVI, tabl. 439)
J Teheran Museum (MDP. XXVI, tabl. 156)
K Louvre Museum (MDP. XVII, tabl. 17)

Fig. 12–3. Proto-Elamite account tablets.

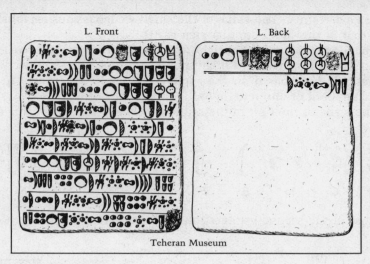

L. Front   L. Back

Teheran Museum

Fig. 12–3 (continued).

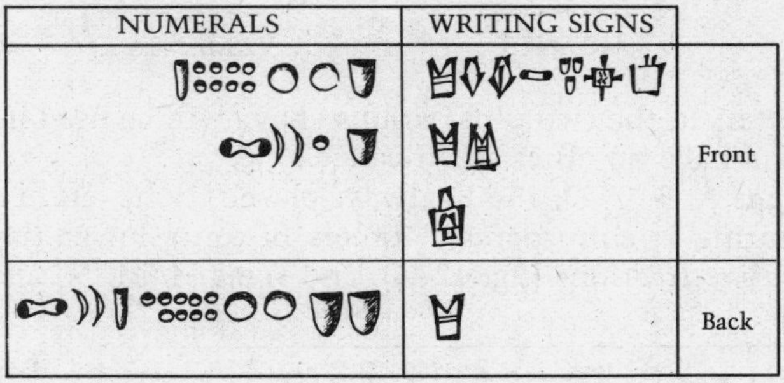

NUMERALS	WRITING SIGNS	
(numerals shown)	(writing signs shown)	Front
(numerals shown)	(writing sign shown)	Back

Fig. 12–4.

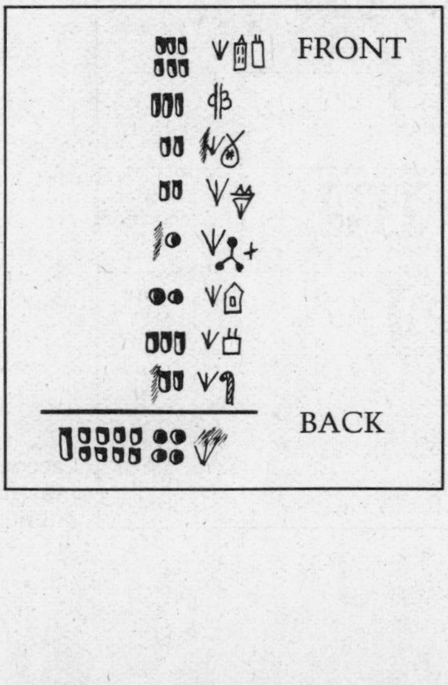

Fig. 12–5.

FRONT

BACK

taken as a certainty. The value of the small wedge is thus established as 1, and that of the small circular sign as 10.

The Elamites wrote from right to left and noted numbers in the same direction, beginning with the highest order of units. They had two written numerations, both based on the additive principle but, except in a few cases, composed of different signs.

The signs of the first numeration are always written in this order:

and those of the second numeration in this order:

(The letters in the two illustrations above will be used in referring to the signs and have no other significance.)

Signs A, B, C, D, and E, always placed to the left of the small wedge representing 1, correspond to orders of units lower than 1, that is, to consecutive fractions (fig. 12–6); and signs H, M, N, and P, as well as I

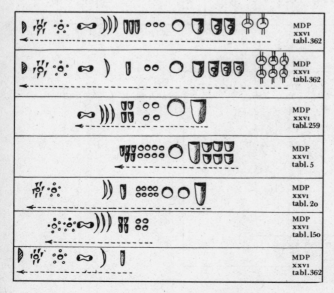

Fig. 12–6. Examples from account tablets showing the first Proto-Elamite numeration.

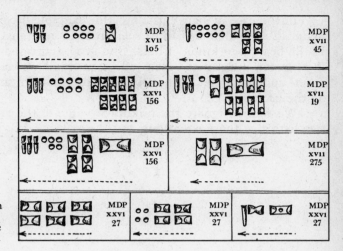

Fig. 12–7. Examples from account tablets showing the second Proto-Elamite numeration.

(or J), K (or L), and O, correspond to orders of units higher than 10, since they are always placed to the right of the small circular sign representing 10 (figs. 12–6, 7).

By taking totals from several tablets, I have obtained the following results (which will be confirmed later):

$$A = \frac{1}{120}; \quad B = \frac{1}{60}; \quad C = \frac{1}{30}; \quad D = \frac{1}{10}; \quad E = \frac{1}{5}$$

Fig. 12–8. Rearranged transcription of tablet K in figure 12–3.

In the case of sign E (the arc), for example, I considered tablet K in figure 12–3, which, as can be seen from its rearranged transcription (fig. 12–8), bears two kinds of lists. One kind, indicated by the writing sign 𝟙 , has ten arcs on the front and two small wedges on the back. The other, indicated by 𝟙 , has five arcs on the front and one small wedge on the back.

We will designate the unknown value of the arc (sign E) as x. The two lists give these equations:

$$x + x + x + x + x + x + x + x + x + x = 2$$

$$x + x + x + x + x = 1$$

Or:

$$10x = 2$$

$$5x = 1$$

And this means that the value of sign E is $1/5$.

Let us now try to evaluate the large circular sign and the large wedge (signs H and M in fig. 12–1). Because they have the values of 3600 and 60, respectively, in the Sumerian system (fig. 11–5), we are at first tempted to assign those same values to them in the Proto-Elamite system. But that would be a mistake. Since the Elamites always wrote numbers from right to left in descending order, sign H would have preceded sign M if their respective values had been 3600 and 60, but in fact sign M preceded sign H, as shown in figure 12–6.

Tablet E in figure 12–3 enables us to decipher the large circle (sign H) without difficulty. A rearranged transcription of that tablet is shown in figure 12–9.

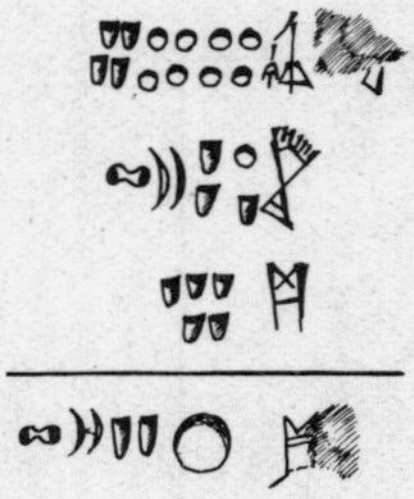

Fig. 12–9.   FRONT / BACK

Disregarding the two arcs and the double circle that appear on both sides, the tablet contains nine small circles and twelve small wedges on the front, and one large circle and two small wedges on the back. Taking account of acquired results and designating the value of the large circle as x, we obtain the following:

Front: $9 \times 10 + 12 = 102$

Back: $1 \times x + 2 = x + 2$

And this gives the equation $x + 2 = 102$, whose solution is $x = 100$.

Now let us consider tablet G in figure 12–3, which contains twenty small circles and two large ones on the front, and one large wedge (sign M) and one large circle on the back.

Giving the large circle the value of 100 that we have just determined, and designating the value of the large wedge as y, we obtain the following:

Front: $20 \times 10 + 2 \times 100 = 400$

Back: $1 \times y + 100 = y + 100$

And this gives the equation $y + 100 = 400$, whose solution is $y = 300$.

We cannot conclude, however, that 100 and 300 are the correct values of signs H and M, respectively, unless we find at least one other tablet in which they give perfectly concordant results. That is the case with tablets H and I in figure 12–3 (see fig. 12–10).

Tablet H, Figure 12–3

Front			3oo + 9×1o	3 9 0
			3oo + 1oo	4 0 0
			2×3oo + 3×1o + 3	6 3 3
				1 4 2 3
Back			4×3oo + 2×1oo + 2×1o + 3	1 4 2 3

Tablet I, Figure 12–3

Front			2×1oo	2 0 0
			3oo + 2×1oo	5 0 0
			3oo	3 0 0
			2×1oo + 4×1o + 4 ...	2 4 4
			3oo + 3×1o	3 3 0
			1oo + 9×1o	1 9 0
				1 7 6 4
Back			5 × 3oo + 2 × 1oo + 6×1o + 4	1 7 6 4

Fig. 12–10.

The results obtained so far, which can be regarded as definitive, are the following:

| $\frac{1}{120}$ | $\frac{1}{60}$ | $\frac{1}{30}$ | $\frac{1}{10}$ | $\frac{1}{5}$ | 1 | 10 | 100 | 300 |

Of the eleven signs in the first Proto-Elamite numeration, nine have been deciphered. The two others are:

N   P

These have been given different interpretations since the beginning of this century. As shown in figure 12–2, for example, values of 600, 1000, 6000, and 10,000 have been attributed to sign N.

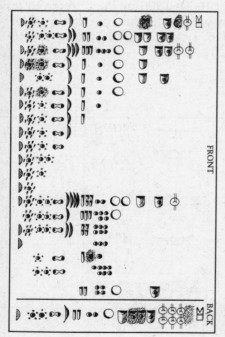

Fig. 12–11. Rearranged transcription of tablet L in figure 12–3.

In an attempt to determine the values of those two signs, we will begin by considering tablet L in figure 12–3 and its rearranged transcription in figure 12–11. According to V. Scheil, the tablet represents an educational exercise in agricultural accounting. To the best of my knowledge, it is the only intact Proto-Elamite document that contains all the signs of the first numeration as well as a general total. On the front there is a series of twenty numerical expressions (probably corresponding to twenty lots of something that seems to be indicated by the writing sign at the beginning of the first line, going from right to left), and on the back is the corresponding general total (preceded by the same writing sign).

We will try adding the numbers on this tablet in several different ways, using the table in figure 12–12, assigning various values to signs N and P, and taking account of previously acquired results.

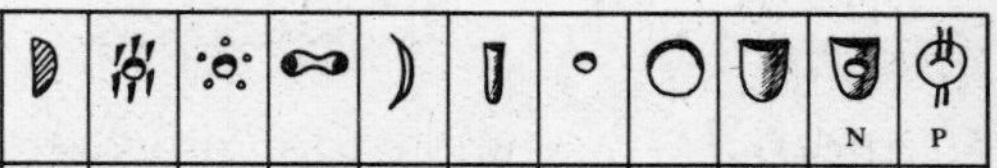

Fig. 12–12. Complete inventory of the numerical signs contained in tablet L, figure 12–3.

											N	P
Number of times that each sign appears	ON THE FRONT	15	15	24	14	19	26	39	11	7	8	5
	ON THE BACK	1	0	2	1	1	2	2	1	1	3	6

First trial. Like V. Scheil (MDP, vol. XXVI), we will assign values of 10,000 and 100,000 to signs N and P respectively.

On the front of the tablet (fig. 12–12), these values give the following result:

$15 \times 1/120 + 15 \times 1/60 + 24 \times 1/30 + 14 \times 1/10 + 19 \times 1/5 + 26 + 39 \times 10 + 11 \times 100 + 7 \times 300 + 8 \times 10,000 + 5 \times 100,000;$

that is:

$$583,622 + {}^{45}/_{120}.$$

On the back (fig. 12–12), they give:

$1 \times 1/120 + 0 \times 1/60 + 2 \times 1/30 + 1 \times 1/10 + 1 \times 1/5 + 2 + 2 \times 10 + 1 \times 100 + 1 \times 300 + 3 \times 10,000 + 6 \times 100,000;$

that is:

$$630,422 + {}^{45}/_{120}.$$

The difference between these two results is 46,800, too great to be attributed to an error on the part of the scribe, so the combination of 10,000 and 100,000 as the respective values of signs N and P must be rejected.

Second trial. Using N = 6000 (V. Scheil in 1923) and P = 100,000 (V. Scheil in 1935), we obtain the following results:

Front: $551,622 + {}^{45}/_{120}$

Back: $618,000 + {}^{45}/_{120}$

Again the difference is too great and the combination of values must be rejected.

Third trial. If we use the combination N = 6000 (V. Scheil in 1923) and P = 10,000 (S. Langdon in 1925), we again fail, because the results are:

$$\text{Front: } 101,622 + {}^{45}/_{120}$$

$$\text{Back: } 78,422 + {}^{45}/_{120}$$

Fourth trial. The combination proposed by R. de Mecquenem in 1949:— N = 1000, P = 10,000—gives these results:

$$\text{Front: } 61,622 + {}^{45}/_{120}$$

$$\text{Back: } 63,422 + {}^{45}/_{120}$$

For a long time, this combination seemed to me the most plausible. It gives relatively satisfactory results, since the difference between the two sides of the tablet is only 1800. Accepting Mecquenem's hypothesis, I assumed that the scribe must have either miscalculated or forgotten to repeat the signs that make up the difference. This would have been understandable enough, considering the large number of signs on the tablet; to err is human.

When I had thought the matter over, however, the value of 1000 for sign N seemed to me "illogical," so to speak, for two reasons. Consider these two numerical expressions, taken from Proto-Elamite tablets:

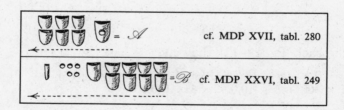

cf. MDP XVII, tabl. 280

cf. MDP XXVI, tabl. 249

On Mecquenem's hypothesis, they have the following values:

$$A = 1 \times 1000 + 6 \times 300 = 2800$$

$$B = 9 \times 300 + 5 \times 10 + 1 = 2751$$

Still on that hypothesis, the following numbers constituted consecutive orders of units:

$$1 \quad 10 \quad 100 \quad 300 \quad 1000 \quad 10,000$$

We may now ask why, if sign N had the value of 1000, the scribes noted 2800 and 2751 as shown above, rather than in these forms:

A = 2800 =
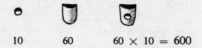

B = 2751 =

Moreover, we know that the Sumerians assigned the value of 10 to the small circle, 60 to the large wedge, and 600 to sign N:

10	60	60 × 10 = 600

Sign N was therefore composed according to the multiplicative principle.

The Elamites assigned the value of 10 to the small circle and 300 to the large wedge. Because of the analogy with Sumerian usage, the value of 300 × 10 = 3000 may seem rather plausible for sign N:

10	300	300 × 10 = 3000?

Those are the reasons that led me to reject Mecquenem's hypothesis.

Fifth trial. Consider the values N = 3000, P = 10,000 (S. Langdon in 1925 and R. de Mecquenem in 1949). They give these results for the two sides of the tablet:

Front: 77,622 + $^{45}/_{120}$

Back: 69,422 + $^{45}/_{120}$

This hypothesis must also be rejected. If the value of 3000 is kept for sign N, another value must be found for sign P.

Examination of the mathematical structure deduced from the values determined for nine other signs of the first Proto-Elamite numeration led me to consider these three values for sign P: 9000, 18,000, and 36,000.

I went on the assumption that the Proto-Elamite fractional system must have been in harmony with the corresponding notational system for whole numbers; that is, there must have been a certain correspondence between a scale of ascending values and a scale of descending values in

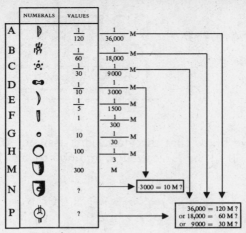

Fig. 12–13.

relation to a reference number. And that is precisely the result obtained if the various known values are expressed in terms of sign M = 300 (fig. 12–13).

Taking N = 3000 as established, we can try it in combination with the three proposed values for P: 9000, 18,000, and 36,000

Sixth trial. The combination N = 3000, P = 9000 must be rejected because it gives implausible results:

> Front: 72,622 + $^{45}/_{120}$
>
> Back: 63,422 + $^{45}/_{120}$
>
> Difference: 9200

Seventh trial. The combination N = 3000, P = 36,000 also gives implausible results:

> Front: 207,622 + $^{45}/_{120}$
>
> Back: 225,422 + $^{45}/_{120}$
>
> Difference: 17,800

Last trial and solution of the problem. The combination N = 3000, P = 18,000, which is in keeping with a coherent mathematical structure (fig. 12–13), gives highly satisfactory results:

> Front: 117,622 + $^{45}/_{120}$ (or 117,622 + $^1/_5$ + $^1/_{10}$ + $^2/_{30}$ + $^1/_{120}$)
> Back: 117,422 + $^{45}/_{120}$ (or 117,422 + $^1/_5$ + $^1/_{10}$ + $^2/_{30}$ + $^1/_{120}$)

But how are we to account for the difference of 200 between the two sides of the tablet? It was simply caused by a careless mistake.

On the back of the tablet, instead of noting the total in this form:

$$\frac{1}{120}+\frac{1}{30}+\frac{1}{30}+\frac{1}{10}+\frac{1}{5} + 1+1+10+10+300+300+3000+3000+3000+ \quad 18,000 \times 6$$

$$117,622 + \frac{1}{5}+\frac{1}{10}+\frac{1}{30}+\frac{1}{30}+\frac{1}{120}$$

the scribe made a large wedge and a large circle instead of two large wedges:

Error

100 300

$$117,422 +\frac{1}{5}+\frac{1}{10}+\frac{1}{30}+\frac{1}{30}+\frac{1}{120}$$

The reason is easy to understand. At this point the scribe held his round stylus in the wrong way: instead of pushing it into the surface of the soft clay at an angle of 30 to 45 degrees, which would have produced a large wedge, he held it perpendicular to the surface, producing a large circle. Concretely, instead of doing this:

300
Result

he did this:

100
Result

In all probability, then, sign N had the value of 3000 and sign P the value of 18,000. All the signs of the first Proto-Elamite numeration (fig. 12–6) are now deciphered.

We have good reason to believe that this system is the older of the two (figs. 12–6, 12–7), since its signs:

1	10	100	300	3000

are the ones that appear on Proto-Elamite account tablets dating from archaic times. They also appear on the earliest numerical tablets and the clay envelopes recently discovered at Susa (figs. 10–9, 10–11, 10–12, 10–13). And finally, they are the signs that, by their shapes, correspond to the tokens contained in those envelopes (figs. 10–5, 10–6, 10–9, 10–10). Because the values of the signs have now been determined, the values of the tokens are also known:

Cylinder Sphere Disk Cone Large punched cone
1 10 100 300 3000

As for the second written numeration, I believe that the Elamites devised it—perhaps in a relatively late period—to note numbers that probably corresponded to objects, commodities, or sizes different from those expressed by the signs of the first system.

I base this hypothesis on analogy with Sumerian usage. During the third millennium B.C., Sumerian scribes had three different systems of numerical notation. The first, which was the oldest and most common, was the one we have already examined; it was used for expressing numbers of people, animals, and objects, as well as measures of weight and length. The second was used for measures of volume and the third for measures of area.

The hypothesis is supported by tablet J in figure 12–3, which bears two clearly differentiated lists (fig. 12–14). The first one is indicated by the writing sign , and the corresponding quantities are expressed in signs of the first Proto-Elamite numeration. The second one seems to be indicated by the still-undeciphered writing signs:

and the corresponding numbers are expressed in signs of the second Proto-Elamite numeration (fig. 12–7).

The numbers given on the back of this tablet correspond to the totals of the first and second lists. Using the values we have already determined,

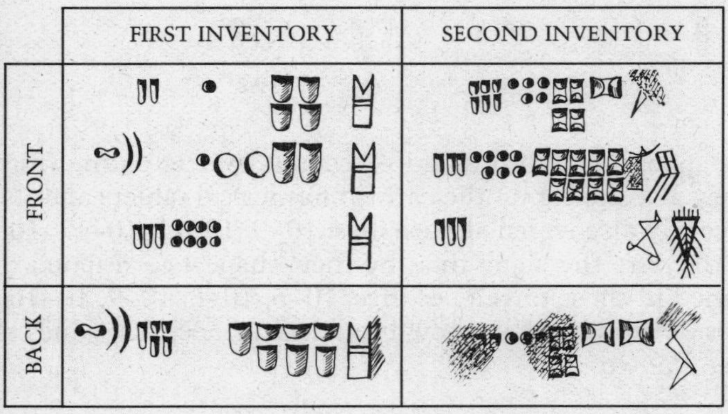
FIRST INVENTORY
SECOND INVENTORY
FRONT
BACK

Fig. 12–14. Rearranged transcription of tablet J in figure 12–3.

we obtain the following by adding the numbers of the first list (fig. 12–14):

Front: $6 \times 300 + 2 \times 100 + 10 \times 10 + 5 + \frac{2}{5} + \frac{1}{10} = 2105 + \frac{2}{5} + \frac{1}{10}$

Back: $7 \times 300 + 5 + \frac{2}{5} + \frac{1}{10} = 2105 + \frac{2}{5} + \frac{1}{10}$

(And this again confirms the accuracy of the values determined.)

Let us now consider the numerical signs of the second list, assigning the value of 1 to the small wedge, 10 to the small circle, 100 to the vertical double wedge, and 1000 to the horizontal double wedge. Addition then gives:

Front: $1000 + 13 \times 100 + 12 \times 10 + 12 = 2432$

Back: $2 \times 1000 + 4 \times 100 + 3 \times 10 + 2 = 2432$

We can therefore establish the values of the signs:

The first value is confirmed by tablet F in figure 12–3, since the corresponding additions give 591 on both sides (fig. 12–15).

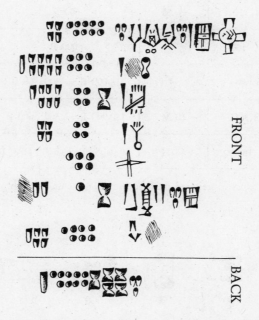

Fig. 12–15. Rearranged transcription of tablet F in figure 12–3.

The foregoing study justifies the tables in figures 12–16 and 12–17 and makes me inclined to believe that the problem posed by Proto-Elamite numerical signs is now practically solved. A question remains, however, concerning sign 0 in figure 12–1. I believe it must have been the sign for $10,000 = 1000 \times 10$, but unfortunately I have not found any document which proves that. The value of the sign cannot be conclusively established until clearer or better-preserved tablets are discovered.

	NUMERALS	VALUES	*A*	*B*
A	〄	$\frac{1}{120}$	$\frac{1}{36,000}$ M	$\frac{1}{2}$ B
B	 〄	$\frac{1}{60}$	$\frac{1}{18,000}$ M	B
C	 〄	$\frac{1}{30}$	$\frac{1}{9000}$ M	2 B
D	 〄	$\frac{1}{10}$	$\frac{1}{3000}$ M	6 B
E	 〄	$\frac{1}{5}$	$\frac{1}{1500}$ M	12 B
F	 〄	1	$\frac{1}{300}$ M	60 B
G	〄	10	$\frac{1}{30}$ M	600 B
H	 〄	100	$\frac{1}{3}$ M	6000 B
M	 〄	300	M	$18,000$ B $= 300 \times 60$ B
N	 〄	3000	10 M	$180,000$ B $= 300 \times 600$ B
P	 〄	18,000	60 M	$1,080,000$ B $= 300 \times 6000$ B

Fig. 12–16. Mathematical structure of the first Proto-Elamite numeration.

F	G	I	J	K	L	O
〄	〄	〄	〄	〄	〄	〄
1	10	100	100	1000	1000	10,000 ?

Fig. 12–17. Values of the signs in the second Proto-Elamite numeration.

Our study has shown the coexistence of two written numerations, probably used for different purposes. The first, obviously "contaminated" by use of the base 60 (fig. 12–16), probably served to express various measures—of volume and area, for example. The second, which seems to have been strictly decimal, was probably used for indicating numbers of people, animals, and objects.

Everything that has just been said would seem to confirm that there were extensive cultural and economic relations between Elam and Sumer in the fourth and third millennia B.C. and that each was influenced by the other.

13

Egyptian Numerals

The language spoken in ancient Egypt has come down to us in writing on stone monuments (temple walls, obelisks, stelae), pottery, limestone plaques, and the fragile, brittle fiber of papyrus. The basic writing system of that language was essentially reserved for stone monuments. It is composed of hieroglyphs, pictorial signs that represent people, animals, edifices, monuments, sacred and profane objects, heavenly bodies, plants, etc. (fig. 13–1).

Worship	Phallus	Quail	Bird in flight	Fish	Uraeus

Woman	Bull	Owl	Scarab	Snake	Flowering reed

Pregnant woman	Hare	Hawk	Bee	Horned viper	Lotus

Fig. 13–1.

"Hieroglyph" usually refers specifically to ancient Egyptian writing, but its meaning has been extended to include any pictograph. The Egyptians regarded their writing signs as "the god's words," and the Greeks called them *grammata hiera* ("sacred letters") or *grammata hierogluphika* ("carved sacred letters") from which our words "hieroglyph" and "hieroglyphic" are derived.

In inscriptions on monuments, hieroglyphs are read either from left to right or from right to left, and either horizontally or vertically (in columns, always read from top to bottom). The orientation of the signs indicates how they are to be read: the people and animals all face the beginning of the line; that is, they look toward the left in a text that is to be read from left to right:

and toward the right in a text that is to be read from right to left:

Egyptian hieroglyphs could signify what they visually represented. This is known as pictography, and it was commonly used in Egyptian writing during all periods. But hieroglyphs also had more extended meanings: a picture of a human leg, for example, could have the meaning not only of "leg" but also of "walk," "run," or "flee," and a picture of the sun could mean "day," "warmth," or "light," or designate the sun god.

Several hundred hieroglyphs were used phonetically. The sounds they expressed were always consonants because vowels were omitted in Egyptian writing (as in some Semitic scripts). In most cases we do not know what the vowels were. To make pronunciation easier, the letter *e* is often inserted between consonants in transliterations of Egyptian words, but that is only a convention.

Hieroglyphs could express one, two, or three consonants. The word for "quail chick" was *wa*, and the sign for that word could be used to indicate the sound *W*. Similarly, the signs for "stool" *(pe)*, "mouth" *(er)*, "hare" *(wen)*, and "scarab" *(kheper)* could be used to indicate the sounds *P*, *R*, *WN* and *KhPR*, respectively.

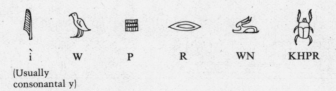

i   W   P   R   WN   KHPR

(Usually consonantal y)

Hieroglyphs fall into three categories, according to the number of consonants they express:

1. Uniliteral signs: one consonant (*W*, *P*, *M*, *R*, etc.). There were twenty-five of them during the Old Kingdom and twenty-four during the Middle Kingdom. The Egyptians could have used that "alphabet" alone to write all the words of their language. It would have spared them the use of a multitude of signs that were both pictorial and phonetic, and it would have avoided redundancies. But that possibility was never exploited. Out of traditionalism, and probably also because writing was an esthetic experience for them, Egyptian scribes preferred to keep a mixed and unnecessarily complicated system.
2. Biliteral signs: two consonants (*WN*, *SW*, etc.).
3. Triliteral signs: three consonants: (*KhPR*, *NFR*, etc.).

With these hieroglyphs, used as ideograms or phonograms, the Egyptians were able to write all the words of their language.

The origin of hieroglyphs: Sumerian or Egyptian?

Egyptian hieroglyphic writing is unquestionably among the oldest of all writing systems, since the earliest known specimens of it, such as King Narmer's mace-head (fig. 13–2), go back to the Thinite period, c. 3000–2850 B.C.

Fig. 13–2. Scene from King Narmer's mace-head (early third century B.C.). Found at Hierakonopolis.

As we have seen, writing seems to have appeared in Sumer one or two centuries earlier: as far as we now know, the oldest Sumerian written documents date from about 3200–3100 B.C., while the first known specimens of Egyptian writing do not go back beyond 3000. Moreover, recent archaeological discoveries have shown that there were regular contacts between Egypt and Mesopotamia in the period from 3300 to 3100. Are we to conclude that the Egyptians borrowed the Sumerian pictographic system and turned it into their own form of writing? No, because Egyptologists have demonstrated that this writing was indigenous to Egypt and developed without foreign influences. "All the plants and animals that appear in the hieroglyphs are from the Nile Valley. . . . The tools and instruments that appear in them had been used in Egypt since the beginning of the fourth millennium B.C. This proves that Egyptian hieroglyphic writing is the product of Egyptian civilization alone, and was born on the banks of the Nile" (Vercoutter).

From the beginning, there were always great differences between Egyptian and Sumerian pictographs, even in the case of signs representing the same object or animal. And while the Sumerians wrote almost exclusively by impressing or inscribing their signs on clay tablets, the Egyptians engraved theirs on stone monuments, using hammer and chisel, or drew them, using a reed with a crushed point dipped in a colored liquid, on pottery, stone chips, or sheets of papyrus.

If the Egyptians borrowed anything from the Sumerians, it can only have been the idea of writing, not Sumerian writing itself.

From drawings to numerals

From the end of the fourth millennium B.C., the Egyptians used a written numeration that was integrated into hieroglyphic writing and enabled them to note whole numbers up to and even beyond 1,000,000. It had a decimal base and employed the additive principle.

There was a special sign for each of the first seven powers of 10: for 1, a vertical line; for 10, a sign with the shape of an upside-down U; for 100, a spiral; for 1000, a lotus blossom; for 10,000, a raised finger, slightly bent at the tip; for 100,000, a tadpole; and for 1,000,000, a kneeling genie with upraised arms (fig. 13–3). When this last sign was first deciphered, it was thought to represent a man frightened by the great size of the number he was indicating. Later analysis showed that it represented a genie holding up the sky and that it could also mean "eternity" or "a

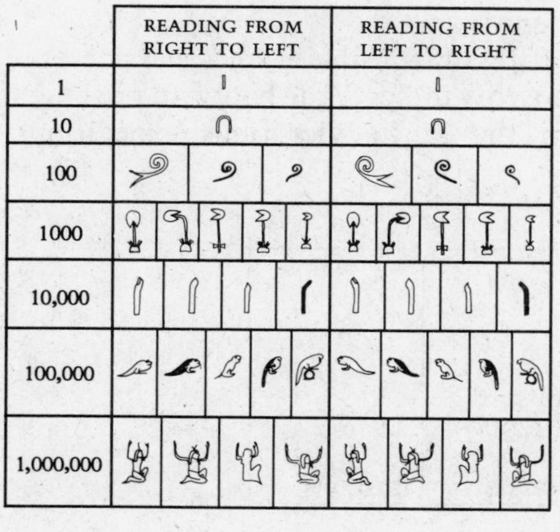

	READING FROM RIGHT TO LEFT			READING FROM LEFT TO RIGHT		
1		ı			ı	
10		∩			∩	
100						
1000						
10,000						
100,000						
1,000,000						

Fig. 13–3. The basic Egyptian hieroglyphic numerals, and their main variants, appearing on stone monuments. Most of them face the beginning of the line and thus indicate the direction in which it is to be read.

million years." It soon lost its numerical value and came to mean only "multitude" or "eternity."

This hieroglyphic numeration was a written version of concrete counting systems using pebbles, sticks, or other material objects. It was based on the additive principle: to represent a number, the sign for each decimal order was repeated as many times as necessary. To make it easier for the reader to count these repeated signs, they were often placed in groups of two, three, or four, arranged vertically, as shown in figure 13–4.

	Units	Tens	Hundreds	Thousands	Ten thousands	Hundred thousands
1						
2						
3						
4						
5						
6						
7						
8						
9						

Fig. 13–4. Representation of the units in each decimal order.

To write a given whole number, the scribe usually began by drawing the units of the largest decimal order contained in the number, then those of the order immediately below it, and so on.

On King Narmer's mace-head (fig. 13–2) there are several hieroglyphic numerals whose totals supposedly indicate the booty, in prisoners and livestock, brought back from the king's victorious expeditions (fig. 13–5).

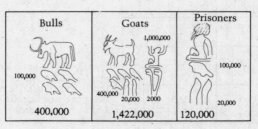

Fig. 13–5.

Are these numbers accurate, or are they an exaggeration intended to glorify King Narmer? Drioton and Vandier say that "as always, the numbers are pure fantasy." But G. Godron believes that they were probably approximations of reality. He points out that the herd counts on Old Kingdom mastabas often give very high numbers for one owner, and that "this inscription concerns a whole country."

Be that as it may, the numbers on the mace-head are generally written in descending order of powers of 10. The notation for 1,422,000, the number of goats, however, seems to be somewhat contrary to the habits of Egyptian scribes, since the sign for 1,000,000 is placed to the right of the goat and on the same line, while the rest of the number is written below. Normally, the number would have been indicated in this form:

Here are other examples, appearing in inscriptions from different periods:

1. On the base of a statue from the early third millennium B.C., found at Hierakonopolis:

47,209

This notation indicates the number of enemies slaughtered by King Khasekhem, in whose honor the statue was erected. It shows the number 47,209 written in an archaic form. On the whole, the drawings and groupings of signs are rather primitive: note the representation of a finger for 10,000 and a lotus for 1000, as well as the alignment of the vertical strokes for 1 and the grouping of the signs for 1000.

2. Inscriptions from the tomb of Sahura, a king who lived during the time when the Pyramids were built, c. twenty-fourth century B.C. (fig. 13–6).

Although in places they have deteriorated with time, these hieroglyphic numerals are perfectly recognizable. Since the heads of the tadpoles are all turned to the left, the notation was meant to be read from left to right. In inscription D, the notation for 200,000 is linear, whereas in inscription B one tadpole is above the other. Thousands are indicated by lotus blossoms connected at the bottom, a practice that was disappearing by the end of the Old Kingdom.

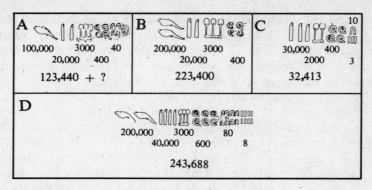

Fig. 13–6.

3. Examples from tomb inscriptions of the First Intermediate Period, late third millennium:

A	B	C	D
77	700	7000	760,000

4. Examples from the center column of the Annals of Thutmose III (1490–1436 B.C.) (fig. 13–7):

276   4622

5. Example from the stela of Ptolemy Philadelphius (282–246 B.C.) at Pithom:

660,000

Simple though it was, the Egyptian hieroglyphic numeration had the drawback of requiring frequent repetitions of signs. To write 98,737, for example, thirty-four signs had to be used. This drawback probably had a great deal to do with the many mistakes made by scribes and stonecutters.

Fig. 13–7. A portion of the Annals of Thutmose III (1490–1436 B.C.), enumerating the booty taken in the twenty-ninth year of his reign. Sandstone bas-relief from Karnak. Louvre Museum, Paris.

Fractions and the dismembered god

To express fractions, the Egyptians generally used the sign for "mouth," which in the context had the meaning of "part," placed above the denominator:

$$\frac{1}{3} \qquad \frac{1}{5} \qquad \frac{1}{6} \qquad \frac{1}{10} \qquad \frac{1}{100}$$

When the whole denominator would not fit under the "mouth" sign, they wrote the excess after it:

$$\frac{1}{249}$$

Some fractions, such as $\frac{1}{2}$, $\frac{2}{3}$, and $\frac{3}{4}$, were represented by special signs. For $\frac{1}{2}$, they used either of two hieroglyphs: ⟝ or ⟝

For $\frac{2}{3}$: ⊓ or ⊓ or ⊕ (literally, "two parts").

And for $\frac{3}{4}$: ⊓ (literally, "three parts").

The last two expressions were not quite conceived as fractions. The Egyptians did not know fractions with a numerator other than 1. To express the equivalent of our fraction $\frac{3}{5}$, for example, they did not put it in the form of $^1/_5$ + $^1/_5$ + $^1/_5$; instead, they put it in the form of $^1/_2$ + $^1/_{10}$: 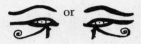. The fraction $^{47}/_{60}$ was expressed as $^1/_3$ + $^1/_4$ + $^1/_5$:

For measures of volume (grain, fruit, liquid), the Egyptians used a curious notation, different from the one above, indicating fractions of the *hekat*, a unit that, according to the traditional estimate given by G. Lefebvre, was equal to about 4.8 liters. This notation employed "Horus-eye fractions" represented by different parts of the eye of the hawk god Horus:

Since the Horus-eye was the eye of both a human being and a hawk, it had the cornea, iris, and eyebrow of the human eye and, below them, the two colored marks characteristic of the peregrine falcon. And since the most commonly used fractions of the *hekat* were $^1/_2$, $^1/_4$, $^1/_8$, $^1/_{16}$, $^1/_{32}$, and $^1/_{64}$, the notation consisted in dividing the Horus-eye into six parts and assigning each of them to one of those six fractions, as shown in figure 13–8. If the line was to be read from right to left, the value of $^1/_2$ was assigned to the right part of the cornea, $^1/_4$ to the iris, $^1/_8$ to the eyebrow, $^1/_{16}$ to the left part of the cornea, $^1/_{32}$ to the oblique colored line, and $^1/_{64}$ to the vertical colored line.

Because the symbolism of the Horus-eye played an important part in rites associated with the Osiris myths, the origin of this notation can probably be found in the myth of the gods Osiris and Horus. It was recorded by several Greek authors, notably Plutarch, in his *Isis and Osiris*.

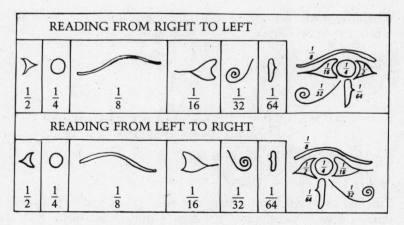

Fig. 13–8. Notation of fractions of the *hekat* for measures of volume.

It is also found, but in fragmentary form, in the magic and religious literature of ancient Egypt.

Nut, the sky goddess, secretly married Geb, the earth god, against the will of Re, the sun god. When Re learned of it, he flew into a terrible rage and cast a spell on Nut to prevent her from giving birth in any month of the year. (According to the legend, at that time the year had only 360 days, divided into twelve months of thirty days each.)

Doomed to eternal barrenness, Nut confided her sorrow to her friend Thoth, the ibis-headed magician god who was not only the supreme master of mathematics, speech, writing, and scribes, but also the protector of the moon and the powerful ruler of time and the calendar. Thoth decided to help Nut. He played dice with the moon and, when he won, made her give him $1/72$ of her light. He used it to make five days and added them to the 360 days of the year as it had existed till then. (It was from this time onward, according to the legend, that the Egyptian year had 365 days divided into twelve months of thirty days each, with the five additional days coming after the end of the last month.)

And so, unknown to Re, Nut gained five days that did not appear in the usual calendar. She quickly took advantage of them to clandestinely give birth to five children, one for each of the days won from the moon by her friend Thoth. It was thus that the gods and goddesses Osiris, Herur, Set, Isis, and Nephthys were born.

In that time, the inhabitants of Egypt were still barbarians. They lived on whatever food they could find and, when they could find nothing else, they ate each other. In short, there was very little they knew how to do; they were barely able to defend themselves against wild beasts. But their lot was about to improve, for a great king was going to educate them, and that king was none other than Osiris, eldest son of the goddess Nut and heir to Geb as the ruler of the earth.

When Osiris came of age, he married his sister Isis, unified the land of Egypt, and became its first sovereign. Under his guidance, the Egyptians ceased living like animals. He revealed the many riches of nature to them and taught them how to cultivate the fruits of the earth, how to distinguish metal from the matter around it, how to work gold and forge brass, and how to make all sorts of tools and weapons. He also gave them laws and, with the aid of Thoth, initiated them into magic and the art of writing. Finally, he taught them to respect the gods and each other. He then traveled over the whole world to civilize it.

Osiris was thus kind and generous. His brother Set, however, was the incarnation of evil: he was jealous, violent, and malicious, and he hated Osiris because everyone else loved him.

Having decided on a cruel plot against Osiris, he acquired seventy-two accomplices. He secretly took Osiris's measurements, then had a superb chest built to the exact size of Osiris's body. It was made of precious wood and lavishly decorated with emeralds, amethysts, and jasper. He had it brought to a feast that he gave in honor of his brother.

The guests all expressed great admiration at the sight of the chest. Set promised to give it to whichever of them could lie inside it and fit it perfectly. They all tried, but Osiris was naturally the only one who succeeded. Set and his sinister accomplices closed the chest with Osiris inside it, nailed it shut, sealed it with molten lead, and threw it into the Nile, which carried it to the sea. The waters of the sea took it to Byblos. Thus perished Osiris, a victim of the spirit of evil.

Isis, Osiris's wife and sister, was overwhelmed with anger and despair. Thoth restored her courage and told her to go off in search of Osiris's body. After many adventures, she found it, brought it back to Egypt, and hid it. But Set discovered the hiding place, broke the body into fourteen pieces, and scattered them in the Nile.

One by one, Isis found the pieces of her husband's body, except for the penis, which had been devoured by the oxyrhynchus, a Nile fish that was an accomplice of the odious Set. She built a sepulcher at each of the places where she had found one of the pieces, which explains why so many Egyptian cities were able to boast that they had the tomb of Osiris.

When she had finished her painful task, Isis, with the help of her sister Nephthys, reconstituted Osiris's body. The two women then called out in heartrending tones, begging Osiris to return to earth. Re, the sun god, heard their cries and was moved by them. He sent Thoth and Anubis, who took the thirteen pieces of Osiris and made them into an immortal body by mummification. The resuscitated Osiris became the god of the dead, immortality of the soul, and plants. And through his penis, which had remained at the bottom of the Nile, the river acquired its life-giving power.

Isis had conceived a posthumous son, Horus, by her husband. For a long time she hid Horus among the tall papyrus plants of a marsh, to keep him from being found by Set. She brought him up with the idea that he would someday avenge the murder of his father. When he felt strong enough, Horus challenged his uncle, Set, and a long struggle began. The fights between Horus and Set were savagely violent. In one of them, Set gouged out Horus's eye, tore it into six pieces, and scattered them throughout Egypt; and in response to this, Horus emasculated him.

The assembly of the gods finally intervened in favor of Horus and put an end to the struggle, which neither adversary had been able to win. The gods then placed Horus on the throne of Egypt so that he could become the god who protected the pharaohs and guaranteed the legitimacy of their reigns. Set was condemned to carry his brother Osiris forever. He became the accursed god of barbarians and the lord of evil.

The divine tribunal ordered Thoth to collect the pieces of Horus's mutilated eye, turn them into a complete and healthy eye by means of his powerful sorcery, and give it back to its owner. The Horus-eye later

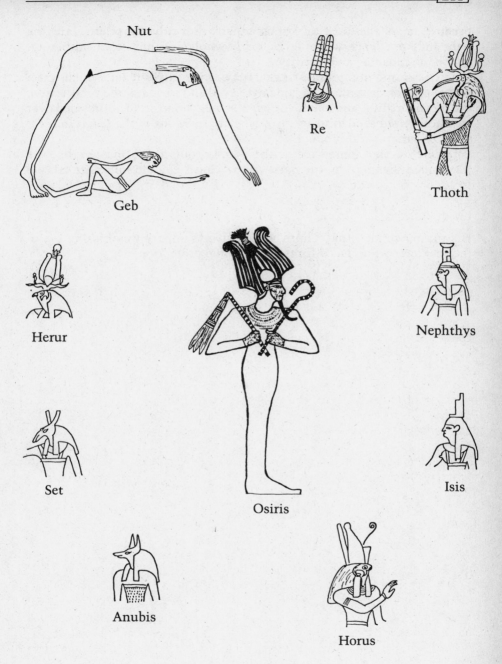

Nut

Re

Thoth

Geb

Herur

Nephthys

Set

Osiris

Isis

Anubis

Horus

Fig. 13–9.

became one of the most important talismans for the Egyptians, bringing light and knowledge, a symbol of bodily wholeness, physical health, total vision, abundance, and fertility.

To commemorate the struggle between Horus and Set, which was a divine sign of the victory of good over evil, and to guarantee total vision, universal fertility, and abundant harvests, Egyptian scribes, presided over by Thoth, used the Horus-eye to designate fractions of the *hekat* in measures of agricultural products.

One day an apprentice scribe pointed out to his master that the fractions designated by the parts of the Horus-eye did not quite add up to 1:

$$\frac{1}{2} + \frac{1}{4} + \frac{1}{8} + \frac{1}{16} + \frac{1}{32} + \frac{1}{64} = \frac{63}{64}$$

His master replied that Thoth would always supply the missing $\frac{1}{64}$ to the reckoner who placed himself under his protection.

14

Kindred Numeration Systems

In this chapter I propose to show that a number of ancient peoples—the Egyptians, Elamites, Cretans, Hittites, Aztecs, Sumerians, Romans, Etruscans, and Greeks, and the peoples of southern Arabia—devised written numerations with closely similar basic concepts.

This illustrates something well known to historians and anthropologists: that different cultures, perhaps widely separated by distance or time, have sometimes taken parallel paths without necessarily having had any contact with each other.

Cretan numerals

From about 2200 to 1400 B.C., the island of Crete was the center of a highly advanced culture now known as Minoan (from the name of the legendary King Minos, who, according to Greek mythology, was one of the first sovereigns of Knossos, the ancient Cretan capital located near the site now occupied by the seaport of Candia). In many respects, this brilliant and original civilization was a precursor of Hellenism. After its fall about 1400 B.C., probably caused by a natural cataclysm or an invasion by the Mycenaeans (a people of Greek origin), the only known traces of it that remained were in the legends gathered by the Greeks of the classical period.

The most spectacular stages in the rediscovery of Minoan civilization, such as the excavation of the famous palace of Knossos, were brought about by the tireless efforts of the British archaeologist Sir Arthur Evans (1851–1941). He was the first to demonstrate that the legends were based on reality, and he proved the existence of one of the oldest known European civilizations.

Since the end of the last century, excavations done mainly at the sites of Knossos and Mallia have brought to light many documents (particularly small stone seals, inscribed vases, and clay bars and tablets) that reveal the existence of a hieroglyphic form of writing in the period from about 2000 to 1600 B.C. (fig. 14–1).

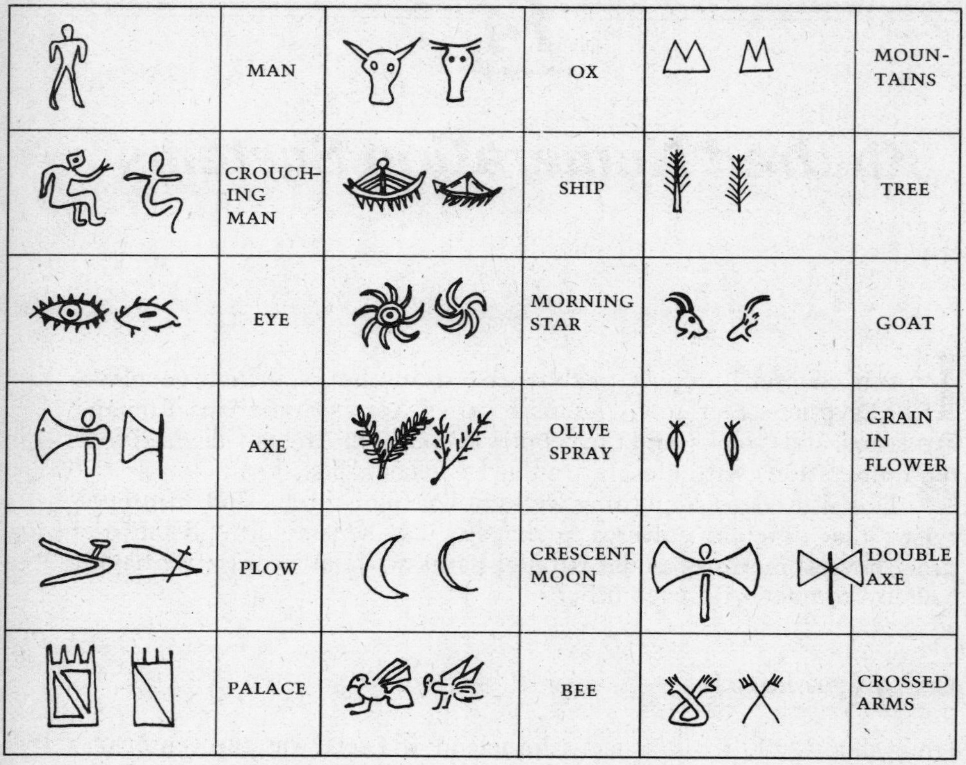

	MAN		OX	MOUN-TAINS
	CROUCH-ING MAN		SHIP	TREE
	EYE		MORNING STAR	GOAT
	AXE		OLIVE SPRAY	GRAIN IN FLOWER
	PLOW		CRESCENT MOON	DOUBLE AXE
	PALACE		BEE	CROSSED ARMS

Fig. 14–1. Cretan hieroglyphs.

Although Cretan hieroglyphic writing has still not been deciphered, some of these documents give evidence of an accounting system, probably used by a bureaucracy that arose in the earliest palaces of Minoan civilization. There are many clay bars and tablets covered with numerals and hieroglyphs whose purpose must have been to note quantities associated with various kinds of commodities. They were probably accounting devices giving details of deliveries or exchanges (fig. 14–2).

The corresponding written numeration was strictly decimal and based on the additive principle. The number 1 was indicated by a small slanted line or a small arc that could be oriented in any direction (Cretan writing sometimes went from left to right, sometimes from right to left, and sometimes in both directions alternately); 10 by a circle or, on clay, by a circular impression made with a round stylus held perpendicular to the surface; 100 by a large slanted line, clearly differentiated from the small line for 1; and 1000 by a diamond-shaped sign:

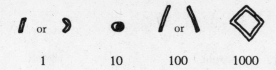

Other numbers were indicated by repeating these numerals as often as necessary.

Minoan civilization had another type of writing that probably evolved from the hieroglyphic system but used schematic signs instead of pictographs. Evans distinguished two forms of it and named them Linear A and Linear B.

Linear A is the older of the two. It was used from the beginning of the second millennium B.C. till about 1400 B.C., more or less conjointly with hieroglyphic writing. "Documents in Linear A have come from a number of sites, notably Haghia Triada, Mallia, Phaestos, and Knossos. The archives of Haghia Triada have yielded many account tablets, but unfortunately they are rather carelessly written [fig. 14–3]. They are inventories, with ideograms and numerals, in the form of small pages. But Linear A also appears on a wide variety of objects: vases (with inscriptions incised, painted, or even written in ink), religious objects (libation tables), clay seals, stamps and labels, large copper ingots, etc. This writing was therefore used not only by administrators but also in sanctuaries, and probably by private individuals" (Masson).

Linear B is later and better known that Linear A. It is usually dated from the period of about 1350 to 1200 B.C. It was during this period that the Mycenaeans conquered Crete and the ancient Minoan civilization spread to mainland Greece, particularly the regions of Mycenae and Tirynthus. The signs of Linear B were usually inscribed on clay tablets, the

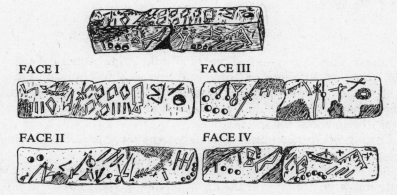

Fig. 14–2. Faces of an inscribed clay bar bearing Cretan numerals and hieroglyphs. Palace of Knossos, first half of second millennium B.C. Arthur Evans, *Scripta Minoa*, vol. 1.

Fig. 14–3. Cretan tablet with signs and numerals of the writing known as Linear A. Haghia Triada, sixteenth century B.C. Godart and Olivier, *Recueil des inscriptions en linéaire A*, vol. 1.

first of which were discovered in 1900 (fig. 14–4). Since then about 5000 have been found in Crete (only at Knossos, but in rather large numbers) and mainland Greece (notably at Pylos and Mycenae). Apparently an altered form of Linear A, Linear B was used to express an archaic Greek dialect, as has been shown by A. Ventris, the British specialist who succeeded in deciphering it after World War II. It is the only Creto-Minoan writing that has so far been deciphered: Linear A and the hieroglyphic system correspond to a language that is still unknown to us.

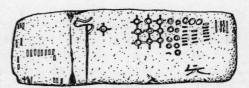

Fig. 14–4. Cretan tablets with signs and numerals of the writing known as Linear B. Arthur Evans, *Scripta Minoa*, vol. 2.

Linear A and Linear B had practically the same numerical signs (fig. 14–5): a vertical line for 1; a horizontal line (or sometimes, in Linear A, a small circular impression) for 10; a circle for 100; a circle with distinguishing marks for 1000; and, in Linear B, the sign for 1000 with a horizontal line inside it for 10,000.

To indicate other numbers, these signs were repeated as often as necessary (fig. 14–6).

The three Cretan numerations of the second millennium B.C. used exactly the same principle as the Egyptian hieroglyphic notation, and this principle was not changed in any way during their existence. Nor were Cretan writing signs and numerals replaced with a cuneiform system, as

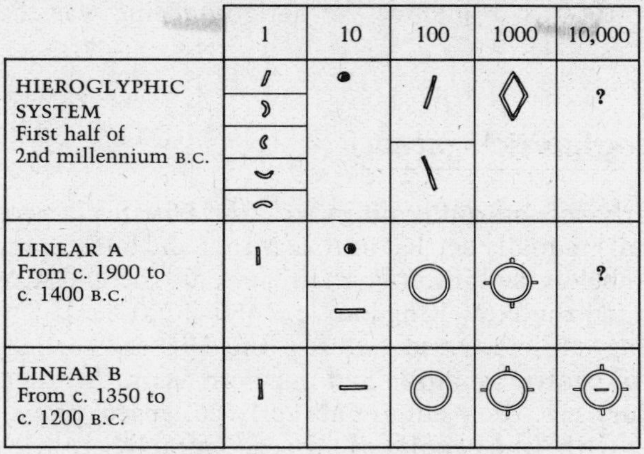

	1	10	100	1000	10,000
HIEROGLYPHIC SYSTEM First half of 2nd millennium B.C.	/ ⟩ ⟨ ◡ ◠	●	/ \	◇	?
LINEAR A From c. 1900 to c. 1400 B.C.	I	● —	○	⊙	?
LINEAR B From c. 1350 to c. 1200 B.C.	I	—	○	⊙	⊖

Fig. 14–5.
Cretan numerals.

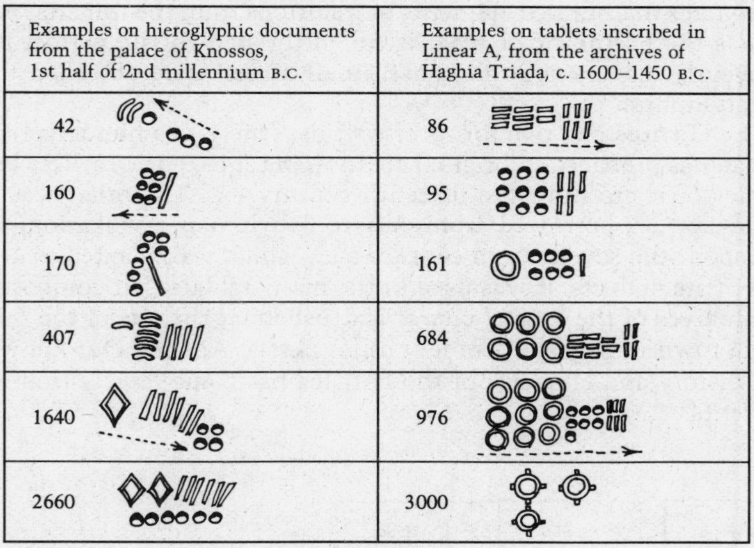

Examples on hieroglyphic documents from the palace of Knossos, 1st half of 2nd millennium B.C.		Examples on tablets inscribed in Linear A, from the archives of Haghia Triada, c. 1600–1450 B.C.	
42		86	
160		95	
170		161	
407		684	
1640		976	
2660		3000	

Fig. 14–6. Principle of Cretan numerations. The examples on the left are from Arthur Evans, *Scripta Minoa*, vol. 1; those on the right are from Godart and Olivier, *Recueil des inscriptions en linéaire A*, vol. 1.

occurred in Mesopotamia. Like the Egyptian hieroglyphic notation, these numerations had a decimal base and used the procedure of juxtaposing signs by addition. And the only numbers with special signs were 1 and the powers of 10.

The sign for 10,000 in Linear B, formed by placing a horizontal line inside the sign for 1000, obviously represented a use of the multiplicative principle (10,000 = 1000 × 10), since the horizontal line was the sign for 10.

The Hittite hieroglyphic system

Beginning in the early second millennium B.C., the Hittites (a people of Indo-European origin) gradually settled in Asia Minor and built a powerful civilization that can be divided into two main periods: the Old Kingdom (c. 1750–1450 B.C.) and the New Kingdom (c. 1450–1200 B.C.).

With alternating successes and failures, the Hittites carried on a policy of conquest in central Anatolia and northern Syria. At the end of the thirteenth century B.C. their empire abruptly collapsed, probably as the result of conflict with "the peoples of the sea," but in about the ninth century B.C. there was a revival in northern Syria, where several small Hittite states maintained elements of traditions from the imperial period. This was the beginning of what is called the neo-Hittite period. But in the seventh century B.C. all those small states were absorbed by the Assyrian Empire.

The Hittites had two forms of writing. One was a hieroglyphic system that was probably an original creation (fig. 14–7); the earliest known examples of it are from the fifteenth century B.C. The other was a cuneiform system borrowed from Assyro-Babylonian civilization, introduced about the seventeenth century B.C., which was adapted to at least three Hittite dialects. It was used in the many tablets that composed the royal archives of the capital city of Hattushah, at the site of the modern Turkish town of Bogazköy, ninety miles east of Ankara. Our knowledge of the history and language of the Hittites has come largely from those documents.

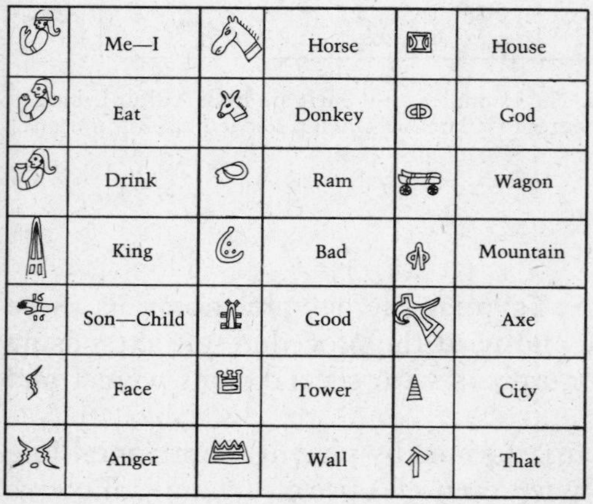

	Me—I		Horse		House
	Eat		Donkey		God
	Drink		Ram		Wagon
	King		Bad		Mountain
	Son—Child		Good		Axe
	Face		Tower		City
	Anger		Wall		That

Fig. 14–7.
Several Hittite hieroglyphs and their meanings.

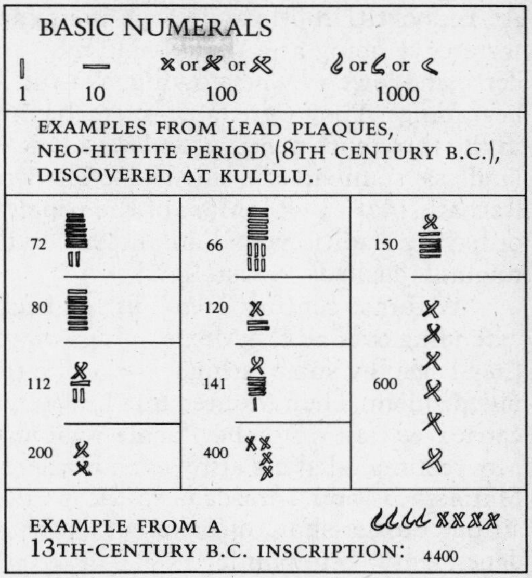

Fig. 14–8. The Hittite hieroglyphic numeration.

After the destruction of the Hittites' empire about 1200 B.C., cuneiform writing fell into disuse, but hieroglyphic writing was still used for cultural, dedicatory, and commercial purposes.

In Hittite hieroglyphic writing, the number 1 was indicated by a vertical line, which was repeated, sometimes in groups of two, three, four, or five to facilitate counting, for other numbers up to 9; 10 was indicated by a horizontal line, 100 by a sign that had roughly the shape of an X, and 1000 by a sign resembling a fishhook (fig. 14–8). These signs were repeated as many times as necessary.

Like the Egyptian hieroglyphic system and the three Cretan numerations, the Hittite hieroglyphic numeration was strictly decimal and additive, and it had special signs only for powers of 10.

The Aztec system

"The Aztecs *(Azteca)* or Mexicans *(Mexica)* were in full control of most of Mexico when the Spanish conquerors penetrated the country in 1519. They had imposed their language and religion from the Atlantic to the Pacific, and from the northern plains to Guatemala, over immense areas. The name of their sovereign, Montezuma, was venerated or feared from one end of that vast territory to the other. Their merchants traveled all over it with caravans of bearers. Their officials collected taxes everywhere. At the borders, Aztec garrisons held unsubdued peoples in check.

At Tenochtitlán (Mexico City), their capital, architecture and sculpture were developing at a remarkable rate, and luxury in clothing, food, gardens, and jewelry was growing. Yet the Aztecs had obscure and difficult beginnings. They were latecomers, having arrived in central Mexico only in the thirteenth century, and for a long time they were regarded as poor, landless, semibarbaric intruders. Their rise dated only from the reign of Itzcoatl (1428–1440). Most of the peoples surrounding them could boast of having traditions and an ancient civilization which those recent immigrants lacked" (Soustelle 1).

Within a century Tenochtitlán became the capital of a vast empire extending over all the Mexican high plateaus. "Under the rule of Itzcoatl, they began by subjugating most of the tribes in the valley that were still independent. Then Montezuma I, who was their ruler from 1440 to 1472, carried warfare into the Puebla region, to the south. Axayacatl, son of Montezuma I, led his army even farther: to Oaxaca. He also attacked the Matlazincas and Tarascans to the west, but the latter took up firm positions on the shore of Lake Pátzcuaro, defeated him, and remained independent" (Lehmann).

The Aztecs' victorious military expeditions all ended with plunder, massacres, the taking of captives, and numerous human sacrifices to their bloodthirsty gods. War was essentially dedicated to the service of the gods. "The Aztecs' civilization, history, society, and arts can be explained only in close correlation with their religion. It was a tyrannical religion containing no elements of hope or even of virtue in the Christian sense" (Lehmann). The Aztecs used war as an instrument of national policy, of

Fig. 14–9. First page of the Codex Mendoza, a document written after the Spanish conquest. Using a number of Aztec hieroglyphs, this illustration summarizes the history of the Aztecs and describes the founding of the city of Tenochtitlán.

course, but its main purpose was to supply the priests with captives for their ritual sacrifices.

In normal times, there were probably about 20,000 of these sacrifices each year. They were believed to be a means of warding off natural disasters, such as earthquakes and droughts, and maintaining the superior energies of the gods. "The mission of the human race in general, and particularly the Aztecs, the people of the sun, consisted in tirelessly repelling the assaults of nonexistence. For that purpose, the sun, the earth, and all the gods had to be supplied with the 'precious water' without which the machinery of the world would cease to function: human blood" (Soustelle 1).

The Aztecs' ritual murders also served a less exalted purpose: they provided food for the local population, as we know from a number of written accounts. "They offered the hearts to the idols, then they cut off the arms and legs and ate them as we eat the meat of animals, and they even sold human flesh in their *tianguis,* or markets" (from Díaz del Castillo's history of the Spanish conquest, cited by Simoni-Abbat).

Below the military aristocracy was a whole caste of artisans and merchants organized into a system of guilds. The largest market of ancient Mexico was at Tlatelolco, the sister city of Tenochtitlán, founded in 1358. Here were assembled all sorts of goods brought by merchants, sometimes from distant places, and the vast amounts of tribute paid by conquered cities. Records of tribute payments were kept by imperial officials, and some of them have come down to us. They show the extraordinary variety of merchandise sold at Tlatelolco: gold, silver, jade, seashells for jewelry, feathers for ceremonial garments, ostentatious clothing, shields, bales of cotton, cacao beans, cloaks, blankets, embroidered fabrics, etc.

It was at the beginning of the sixteenth century, during the reign of Montezuma II, that this civilization collapsed. In 1519 Cortez and a handful of men armed with guns landed at the present site of Veracruz and headed for the high plateau. Tribes that were enemies or vassals of the Aztecs provided them with soldiers and food. After a fierce struggle they captured Tenochtitlán on August 13, 1521, and destroyed the Aztecs' empire forever.

Aztec writing, at least at the time of the Spanish conquest, was a kind of compromise between an ideographic system and a phonetic notation (fig. 14–9). Its signs were drawings with varying degrees of realism. Some of them designated people, animals, objects, or ideas, and others (or the same ones) expressed sounds. In particular, proper nouns were written according to the principle of the rebus, an example of which would be combining two drawings, one of a fire and the other of a stone, to represent the name Firestone. (Many of the rebuses are only approximate, however, because word endings were often disregarded.) Thus the name of the city of Coatlan ("Place of Snakes") was indicated by a drawing of a snake *(coatl)* and the sign for "tooth" *(tlan).* For Itztlan ("Place of

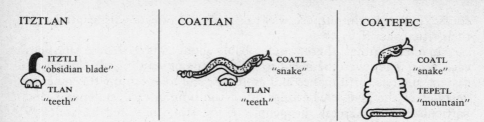

ITZTLAN   COATLAN   COATEPEC

ITZTLI "obsidian blade" / TLAN "teeth"   COATL "snake" / TLAN "teeth"   COATL "snake" / TEPETL "mountain"

Fig. 14–10. Examples of Aztec names written according to the principle of the rebus.

Obsidian"): an obsidian blade *(itztli)* and a tooth *(tlan)*. For Coatepec ("Place of the Mountain of Snakes"): a snake *(coatl)* and a mountain *(tepetl)* (fig. 14–10).

The Aztecs undoubtedly developed their writing independently of any Old World peoples. It is known to us from manuscripts called codices (codex in the singular), some of which were written after the Spanish conquest. Several of them deal with religion, divination, and magic; others relate mythical or historical events, such as tribal migrations, the founding of cities, and the origins and histories of dynasties; still others are records of the enormous amounts of tribute that imperial officials collected from subjugated cities for the notables of the city of Tenochtitlán. (figs. 14–11, 14–12).

A codex is a long strip of parchment made of pounded plant fiber, strengthened with a gum substance and coated on both sides with white lime. The outlines of characters were painted on it in black, then filled in with color. Finally it was folded, accordion fashion, and placed between wooden or leather covers that gave it the general appearance of a modern book.

Fig. 14–11. A page of the Codex Mendoza showing the tribute that seven Mexican cities were to pay to the notables of the city of Tenochtitlán.

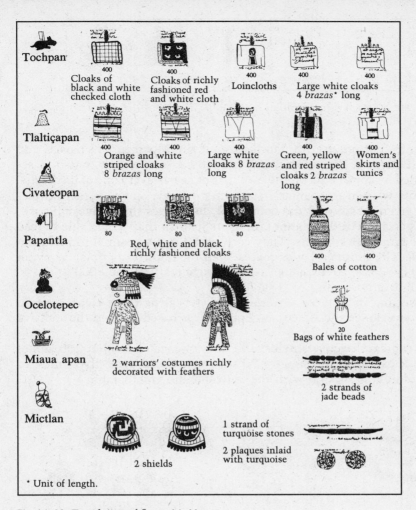

Fig. 14–12. Translation of figure 14–11.

The Codex Mendoza is one of the most remarkable of those precious documents. "By order of Viceroy Antonio de Mendoza it was compiled from ancient Mexican documents in around ten days ('as the fleet was going to leave') for sending to the Spanish court, in European form but purely indigenous style. It is in three sections which deal with the conquests of the Aztecs, the tribute they received from each conquered town, and the life cycle of the Aztecs, beginning from birth and including education, punishment, amusements, warriors' insignia, battles, genea-

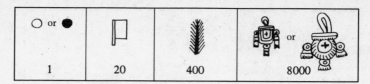

Fig. 14–13. Aztec numerals.

logical information on royal families, a plan of Moctezuma's palace, and so on" (Peterson). "Its special importance lies in the fact that it is accompanied by comments on the meaning of nearly every one of its details, comments written in Spanish by a *contemporary*, on the basis of explanations given *directly* by the council of the Aztecs themselves" (Ross).

It is because of those annotations in Spanish that we are able to read the Aztec number symbols. They correspond to a vigesimal numeration (base 20). The number 1 was represented by a dot or a circle, 20 by an axe, 400 (= 20 × 20) by a sign resembling a feather, and 8000 (= 20 × 20 × 20) by a purse (fig. 14–13).

Other numbers were represented by repeating these signs as often as necessary, since the Aztec numeration was based only on the additive principle.

To indicate twenty shields, 100 bags of cacao beans, or 200 jars of honey, for example, the sign for 20 was reproduced once, five times, or ten times in conjunction with the corresponding pictograph:

20 shields

100 bags of
cacao beans

200 jars
of honey

Similarly, for 400 decorated cloaks, 800 deer hides, or 1600 cacao beans, the sign for 400 was reproduced once, twice, or four times:

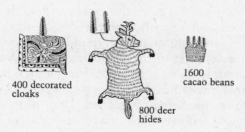

400 decorated
cloaks

800 deer
hides

1600
cacao beans

The Codex Telleriano-Remensis, a manuscript written after the Spanish conquest, states that in 1487, for the consecration of a building in Tenochtitlán, the Aztecs sacrificed 20,000 people from subjugated regions. The scribe wrote the number in the following form (fig. 14–14):

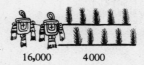

16,000   4000

Fig. 14–14. Detail of a page of the Codex Telleriano-Remensis.

The Spanish commentator, who added notes in his own language to help European readers understand the manuscript, knew the sign for 400 but not the one for 8000, with the result that he attributed a value of only 4000 to the entire numerical expression.

Greek acrophonic systems

We will now examine the written numerations used by the Greeks in their monument inscriptions during the second half of the first millennium B.C.

One of these systems, the Attic (or Athenian) numeration, had a special sign for each of the numbers 1, 5, 10, 50, 100, 1000, 5000, 10,000, and 50,000, and was based primarily on the additive principle (fig. 14–15).

Mathematically speaking, this system is quite similar to those of the Romans and Etruscans. But it differs in one way from all the other systems we have so far considered: with the exception of the vertical bar for 1, its numerals are the initials of the Greek words for the corresponding numbers, or combinations of them. These are known as acrophonic numerals. The signs for 5, 10, 100, 1000, and 10,000 are shown in figure 14–16.

1 I	100 H	10,000 M
2 II	200 HH	20,000 MM
3 III	300 HHH	30,000 MMM
4 IIII	400 HHHH	40,000 MMMM
5 Γ	500 ⌐Н	50,000 ⌐Н
6 ΓI	600 ⌐НH	60,000 ⌐Н M
7 ΓII	700 ⌐НHH	70,000 ⌐Н MM
8 ΓIII	800 ⌐НHHH	80,000 ⌐Н MMM
9 ΓIIII	900 ⌐НHHHH	90,000 ⌐Н MMMM
10 Δ	1000 X	
20 ΔΔ	2000 XX	
30 ΔΔΔ	3000 XXX	
40 ΔΔΔΔ	4000 XXXX	
50 ⌐Δ	5000 ⌐Н	
60 ⌐ΔΔ	6000 ⌐Н X	
70 ⌐ΔΔΔ	7000 ⌐Н XX	
80 ⌐ΔΔΔΔ	8000 ⌐Н XXX	
90 ⌐ΔΔΔΔ	9000 ⌐Н XXXX	

Fig. 14–15. System of numerical notation used in Attic inscriptions from the fifth century B.C. to the beginning of the Christian era.

NUMERICAL SIGN	VALUE	INITIAL OF THE WORD:
Γ (letter *pi*, archaic form of π)	5	Πεντε (*pente*, "five")
Δ (letter *delta*)	10	Δεκα (*deka*, "ten")
H (letter *eta*)	100	Ηεκατον (*hekaton*, "hundred")
X (letter *khi*)	1000	Χιλιοι (*khilioi*, "thousand")
M (letter *mu*)	10,000	Μυριοι (*murioi*, "ten thousand")

Fig. 14–16.

Numbers 50, 500, 5000, and 50,000 are indicated by combinations of these signs, in accordance with the multiplicative principle. When one of the acrophonic numerals Δ, H, X, and M is placed inside the letter Γ, the sign for 5, its value is quintupled (fig. 14–17).

Only cardinal numbers were noted by this system. (Originally, ordinal numbers were always noted by writing the words for them, but beginning in the fourth or fifth century B.C. they were noted by another method, which we will consider later.) It was used for indicating weights, measures, and sums of money. This was the system for which Greek

50	⌐Δ = Γ . Δ	5 × 10
500	⌐H = Γ . H	5 × 100
5000	⌐X = Γ . X	5 × 1000
50,000	⌐M = Γ . M	5 × 10,000

Fig. 14–17.

counting boards of the type represented by the Salamis abacus were designed.

To note sums expressed in drachmas, the Athenians repeated their acrophonic numerals as often as necessary and replaced the usual sign for 1 with the symbol Ͱ, which represented the drachma. For example:

$$\text{XXX} \;\lceil\!\!\rceil\text{H} \;\Delta\Delta\Delta \;\text{ͰͰͰ}$$

3000   500   100   30   3

------------------------ →

3633 DRACHMAS

For multiples of the talent, which was worth 6000 drachmas, they used their ordinary numerals with the addition of the letter *T*, the initial of *talanton*:

$$\lceil\!\!\rceil \;\lceil\!\!\rceil\!\!\text{Δ} \;\text{ΔΔΔΔ} \;\lceil\!\!\rceil \;\text{T T T}$$

500   50   40   5   3

------------------------ →

598 TALENTS

Submultiples of the drachma (the obol, the half-obol, the quarter-obol, and the khalkos) were indicated as in the figure 14–18.

1 KHALKOS (or ⅛ obol)	**X**	X: initial of ΧΑΛΚΟΥΣ
1 QUARTER-OBOL	Ͻ or **T**	T: initial of ΤΕΤΑΡΤΗΜΟΡΙΟΝ
1 HALF-OBOL	**C**	
1 OBOL (⅙ drachma)	**I** or **O**	O: initial of ΟΒΟΛΙΟΝ

Fig. 14–18.

Fig. 14–19. Fragment of a Greek inscription, Athens, fourth century B.C. In the third line is a notation for the sum of 3 talents and 3935 (+ *x*?) drachmas in this form:

$$\text{T T T XXX} \;\lceil\!\!\rceil\text{HHHH} \;\Delta \Delta \Delta \;\lceil\!\!\rceil$$

3   3000   500   400   30   5

TALENTS   DRACHMAS

With these signs, the Athenians had no great difficulty in expressing the monetary sums they used in everyday life (and they had similar conventions for weights and measures), as can be seen from the examples in figure 14–20.

Acrophonic numerations analogous to the Attic system were used in the many different states of the ancient Greek world during this same period (figs. 14–21, 14–22).

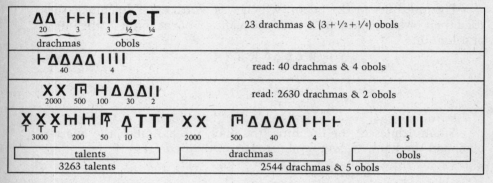

Fig. 14–20.

1 ⊢1 drachma	10 Δ	100 H	1000 X
2 ⊢⊢	20 ΔΔ	200 HH	2000 XX
3 ⊢⊢⊢	30 ΔΔΔ	300 HHH	3000 XXX
4 ⊢⊢⊢⊢	40 ΔΔΔΔ	400 HHHH	4000 XXXX
5 ⊢⊢⊢⊢⊢	50 Γᴰ	500 Γ⁻	5000 Γˣ⁻
6 ⊢⊢⊢⊢⊢⊢	60 Γᴰ Δ	600 Γ⁻ H	6000 Γˣ⁻ X
7 ⊢⊢⊢⊢⊢⊢⊢	70 Γᴰ ΔΔ	700 Γ⁻ HH	7000 Γˣ⁻ XX
8 ⊢⊢⊢⊢⊢⊢⊢⊢	80 Γᴰ ΔΔΔ	800 Γ⁻ HHH	8000 Γˣ⁻ XXX
9 ⊢⊢⊢⊢⊢⊢⊢⊢⊢	90 Γᴰ ΔΔΔΔ	900 Γ⁻ HHHH	9000 Γˣ⁻ XXXX

Ex: Γˣ⁻ Γ⁻ HH ΔΔΔ ⊢⊢⊢⊢⊢⊢⊢⊢⊢

   5000 500 200   40   9

- →

5749 DRACHMAS

Fig. 14–21. Numerical system used in Greek inscriptions on the island of Kos, third century B.C.

1 drachma	⊢ 'or I ¨	
5	Γ ·	Π , initial of Πεντε , "five"
10	▷ ¨or Δ˙	Δ , initial of Δεκα , "ten"
50	ΓE or Γᵉ·	ΠE, abbreviation of Πεντεδεκα , "fifty"
100	⊢E	HE , abbreviation of Ηεκατον , "hundred"
300	⊤E·	T.HE , abbreviation of Τριακοσιοι , "three hundred"
500	Π⊢E or Π⊢E	Π.HE , abbreviation of Πεντακοσιο , "five hundred"
1000	Ψ	Old Boeotian form of X, initial of Χιλιοι , "thousand"
5000	⌐	Π.X , abbreviation of Πενταχιλιοι , "five thousand"
10,000	M	M , initial of Μυριοι , "ten thousand"
	*Sign attested only in Thespiae.	**Sign attested only in Orchomenus.

Fig. 14–22. Numerical system used in Greek inscriptions in Orchomenus and Thespiae, third century B.C.

Although the Attic system (the oldest known Greek acrophonic numeration) gradually spread during the time of Pericles, when Athens established dominance over a number of Greek states, differences still remained. Each state had its own system of weights and money; the idea of a standard international system was still foreign to the Greek mind. But comparison of the Greek acrophonic numerations reveals their common origin (fig. 14–23).

Numerals of the Kingdom of Saba

A system similar to those we have just seen was used from the fifth to the second or first centuries B.C. by the peoples of southern Arabia, notably the Minaeans and the Sebaeans (inhabitants of Saba, whose biblical name is Sheba).

It appears in some of the many South Arabic inscriptions that these peoples have left to us, inscriptions concerning a wide range of subjects: construction of multistoried buildings, irrigation projects with large dams, offerings to the gods, animal sacrifices, conquests, booty, etc. The writing used in them recorded Semitic languages related to Arabic and was probably derived, with notable alterations, from the writing of the Phoenicians. It had an alphabet of twenty-nine letters, all consonants.

The corresponding written numeration is based on the additive principle and has special signs not only for the powers of 10 but also for the numbers 5 and 50 (fig. 14–24). It is acrophonic: except in the case of 1 and 50, all its signs are letters of the alphabet, and these letters are the initials of the Semitic words for 5, 10, 100, and 1000 (fig. 14–25).

Fig. 14–23. Table of Greek numerical signs used to indicate sums of money (drachmas, in the case of the signs shown here) in the second half of the first millennium B.C., revealing the common origin of all the Greek acrophonic numerations employed during that period.

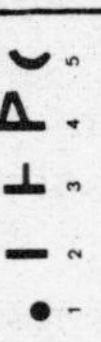

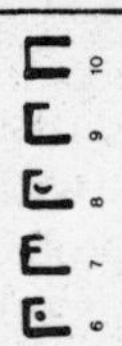

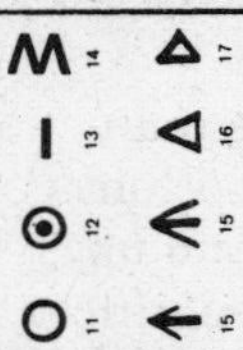

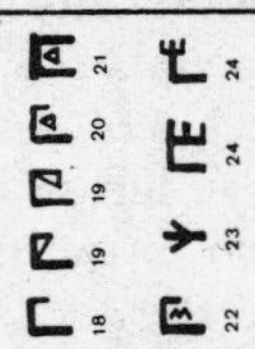

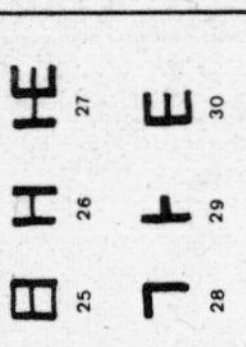

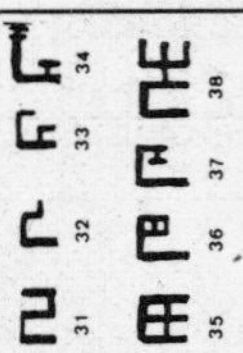

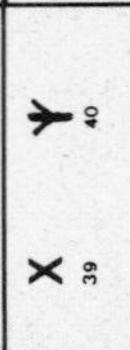

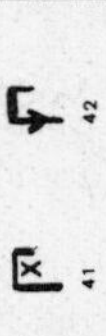

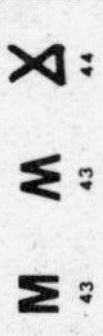

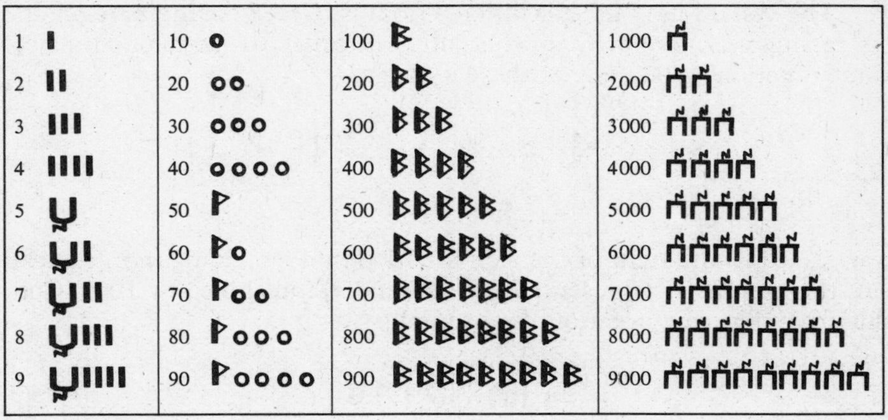

1	**I**	10	**o**	100	**B**	1000	**ň**
2	**II**	20	**oo**	200	**BB**	2000	**ňň**
3	**III**	30	**ooo**	300	**BBB**	3000	**ňňň**
4	**IIII**	40	**oooo**	400	**BBBB**	4000	**ňňňň**
5	**Y**	50	**P**	500	**BBBBB**	5000	**ňňňňň**
6	**YI**	60	**Po**	600	**BBBBBB**	6000	**ňňňňňň**
7	**YII**	70	**Poo**	700	**BBBBBBB**	7000	**ňňňňňňň**
8	**YIII**	80	**Pooo**	800	**BBBBBBBB**	8000	**ňňňňňňňň**
9	**YIIII**	90	**Poooo**	900	**BBBBBBBBB**	9000	**ňňňňňňňňň**

Fig. 14–24. Signs of the South Arabic numeration. This system is known only for the period from the fifth to the second or first centuries B.C. In inscriptions after the beginning of the Christian era, numbers were apparently always expressed by spelling out their names.

The development of this system may have been influenced by the Greeks, since they are known to have had contacts with the Sabaeans and Minaeans, at least during the second half of the first millennium B.C. But that is only a conjecture.

In Minaean and Sabaean inscriptions, numerical expressions are usually placed between the signs ▌ ▌ to avoid confusion between alphabetic numerals and ordinary letters (figs. 14–26, 14–27). The numerals often change their orientation within a single inscription, since the lines in South Arabic writing were read alternately from left to right and from right to left.

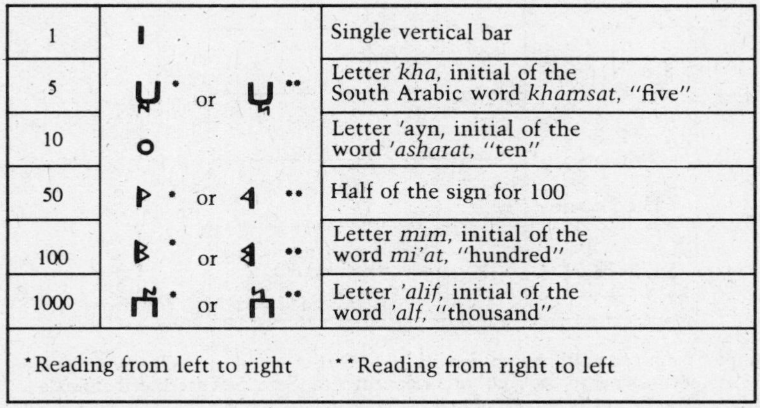

1	**I**	Single vertical bar
5	**Y** ˙ or **Y** ˙˙	Letter *kha*, initial of the South Arabic word *khamsat*, "five"
10	**o**	Letter *'ayn*, initial of the word *'asharat*, "ten"
50	**P** ˙ or **4** ˙˙	Half of the sign for 100
100	**B** ˙ or **ᘓ** ˙˙	Letter *mim*, initial of the word *mi'at*, "hundred"
1000	**ň** ˙ or **ň** ˙˙	Letter *'alif*, initial of the word *'alf*, "thousand"
˙Reading from left to right ˙˙Reading from right to left		

Fig. 14–25.

The system used in Saba differed from its Greek counterparts in one interesting way: it contained what might be called the germ of our place-value notation. When one of these signs:

$$\text{o} \qquad \text{◁} \quad \text{or} \quad \text{ᚹ} \qquad \text{◁|} \quad \text{or} \quad \text{ᛒ}$$

$$10 \qquad\qquad 50 \qquad\qquad\qquad 100$$

was placed to the right of the sign for 1000 in a line going from right to left, the number it represented was (mentally) multiplied by 1000. Consider the following notation, for example:

$$\text{ᚼᚼ ooo ◁ ◁ ◁|}$$

$$2000 \quad 30 \quad 50 \quad 200$$

We would be inclined to interpret it as:

$$200 + 50 + 30 + 2000 = 2280$$

according to the traditional use of the additive principle, but to the Sabaeans it had the value of:

$$(200 + 50 + 30) \times 1000 + 2000 = 282{,}000.$$

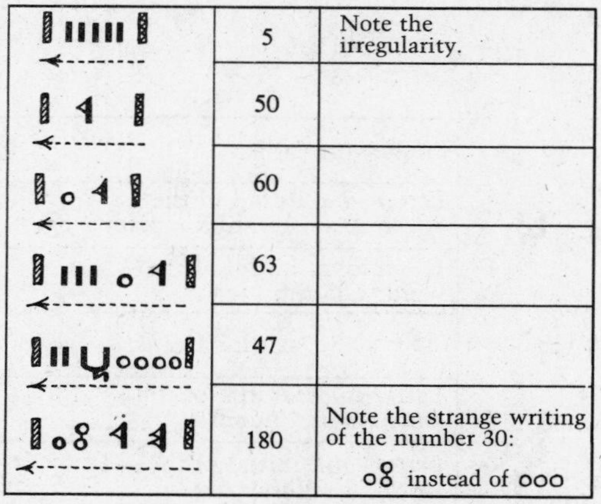

⎸ ‖‖‖‖ ⎸	5	Note the irregularity.
⎸ ◁ ⎸	50	
⎸ o ◁ ⎸	60	
⎸ ‖‖ o ◁ ⎸	63	
⎸‖ ⨆ ooooo⎸	47	
⎸ o⦚ ◁ ◁| ⎸	180	Note the strange writing of the number 30: o⦚ instead of ooo

Fig. 14–26. Examples from Minaean inscriptions, third to first centuries B.C. The numbers refer to the sizes of containers offered to the gods, other offerings to them, and sacrificed animals. (Personal communication from Christian Robin.)

Similarly, when one of those signs was placed to the left of the sign for 1000 in a line going from left to right, its value was multiplied by 1000. The notation:

$$ℬℬℾ \circ \circ \circ \, \text{ⴌⴌ}$$

200 50 30   2000

- - - - - - - - - - - - →

thus stood for 282,000, not 2280.

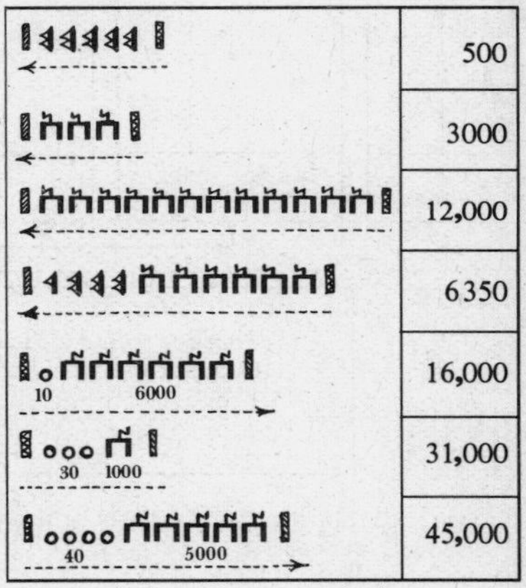

(symbols)	500
(symbols)	3000
(symbols)	12,000
(symbols)	6350
(symbols)	16,000
(symbols)	31,000
(symbols)	45,000

Fig. 14–27. Examples from Sabaean inscriptions, fifth century B.C., found mainly at the site of Sirwah. They describe conquests and enumerate military forces, material resources, booty, prisoners, etc. (Personal communication from Christian Robin.)

Since this practice might cause confusion, however, the Sabaeans developed the habit of writing the word for the corresponding number in front of each notation. Centuries later, that precaution enabled specialists to decipher the notations without ambiguity.

A significant improvement

The numerations we have examined so far have one thing in common: each is a written version of a concrete system in which a given number is represented by grouping as many objects, with numerical values assigned to them, as necessary. They belong to the category of additive numerations because each of them has a limited number of numerical signs, entirely independent of each other, whose juxtaposition implies the sum of the corresponding values, and the value of a sign is unrelated to its position. The additive principle is thus sharply distinct from the place-value principle of our modern numeration, since according to the

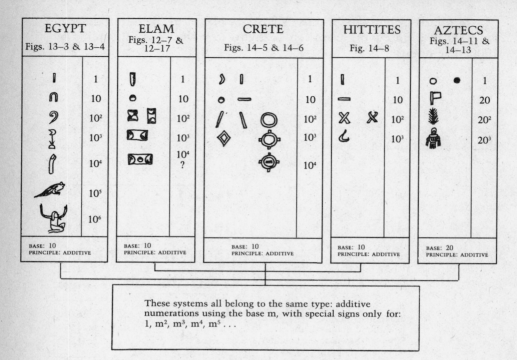

These systems all belong to the same type: additive numerations using the base m, with special signs only for: 1, m², m³, m⁴, m⁵ . . .

Fig. 14–28. Additive numerations of the first type.

latter principle each numeral has a variable value that depends on its position in relation to others.

Some additive numerations use the base 10 and have a special sign only for 1 and each power of 10. Examples: the two systems used in Elam during the first half of the third millennium B.C. (figs. 12–7, 12–17), the Egyptian hieroglyphic system (figs. 13–3, 13–4), the Cretan systems (figs. 14–5, 14–6), and the Hittite hieroglyphic system (fig. 14–8). These systems are all of the same type: written numerations using the base m, with special signs for numbers 1, m, m^2, m^3, m^4, etc.

The Aztec system is also of this type, since it is an additive numeration of base 20, with special signs for 1, 20, 400 (= 20^2), and 8000 (= 20^3). It differs from those mentioned above only in its base and the forms of its signs (fig. 14–28).

These numerations use the simple procedure of expressing a given number by repeating signs for powers of the base as many times as necessary. To express the number 7699 in the Egyptian hieroglyphic system, for example, the sign for 1000 was reproduced seven times, the sign for 100 six times, the sign for 10 nine times, and the sign for 1 nine times, making a total of thirty-one signs:

7699

It was to avoid such tedious repetitions that some peoples who had a system of this kind added several extra signs to their original list, which naturally meant that they had to use fewer symbols in writing numbers.

This was done by the Greeks and the peoples of southern Arabia, who simplified their original systems by introducing a sign for 5, then another for 5 × 10, then another for 5 × 100, and so on. (Compare fig. 14–29 with fig. 14–31), which shows these signs as they were introduced into the system used in the Greek republic of Epidaurus.) To write the number 7699, which can be broken down into 5000 + 1000 + 1000 + 500 + 100 + 50 + 10 + 10 + 10 + 10 + 5 + 1 + 1 + 1 + 1, they now needed only fifteen signs, rather than the thirty-one needed before:

Ⴊ XX ႮႬ Ⴊ∆∆∆∆ႫIIII

| 5000 | 2000 | 500 | 100 | 50 | 40 | 5 | 4 |

→

7699

• 1 drachma	— 10 drachmas	ᗭ 100 drachmas	X 1000 drachmas
2 ⋮	20 =	200 ᗭᗭ	2000 XX
3 ⫶•	30 =−	300 ᗭᗭᗭ	3000 XXX
4 ⫶⫶	40 ==	400 ᗭᗭᗭᗭ	4000 XXXX
⫶⫶• 5 dr.	==− 50 dr.	ᗭᗭᗭᗭᗭ 500 dr.	XXXXX 5000 dr.
6 ⫶⫶⫶	60 ===	600 ᗭᗭᗭᗭᗭᗭ	6000 XXXXXX
7 ⫶⫶⫶•	70 ===−	700 ᗭᗭᗭᗭᗭᗭᗭ	7000 XXXXXXX
8 ⫶⫶⫶⫶	80 ====	800 ᗭᗭᗭᗭᗭᗭᗭᗭ	8000 XXXXXXXX
9 ⫶⫶⫶⫶•	90 ====−	900 ᗭᗭᗭᗭᗭᗭᗭᗭᗭ	9000 XXXXXXXXX

ᗭ. old form of the letter H; initial of HEKATON "hundred"
X. initial of χιλιοι "thousand"

EXAMPLE   X ᗭᗭᗭᗭᗭᗭᗭᗭᗭ ===− ⫶⫶⫶•
          1000   900   70   9
→
1979 DRACHMAS

Fig. 14–29. Written numeration used in Greek inscriptions in Epidaurus, early fourth century B.C. It has the same structure as the Cretan numeration. It is acrophonic only for 100 and 1000 and has no special sign for 5, 50, 500, or 5000.

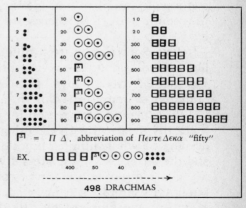

1 •	10 ⊙	10 ᗭ
2 ⋮	20 ⊙⊙	20 ᗭᗭ
3 ⫶•	30 ⊙⊙⊙	300 ᗭᗭᗭ
4 ⫶⫶	40 ⊙⊙⊙⊙	400 ᗭᗭᗭᗭ
5 ⫶⫶•	50 Ⴊ	500 ᗭᗭᗭᗭᗭ
6 ⫶⫶⫶	60 Ⴊ⊙	600 ᗭᗭᗭᗭᗭᗭ
7 ⫶⫶⫶•	70 Ⴊ⊙⊙	700 ᗭᗭᗭᗭᗭᗭᗭ
8 ⫶⫶⫶⫶	80 Ⴊ⊙⊙⊙	800 ᗭᗭᗭᗭᗭᗭᗭᗭ
9 ⫶⫶⫶⫶•	90 Ⴊ⊙⊙⊙⊙	900 ᗭᗭᗭᗭᗭᗭᗭᗭᗭ

Ⴊ = Π Δ . abbreviation of Πεντε Δεκα "fifty"

EX.   ᗭ ᗭ ᗭ ᗭ Ⴊ⊙ ⊙ ⊙ ⊙ ⫶⫶⫶⫶
      400   50   40   8
→
498 DRACHMAS

Fig. 14–30. Written numeration used in Greek inscriptions in Nemea, 4th century B.C. It is a decimal system with only one extra sign, for 50.

• 1 drachma	— 10 drachmas	日 or H 100 drachmas	X 1000 drachmas	M 10,000 drachmas
2 ••	20 =	200 HH	2000 XX	20,000 MM
3 •••	30 =—	300 日日日	3000 XXX	30,000 MMM
4 ••••	40 ==	400 HHHH	4000 XXXX	40,000 MMMM
Γ˙ 5 dr.	Γ² or Γ² 50 dr.	Γ日 or Γ日 or Γ 500 dr.	Γˣ 5000 dr.	?
5 Γ˙.	50 Γ²—	600 Γ日日	6000 Γˣ X	
7 Γ˙..	70 Γ²=	700 Γ日HH	7000 Γˣ XX	
8 Γ˙...	80 Γ²=—	800 Γ日日日日	8000 Γˣ XXX	
9 Γ˙....	90 Γ²==	900 Γ日HHHH	9000 Γˣ XXXX	

Fig. 14–31. Later numeration (from the end of the fourth century B.C. to the middle of the third) used in Epidaurus.

In this way, they devised additive numerations that used the base m and had special signs for numbers 1, k, m, km, m^2, km^2, m^3, etc., the number k being a divisor of the base and functioning as an auxiliary base.

The Etruscans and Romans, Tuscan peasants, Swiss and Dalmatian shepherds, and the Zuñi Indians of North America (whose traditions are 2000 years old) achieved the same result independently and in a different way, starting from the ancient practice of recording numbers with notches in wood or bone, using a single vertical notch for 1, a V-shaped notch for 5, an X-shaped notch for 10, and so on, combining a limited number of symbols. This method enables its users to express relatively large numbers, as large as they are likely to need, without ever having to consider a group of more than four identical signs, and it necessarily corresponds to a count with these successive stages: 1, 5, 10, 5 × 10, 10^2, 5 × 10^2, etc.

The Sumerian written numeration used from the late fourth millennium till about the eighteenth century B.C. is quite similar to the systems described above. This may not be obvious at first sight: the Sumerian system is based on the additive principle, but it is sexagesimal!

In explaining my reason for concluding that this system is mathematically identical with the others, I will continue using the letter m to designate a whole number functioning as the base of a numerical system, and the letter k to designate a divisor of that base.

An additive numeration is of the same type as the Roman numeration, for example, if it has a special sign for each of these numbers: 1, k, m, km, m^2, km^2, m^3, etc. (fig. 14–32)—that is, if it has a special sign not only for each power of its base (1, m, m^2, m^3, m^4, etc.) but also for the product of each of them and the number k (k, km, km^2, km^3, etc.). The Roman and Etruscan systems, as well as the Greek and South Arabic acrophonic numerations, meet these requirements, since their successive signs correspond to:

$$1 \quad 5 \quad 10 \quad 5 \times 10 \quad 10^2 \quad 5 \times 10^2 \quad 10^3...$$
$$\updownarrow \quad \updownarrow \quad \updownarrow \quad \updownarrow \quad \updownarrow \quad \updownarrow$$
$$k \quad m \quad k \times m \quad m^2 \quad k \times m^2 \quad m^3...$$

In the Sumerian numeration, the signs correspond to:

$$1 \quad 10 \quad 60 \quad 10 \times 60 \quad 60^2 \quad 10 \times 60^2 \quad 60^3$$
$$\updownarrow \quad \updownarrow \quad \updownarrow \quad \updownarrow \quad \updownarrow \quad \updownarrow$$
$$k \quad m \quad k \times m \quad m^2 \quad k \times m^2 \quad m^3$$

This shows that it is of the same mathematical type as the Roman numeration and others like it.

Another way of considering the matter will make it even clearer. Instead of starting with an examination of the Roman system and going from there to the Sumerian system, I will begin by examining the latter.

The first thing to be noted is that in the Sumerian system the sequence of numbers represented by a special sign can be written in this form:

$$1$$
$$10$$
$$10 \times 6$$
$$10 \times 6 \times 10$$
$$10 \times 6 \times 10 \times 6$$
$$10 \times 6 \times 10 \times 6 \times 10$$
$$10 \times 6 \times 10 \times 6 \times 10 \times 6$$

This is not surprising, since we already know that the Sumerian numeration was built on a compromise between two complementary divisors* of the base (10 and 6) which function as alternating bases.

Next, an additive numeration of base m is of the same type as the Sumerian system if its sequence of special signs corresponds to the following, in which a and b are complementary divisors of the base:

$$1$$
$$a$$
$$a \times b$$

$a \times b \times a$	$(= a^2b)$
$a \times b \times a \times b$	$(= a^2b^2)$
$a \times b \times a \times b \times a$	$(= a^3b^2)$
$a \times b \times a \times b \times a \times b$	$(= a^3b^3)$
$a \times b \times a \times b \times a \times b \times a$	$(= a^4b^3)$

This is precisely the case with the Greek, Etruscan, and Roman systems, whose sequences of signs correspond to:

*The term "complementary divisors" is used here to designate a pair of divisors, a and b, of a number m, such that $a \times b = m$. Thus while 6 and 10 are complementary divisors of 60, 6 and 20 are not.

$$1$$
$$5$$
$$5 \times 2$$
$$5 \times 2 \times 5$$
$$5 \times 2 \times 5 \times 2$$
$$5 \times 2 \times 5 \times 2 \times 5$$
$$5 \times 2 \times 5 \times 2 \times 5 \times 2$$

In other words, the sequence of numbers associated with the basic signs of the Roman numeration, for example, is generated by successive multiplications, alternately by 5 and 2. It is therefore a system built on a compromise between the complementary divisors 5 and 2 of the base 10, these two numbers functioning as alternating bases. So the foregoing numerations are of the same type, mathematically speaking, which amounts to saying that $1, k, m, km, m^2, km^2, m^3 \ldots$, in which k is a divisor of m, is equivalent to $1, a, ab, a^2b, a^2b^2, a^3b^2, a^3b^3 \ldots$, in which a and b are complementary divisors of m. We can pass from one to the other by a simple change of notation: from the first to the second by positing $a = k$ and $b = \frac{m}{k}$, and from the second to the first by positing $k = a$ and $m = ab$.

But while the Sumerian system is mathematically identical with others of its type, it differs from them in its historical development. As we have seen, Sumerian numerals were originally fashioned on the basis of the clay tokens used for numerical purposes. That is, the structure of the Sumerian written numeration, which was preserved all through the history of the Sumerian people, was derived from that of an older concrete numeration, which was in turn derived from a still older oral system (fig. 14–33).

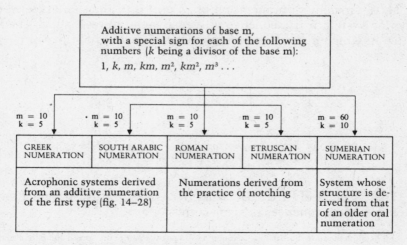

Additive numerations of base m, with a special sign for each of the following numbers (k being a divisor of the base m):

$1, m, km, m^2, km^2, m^3 \ldots$

m = 10 k = 5	m = 10 k = 5	m = 10 k = 5	m = 10 k = 5	m = 60 k = 10
GREEK NUMERATION	SOUTH ARABIC NUMERATION	ROMAN NUMERATION	ETRUSCAN NUMERATION	SUMERIAN NUMERATION
Acrophonic systems derived from an additive numeration of the first type (fig. 14–28)		Numerations derived from the practice of notching		System whose structure is derived from that of an older oral numeration

Fig. 14–32. Additive numerations of the second type.

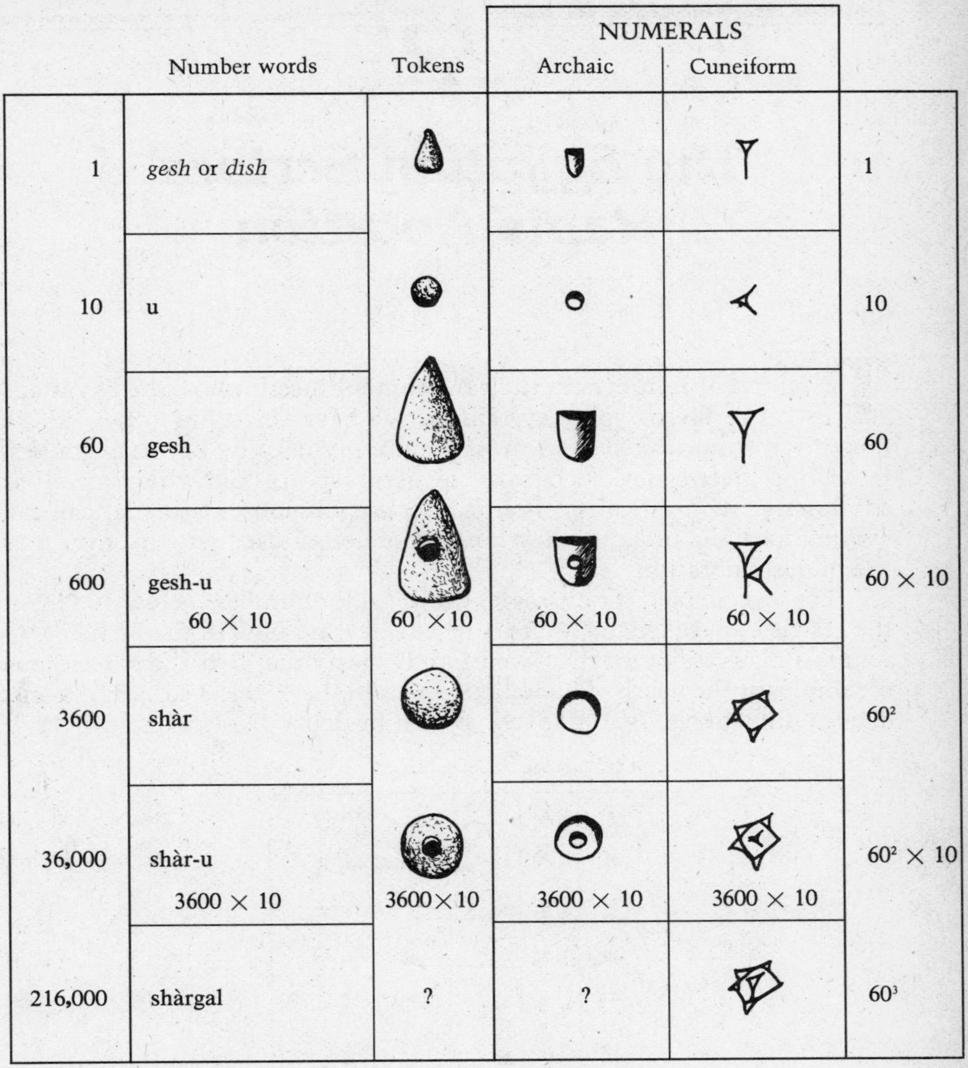

	Number words	Tokens	NUMERALS		
			Archaic	Cuneiform	
1	*gesh* or *dish*				1
10	u				10
60	gesh				60
600	gesh-u 60×10	60×10	60×10	60×10	60×10
3600	shàr				60^2
36,000	shàr-u 3600×10	3600×10	3600×10	3600×10	$60^2 \times 10$
216,000	shàrgal	?	?		60^3

Fig. 14–33. Sumerian numerations.

15

The Egyptian Scribes'
Rapid Notation

To represent numbers in their monument inscriptions, the Egyptians used the hieroglyphic system, as we have seen (figs. 13–3, 13–4). But this was not the system most commonly used by Egyptian scribes. In writing inventories, accounts, censuses, reports, and wills, as well as administrative, juridical, economic, literary, religious, mathematical, and astronomical documents, they much more often used what is known as the hieratic notation.

Let us consider, for example, the signs for numbers below 10,000 in the Harris Papyrus (named for its discoverer and now in the British Museum), a manuscript for the Twentieth Dynasty that lists the possessions of temples at the death of Pharaoh Ramses III (1192–1153 B.C.). The signs, disregarding graphic variants, are shown in figure 15–1.

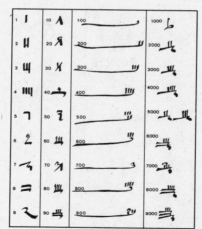

Fig. 15–1.

Most of these signs seem to have nothing in common with their hieroglyphic counterparts. Although the signs for numbers 1 to 4 obviously have a direct visual reference, the others have none.

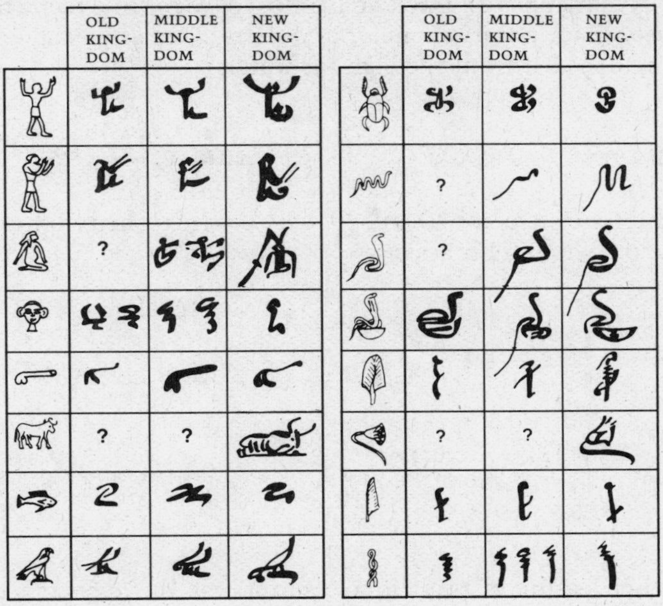

Fig. 15–2. Egyptian hieratic signs, with their hieroglyphic models. In most cases the hieratic version is so different from the original hieroglyph that no connection between them is discernible.

Are we to conclude that these numerals did not evolve from their hieroglyphic counterparts, that they were a kind of shorthand artificially created by scribes as a means of writing faster? The discussion below will enable us to answer that question.

Hieratic script

Since Egyptian hieroglyphs, which were carefully inscribed on stone monuments for commemorative or decorative purposes, were not well suited to rapid writing, scribes of the early dynasties began a process of gradually simplifying them, which led to the cursive signs we call hieratic. The pictorial details of the original hieroglyphs became less numerous and their outlines were stylized (fig. 15–2). In the hieratic sign for water, for example, the undulations that show clearly in the hieroglyph disappeared and were replaced by a thick and almost straight line. Similarly, the spiral of the sign for 100 was transformed into a long and only slightly curved line.

To avoid the confusions that were always possible in that kind of writing, scribes often added certain characteristic details of the person, animal, or object represented, and sometimes added one or more marks to a sign. A mark above a thick and almost straight line identified it with the hieroglyph for negation (a pair of outstretched arms and open hands) and distinguished it from the sign for water:

And the hieratic signs for 20 and 30 were distinguished from the sign for 10 by one or two extra marks:

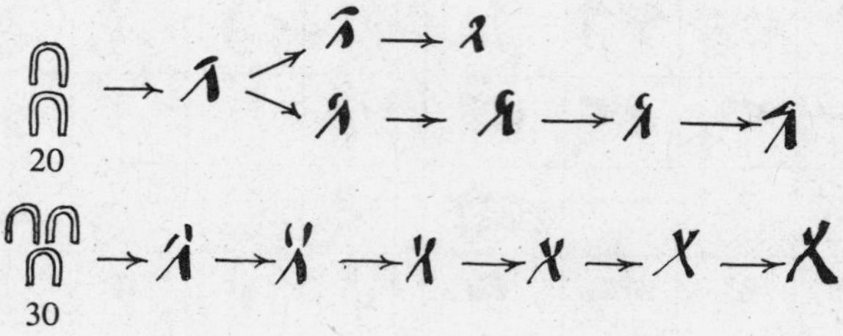

Another feature of this cursive script was the appearance of many ligatures (connecting strokes made without lifting the brush), not only within signs but also between them. This explains why groupings of five, six, seven, eight, or nine vertical strokes eventually took on forms that had no direct visual reference to the numbers they represented (fig. 15–3).

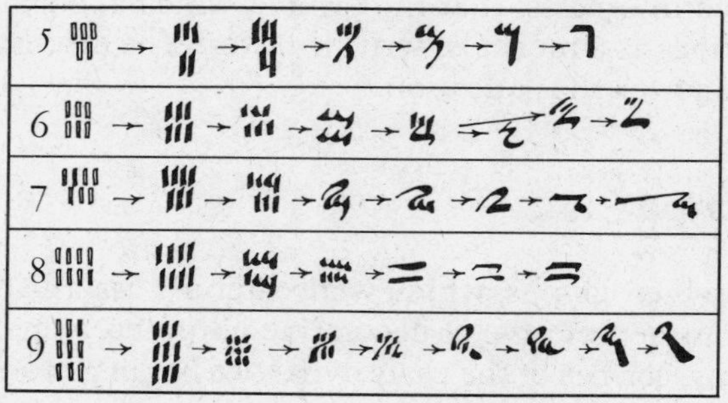

Fig. 15–3.

"Taking account of the possibilities and limitations of the reed pen, the goal was to save time, to write faster. The nature of the reed pen strongly influenced Egyptian handwriting. Scribes tried to write as much as possible in uninterrupted motions, sometimes using quick little strokes and sometimes a single larger stroke. . . . Some hieroglyphs were drastically changed; the cursive forms bore only a vague resemblance to their prototypes. The complex hieroglyphs, however—the wasp, the grasshopper, the crocodile, and others—usually remained quite recognizable in hieratic" (Sainte-Fare Garnot).

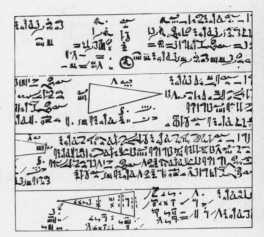

Fig. 15–4. Detail of the Rhind Papyrus, an important mathematical document written in hieratic characters, dating from about the seventeenth century B.C. It is a copy of an earlier treatise that probably went back to the Twelfth Dynasty (1991–1786 B.C.). British Museum.

Hieratic script therefore cannot be regarded as an ancient version of modern shorthand because there is one important difference: modern shorthand is composed of signs that were artificially created for the purpose of rapid writing and were not derived from the letters of our alphabet to any significant extent, whereas Egyptian hieratic signs were developed from pictorial prototypes. Their development was accentuated by a change of writing materials: a reed pen and papyrus instead of a chisel and stone (figs. 15–4, 15–5).

But "unlike what happened among most other peoples, this cursive writing never eliminated hieroglyphs in decorations on buildings, inscriptions in stone, etc., and it never even had any appreciable influence on the shapes of hieroglyphs" (Février). It was developed and used conjointly with its hieroglyphic model, just as the letters of our ordinary handwriting coexist with the capitals often used on our monuments. (Imagine the bewilderment of a Chinese or an Arab, for example, who did not know our alphabet and tried to match our handwritten letters with our printed ones.)

A   B   C   D   E   F   K   R   S

col. 4　　　　col. 3　　　　col. 2　　　　col. 1

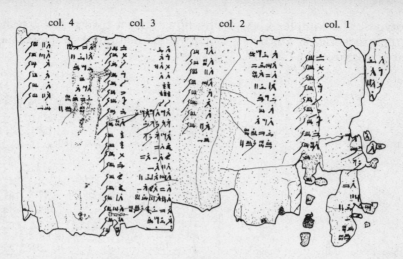

Fig. 15–5. A mathematical manuscript written on leather in hieratic characters. It is a table for converting fractions into sums of fractions with a numerator of 1, often used by scribes who had to make calculations.

For nearly 2000 years, between the third and first millennia B.C., Egyptian hieratic script was commonly used in administration, legal matters, commerce, education, magic, literature, science, and private correspondence. Beginning in the twelfth century B.C. it was gradually replaced in its ordinary uses by another cursive script known as demotic. It did not disappear entirely, however, because it was still used (till the third century A.D.) for religious texts. This explains the name given to it by the Greeks: the word *hieratikos* means "sacred."

A remarkable simplification

Like other hieratic signs, Egyptian hieratic numerals were simplifications of hieroglyphs. Still similar to their hieroglyphic originals in the third millennium B.C., they gradually underwent drastic alterations and then evolved independently of their models. They were further changed by the use of ligatures and the introduction of distinguishing marks, until they had only a very vague resemblance to their prototypes and sometimes became signs with no direct visual reference (fig. 15–6).

All this shows that the development of hieratic numerals from their hieroglyphic models took place without a sharp break, and for practical reasons.

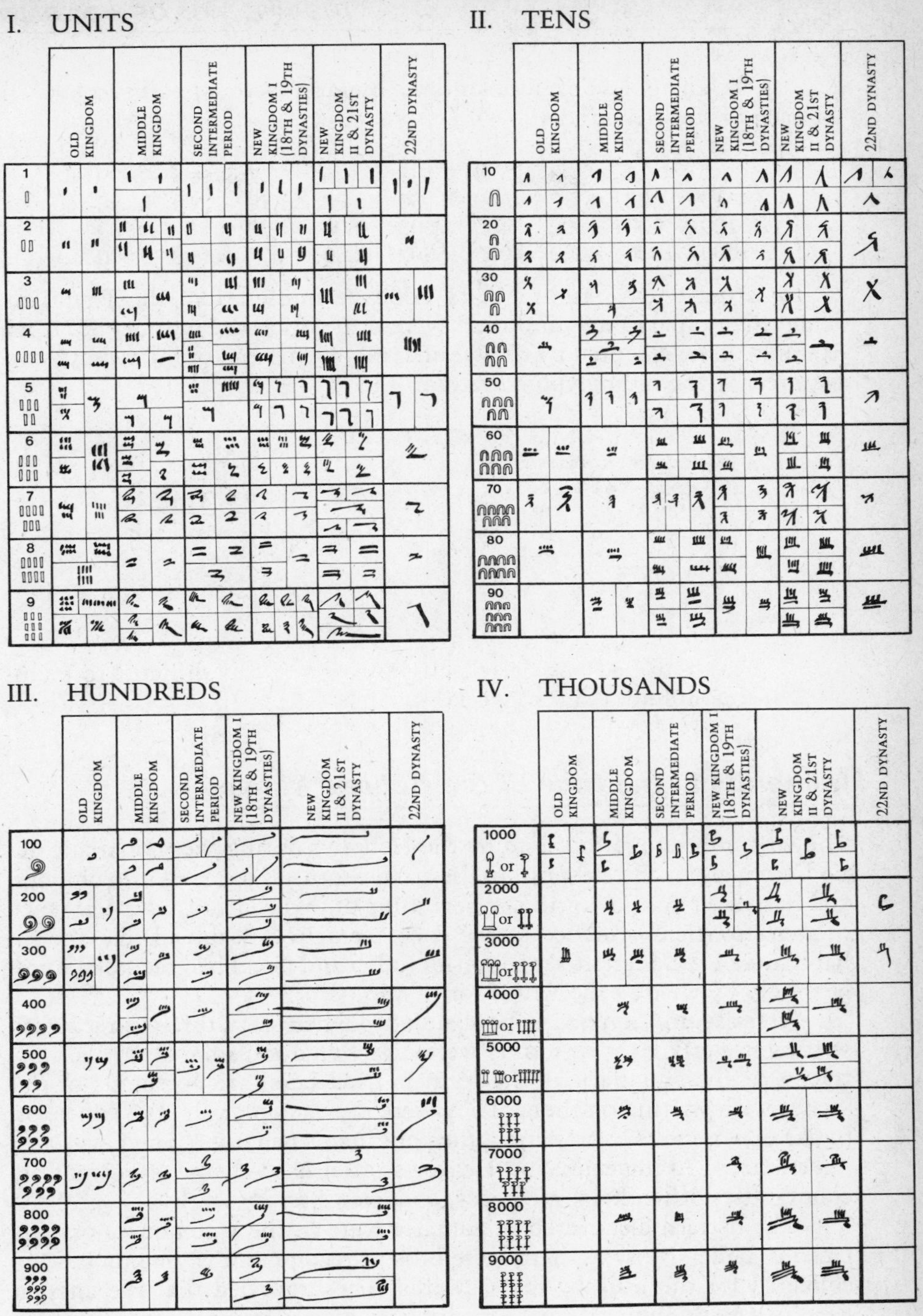

I. UNITS

II. TENS

III. HUNDREDS

IV. THOUSANDS

Fig. 15–6. Egyptian hieratic numerals.

By the end of its development, the hieratic system had a special sign for each of these numbers:

1	2	3	4	5	6	7	8	9
10	20	30	40	50	60	70	80	90
100	200	300	400	500	600	700	800	900
1000	2000	3000	4000	5000	6000	7000	8000	9000

Starting with a rudimentary additive numeration, the Egyptians achieved a remarkably simplified system in which the number 3577, for example, is represented by only four symbols, instead of the twenty-two required by the hieroglyphic system:

HIEROGLYPHIC NOTATION

7   70   500   3000

3577

HIERATIC NOTATION

7   70   500   3000

3577

This system had one drawback, however: it was difficult to memorize all the different signs it used.

Numerals in the time of the Hebrew kings

The written numeration used by the Hebrews during the monarchic period (from about the eleventh to the sixth centuries B.C.) fell into oblivion and remained unknown till the beginning of this century. In the absence of archaeological evidence to the contrary, it was assumed that the inhabitants of the ancient kingdoms of Judah and Israel had recorded numbers only by writing the words for them.

This assumption had to be abandoned when sixty-three ostraca with writing in Early Hebrew letters were discovered in Samaria. Ostraca (ostracon in the singular), as the term is used here, are potsherds or flat stones with writing on them. They were the "scrap paper" of scribes, or, to use G. Posener's expression, "the poor man's papyrus." The Egyptians, Phoenicians, Aramaeans, and Hebrews often used them for current accounts, lists of workers, messages, and even literary works.

The ostraca discovered in Samaria were in the warehouses of King Omri of Israel. Most of them were bills or receipts for payments in kind collected for the king by his stewards. They revealed that the ancient Israelites could indicate numbers not only by writing the words for them but also by means of genuine numerical signs.

This was later confirmed by other archaeological discoveries:

- About 20 Early Hebrew ostraca found at Lachish in 1935, most of them from the last months before the city was captured by Nebuchadnezzar II in 587 B.C. They were messages from a military commander to his subordinates.
- More than 100 Early Hebrew ostraca from the monarchic period discovered at Arad, a city in the eastern Negev near the border of the ancient kingdom of Juda, on the main road to Edom (fig. 15–7).
- Many Israelite inscribed weights.
- Similar discoveries at Kadesh-barnea and the Ophel Hill in Jerusalem.

FRONT   BACK

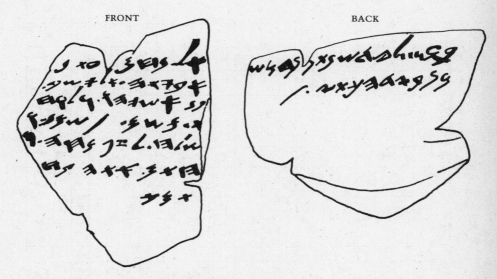

Fig. 15–7. Ostracon with Early Hebrew writing, sixth century B.C., found at Arad. On the back, it has the number 24 written in this form: ᵐᵐᵉ𝄪 .

It can easily be seen from figure 15–8 that the Early Hebrew numerals are simply the Egyptian hieratic numerals in the form they had during the New Kingdom (fig. 15–6).

This borrowing confirms the importance of cultural relations between Palestine and Egypt, as well as Egyptian influence on Israelite administration during the monarchic period, several other aspects of which have already been pointed out by historians.

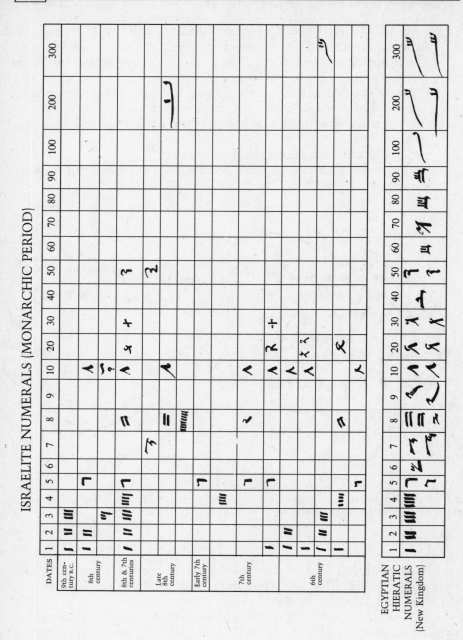

Fig. 15–8. Table showing that the numerals used in Palestine during the Israelite monarchic period are the numerals of the Egyptian hieratic system. This has been further confirmed by recent archaeological excavations in Israel: Early Hebrew ostraca discovered at Kadesh-barnea in 1978 have yielded a nearly complete series of hundreds and thousands in their Egyptian hieratic forms. (Personal communication from André Lemaire.)

PART IV

Numerals
and Letters

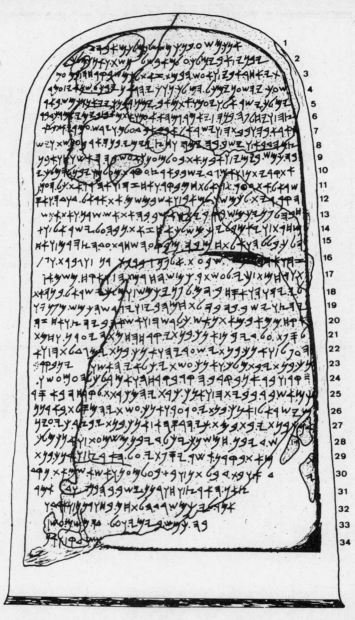

Fig. 16–1. Stele of King Mesha of Moab, a contemporary of the Israelite sovereigns Ahab (874–853 B.C.) and Jehoram (851–842 B.C.). This is one of the oldest known inscriptions in Early Hebrew characters. The language recorded is Moabite, a Canaanite dialect closely related to Hebrew. King Mesha had the stele erected in 842 B.C. at Dibon-gad, his capital. It gives us information on relations between Moab and Israel, and is the only known document of its time, found outside of Palestine, in which the name of Yahweh is explicitly mentioned.

16

Numerical Use of the Hebrew Alphabet

To indicate dates in the Hebrew calendar, to number chapters and verses of the Old Testament, and to paginate some Hebrew books, Jews still use a notational system whose numerals are letters of the Hebrew alphabet.

Hebrew writing, as it appears in manuscripts and printed editions of the Old Testament, uses the alphabet known as Square Hebrew, which has twenty-two letters (fig. 16–3). The order of these letters is the same as that of the Phoenician letters. This order is given by the Bible itself in many poetic passages that present Hebrew letters in the form of acrostics (Psalms 9, 10, 25, 34, 37, 111, 112, etc.).

Because of the great activity of Aramaean traders—who were, so to speak, a "landfaring" version of the earlier seafaring Phoenicians—their language (Aramaic) and writing spread all over the Near East, beginning in the fifth century B.C. The Israelites adopted them at about that time and began gradually changing the Aramaic letters into the forms that became Square Hebrew near the beginning of the Christian era.

Like most alphabetic scripts now in use, Aramaic writing came from the Phoenician alphabet. The script known as Early Hebrew (fig. 16–1), which was commonly used during the first half of the first millennium B.C. in the ancient kingdoms of Judah and Israel and was found sporadically in Palestine till the second century A.D., came directly from the Phoenician alphabet (fig. 16–2). (The Samaritans, faithful to ancient Jewish traditions, still use an alphabet that is a variant of Early Hebrew.)

As in most Semitic scripts, Hebrew letters are written and read from right to left, but they are always separated from each other (like printed letters of the Roman alphabet) and usually have the same form whether they are at the beginning or the end of a word. Five of them, however, have a different form at the end of a word; in most cases the final form is derived from the ordinary form by prolonging the lower stroke. (See next page.)

On the whole, Square Hebrew letters are rather simple and well balanced, but they must be written carefully because some of them have

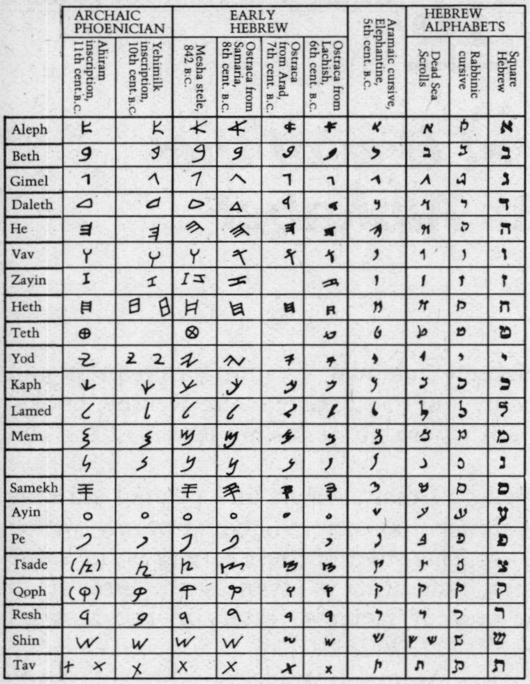

| | ARCHAIC PHOENICIAN | | EARLY HEBREW | | | | Aramaic cursive, Elephantine, 5th cent. B.C. | HEBREW ALPHABETS | | |
	Ahiram inscription, 11th cent. B.C.	Yehimilk inscription, 10th cent. B.C.	Mesha stele, 842 B.C.	Ostraca from Samaria, 8th cent. B.C.	Ostraca from Arad, 7th cent. B.C.	Ostraca from Lachish, 6th cent. B.C.		Dead Sea Scrolls	Rabbinic cursive	Square Hebrew
Aleph										א
Beth										ב
Gimel										ג
Daleth										ד
He										ה
Vav										ו
Zayin										ז
Heth										ח
Teth										ט
Yod										י
Kaph										כ
Lamed										ל
Mem										מ
										נ
Samekh										ס
Ayin										ע
Pe										פ
Tsade										צ
Qoph										ק
Resh										ר
Shin										שׁ
Tav										ת

Fig. 16–2. Ancient Semitic alphabets compared with modern Hebrew.

LETTER	Kaph	Mem	Nun	Pe	Tsade
ORDINARY FORM	כ	מ	נ	פ	צ
FINAL FORM	ך	ם	ן	ף	ץ

confusing resemblances to each other that may cause a neophyte to misread them. This is particularly true of the letters below, which are grouped to make comparison easier.

The Hebrew written numeration uses the twenty-two letters of the Hebrew alphabet, taken in the order of the Phoenician letters from which

ב כ פ	ד ר ך	מ ט	ו ז
B K P	D R K final	M T	V Z
ג נ	ת ח ה	ס ם	ע צ
G N	H Ḥ T	S M final	ʿ TS

they are derived. The first nine (from aleph to teth) are associated with numbers 1 to 9, the next nine (from yod to tsade) with the nine tens, and the last four (from qoph to tav) with the first four hundreds (fig. 16–4).

MODERN HEBREW ALPHABET

א Aleph	ו Vav	כ Kaph	ע Ayin	ש Shin
ב Beth	ז Zayin	ל or ל Lamed	פ Pe	ת Tav
ג Gimel	ח Heth	מ Mem	צ Tsade	
ד Daleth	ט Teth	נ Nun	ק Qoph	
ה He	י Yod	ס Samekh	ר Resh	

Fig. 16–3.

Hebrew letters	Names and sounds of the letters		Numerical values	Hebrew letters	Names and sounds of the letters		Numerical values
				כ	KAPH	k	20
				ל	LAMED	l	30
א	ALEPH	'a	1	מ	MEM	m	40
ב	BETH	b	2	נ	NUN	n	50
ג	GIMEL	g	3	ס	SAMEKH	s	60
ד	DALETH	d	4	ע	AYIN	'	70
ה	HE	h	5	פ	PE	p	80
ו	VAV	v	6	צ	TSADE	ts	90
ז	ZAYIN	z	7	ק	QOPH	q	100
ח	HETH	ḥ	8	ר	RESH	r	200
ט	TETH	ṭ	9	ש	SHIN	sh	300
י	YOD	y	10	ת	TAV	t	400

Fig. 16–4. Hebrew alphabetic numeration.

To write a number that cannot be represented by a single letter, the letters indicating the orders of units contained in it are juxtaposed, beginning with the highest order and going from right to left (figs. 16–5, 16–6). Numbers can thus be rather easily expressed in Hebrew writing. But if letters are used to express numbers, how can a group of ordinary letters be distinguished from a group of alphabetic numerals?

Translation:

THIS IS THE MONUMENT OF ESTHER, DAUGHTER OF ADAIO, WHO DIED IN THE MONTH OF SHEVAT IN YEAR 3 [א] OF THE SHEMITA. YEAR THREE HUNDRED 46 [שׁמ] AFTER THE DESTRUCTION OF THE TEMPLE [of Jerusalem].*
PEACE, PEACE UPON HER.

*346 + 70 = A.D. 416

Fig. 16–5. Jewish gravestone inscriptions (in Aramaic) dated A.D. 416, found near the southwestern coast of the Dead Sea. Amman Museum, Jordan.

Translation:

HERE LIES AN INTELLIGENT WOMAN, PROMPT IN ALL THE PRECEPTS OF THE FAITH, WHO FOUND THE FACE OF GOD, THE MERCIFUL, IN THE TIME WHICH COUNTS [?]. WHEN HANNA DEPARTED, SHE WAS 56 YEARS OF AGE.

$$נ\ ו\ =\ ו\ נ$$
$$\underset{6\quad 50}{}$$
$$56$$

Fig. 16–6. Part of a bilingual (Hebrew-Latin) inscription on a gravestone found at Oria (southern Italy), dating from the seventh or eighth century A.D.

When a number is represented by a single letter, it is customary to mark the letter with a slightly inclined accent placed to the left of its upper part:

ל״ה   שנ״ב

2 50 300   5 30

When a number is represented by two or more letters, the accent is usually doubled and placed between the last two letters on the left:

ש׳   פ׳   ל׳   ג׳   א׳

300   80   30   3   1

to notify the reader that the group of letters does not constitute a word (fig. 16–7). But since these accents also serve as signs of abbreviation, other systems have been used, such as placing a dot above each letter or drawing a line above all the letters in the group (fig. 16–8).

Fig. 16–7. Page of a Hebrew codex, A.D. 1311, reproducing Psalms 117 and 118, with the numbers indicated in the right-hand margin by means of Hebrew alphabetic numerals.

קנ	150	כ״ה	25
קע״ה	175	כ״ז	27
קצ״ו	196	כ״ח	28
רי״ט	219	ל״ב	32
שי״ב	312	מ״ד	44

Fig. 16–8. Numerical notations in medieval Hebrew documents and inscriptions.

The number 400 is the largest that can be written with a single letter. For numbers from 500 to 900 (fig. 16–9), the usual practice is to combine the letter tav (which has the value of 400) with the letters that represent the remaining hundreds:

תתק	תת	תש	תר	תק
100 400 400	400 400	300 400	200 400	100 400
900	800	700	600	500

Fig. 16–9. Numerical notations on Jewish gravestones in Spain.

Numbers 500, 600, 700, 800, and 900 can also be represented by the final form of the letters kaph, mem, nun, pe, and tsade, respectively:

ץ	ף	ן	ם	ך
TS final	P final	N final	M final	K final
900	800	700	600	500

This latter system, however, is accepted only for Cabalistic calculations; in common usage, the final forms of these five letters have the same numerical values as their ordinary forms.

To represent thousands, two dots are usually placed above the letter for the corresponding units, tens, or hundreds. In other words, when a Hebrew alphabetic numeral has two dots above it, its value is multiplied by 1000:

For example, the year 5739 of the Jewish calendar,* which corresponds to the period from October 2, 1978, to September 21, 1979, in the Gregorian calendar, can be indicated as follows:

This rule has not always been strictly followed, however; simplifications have often been used. A Jewish gravestone inscription in Barcelona, dating from A.D. 1299 or 1300, expresses the year 5060 in the Jewish calendar in this form:

$$\underset{\substack{60 \quad 5}}{\text{הֹ סֹ}} \qquad (= 5 \times 1000 + 60)$$

Here, the purpose of the dots is to avoid confusion between ordinary letters and numeral letters. The notation is unambiguous because numbers are written from right to left, beginning with the highest numeral letter, so that any numeral letter always has a higher value than the one to the left of it. In this example, since the letter he, which ordinarily has two possible values, 5 and 5000, is to the right of the letter samekh, which has the value of 60, it can only have the value of 5000.

Two other examples:

5109	הֹקֹטֹ 9 100 5	Gravestone inscription at Toledo, Spain, dated A.D. 1349
5156	הֹקֹנֹוֹ 6 50 100 5	Manuscript dated A.D. 1396

Thousands are sometimes omitted from expressions of dates, as in the inscription in figure 16–10, where the year 4804 is written 804.

---

*"In its present form, this calendar goes back to the fourth century A.D. . . . The starting point of the calculations was the new moon on Monday, September 24, 344, the first of Tishri in the Jewish calendar, and the beginning of a year. Having reckoned that 216 Metonic cycles, or 4104 years, were enough to contain their past, Jewish chronologists placed the first new moon of Creation at Monday, October 7, 3761 B.C." (Couderc).

שְׁמוּאֵל בַּר חֲלַאבּוּ ...

... בִּשְׁנַת תתד
804

"... SAMUEL SON OF HALABU ...
IN THE YEAR 804"

Fig. 16–10. Fragment of a Jewish gravestone inscription found at Barcelona, dated in the year 804 (= 4804) of the Jewish calendar (4804 − 3760 = A.D. 1044).

To indicate 5845, the number of verses in the Torah, some medieval Jewish scholars represented thousands and hundreds by the numeral letters for the corresponding units:

הח"מה

He   Mem   Heth   He

5 . 40 . 8 . 5

◀ − − − − − − − − − −

This notation is unambiguous, because of the rule described above. The letter heth, for example, normally has the value of 8 but cannot have that value here because it is to the right of the letter mem, whose value is 40, and neither can its value be 8000, since it is to the left of the letter he, which represents 5000; its value must therefore be 800.

The origin of the notation is easily explained. In Hebrew, the number 5845 is enunciated as follows:

ḥamishat   alafim   shmoneh   meot   arba'im   ve   ḥamisha
five   thousand   eight   hundred   forty   and   five

and can be written in this way:

ה' אלפים ח'מאות מ"ה
5   40   hundred   8   thousand   5

◀ − − − − − − − − − − − − − − − − − −

If the expression is shortened by omitting the words for "thousand" and "hundred," what remains is the notation in question. Such mixed expressions are found, for example, in Jewish gravestone inscriptions in Spain and in certain medieval manuscripts.

Another irregularity arises from the fact that, because of religious scruples, Jews generally avoid representing the number 15 in its regular form:

Instead, they use the following combination (also found in Jewish grave-stone inscriptions in Spain as well as in modern bilingual calendars, in which the days of each month are numbered regularly except for the fifteenth and sixteenth days):

The same irregularity extends to the number 16. Rather than being written in its regular form:

it is usually written as:

(The rule seems to have been strictly followed for 15, but not always for 16. On the title page of a fifteenth-century copy of Maimonides' *Mishneh Torah*, made in Portugal, the number 15 appears in the form of 9 + 6, but 16 is noted as 10 + 6.)

These anomalies stem from the taboo imposed by the Jewish religion on writing the divine name YHWH (Yahweh):

יהוה

In Jewish tradition this four-letter name, often designated as the Tetragrammaton, is regarded as the True Name of the God of Israel, and must never be written or spoken. However, abbreviated forms of it (YH, YW, HW, YHW, etc.) are sometimes written:

יה ין הן יהן ...

The prohibition against writing the Divine Name itself is therefore extended to the combinations 10 + 5 (YH) and 10 + 6 (YW).

ARCHAIC PHOENICIAN ALPHABET		GREEK ALPHABETS							CLASSIC GREEK ALPHABET		
		ARCHAIC Thera		EASTERN Miletus	Corinth		WESTERN Boeotia				
Aleph	�片 ⟨	𐤠 A		A		A A	⟨ ⟨		⟨	A α	Alpha
Beth	9 𐤔		𐤩	B		𐤔ᴜ⫭		B		B β	Beta
Gimel	𐤂 ⋀	Γ		Γ		< ⊂ι		⋀		Γ γ	Gamma
Daleth	△ ◁	Δ		Δ		Δ		D		Δ δ	Delta
He	∃ ∃	F		ε		B B		F E		E ε	Epsilon
Vav	Y Y	ϝ				F		F Ϲ		Ϲ Ϲ	Digamma*
Zayin	I I					I				Z ζ	Zeta
Heth	⊞ ⊟ ⊟	⊟		⊟ H		⊟		⊟ H		H η	Eta
Teth	⊕	⊕		⊗		⊗		⊕ ⊞ ⊖		Θ θ	Theta
Yod	⨕ 2	⨕		ι		⟨ ⟨		ι		I ι	Iota
Kaph	⩔ ⩔	K		K		K		K		K κ	Kappa
Lamed	⟨ ℓ	⋀		⋀ ⋀		Γ ⋀		ℓ		Λ λ	Lambda
Mem	⟨ ⟨	M [m]		M		M		M M		M μ	Mu
Nun	⟨ ⟨	N		N N		N		N		N ν	Nu
Samekh	⊞	⊞ [z]		⊞ [ks]		⊞				Ξ ξ	Xi
Ayin	O °	O ⊙		o		O		O ◇		O o	Omicron
Pe	⟩ ⟩	⟨		⟨		⟨		⟨ ⟨ ⊓		Π π	Pi
Tsade	⊬	M [s]		[s]		M				⋔ ⟨	San*
Qoph	𐤒	Φ 𐤒				φ		φ φ		ϙ φ	Koppa*
Resh	⟨ 𐤓	⟨		⟨		⟨		⟨ R		P ρ	Rho
Shin	W w	⟨		⟨ Σ				⟨ ⟨		Σ σ	Sigma
Tav	⨉ ✗	⊤		⊤		⊤		⊤		T τ	Tau
		Y [u]		Y		V		Y V		Y υ	Upsilon
		⊙				Φ		Φ		Φ φ	Phi
				✗ [kh]		X ✛		✛ [ks]		X χ	Khi
		V Y [ps]				Y		Y ⩔ [kh]		Ψ ψ	Psi
		Ω								Ω ω	Omega

Fig. 17–1. Greek alphabets compared with the archaic Phoenician alphabet. (Letters marked with an asterisk occurred in early forms of Greek and gradually became obsolete.)

17

Greek Alphabetic Numerals

A Greek papyrus scroll dating from the last quarter of the third century B.C., and now in the Cairo Museum, will serve to illustrate the principle of the Greek alphabetic numeration. The contents of the papyrus have been published by Guéraud and Jouguet. "It obviously served an educational purpose," they write. "It was a kind of textbook that gave the child practice in reading and calculating, as well as various notions useful to his education. . . . While he was learning to read, he also learned to know numbers. The sequence of numbers comes after tables of syllables, and this arrangement is natural enough, since Greek letters also had numerical values. It is logical that after learning how letters were combined to express syllables, the pupil learned how they were combined to express numbers."

The numeration in question uses the twenty-four letters of the classical Greek alphabet, plus the three obsolete letters digamma, koppa, and san (fig. 17–1). These twenty-seven signs are divided into three groups. The first, for units, consists of the first eight letters of the classical alphabet, with digamma for the number 6; the second, for tens, contains the next eight classical letters, with koppa for 90; the third, for hundreds, contains the last eight classical letters, with san for 900 (fig. 17–2).

UNITS				TENS				HUNDREDS			
A	α	alpha	1	I	ι	iota	10	P	ρ	rho	100
B	β	beta	2	K	κ	kappa	20	Σ	σ	sigma	200
Γ	γ	gamma	3	Λ	λ	lambda	30	T	τ	tau	300
Δ	δ	delta	4	M	μ	mu	40	Y	υ	upsilon	400
E	ϵ	epsilon	5	N	ν	nu	50	Φ	ϕ	phi	500
Ϝ	Ϛ	digamma	6	Ξ	ξ	xi	60	X	χ	khi	600
Z	ζ	zeta	7	O	o	omicron	70	Ψ	ψ	psi	700
H	η	eta	8	Π	π	pi	80	Ω	ω	omega	800
Θ	θ	theta	9	Ϙ	ϙ	koppa	90	Ϡ	ϡ	san	900

Fig. 17–2. The Greek alphabetic numeration, whose principle is similar to that of the Hebrew numeral letters (cf. fig. 16–4).

Other numbers are represented by means of the additive principle. For numbers 11 to 19, for example, letters indicating units are placed to the right of the letter iota, which has the value of 10.

A horizontal line was commonly drawn above these numeral letters to distinguish them from ordinary letters. (In modern books they are sometimes distinguished by an accent to the upper right of each letter, but that was not an ancient Greek practice; the accent had a different function.)

At the beginning of the papyrus scroll mentioned above is a partially intact sequence of written numbers, transcribed and translated in figure 17–3.

H̄	8		K̄	20
Θ̄	9		K̄Ā	21
Ī	10		K̄B̄	22
ĪĀ	11		K̄Γ̄	23
ĪB̄	12		K̄Δ̄	24
ĪΓ̄	13		K̄ε̄	25

Fig. 17–3.

"We are struck by the elementary character of the list of numerals. It does not even contain all the symbols the pupil would need in order to understand the table of squares at the end of the 'textbook' (figs. 17–4, 17–5). But this table itself, besides giving a few notions of elementary reckoning, provided an opportunity to show the pupil a series of symbols

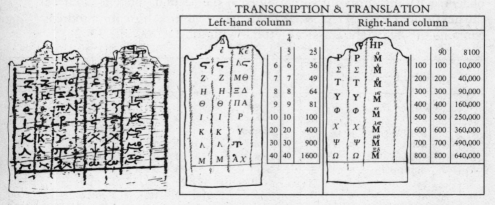

TRANSCRIPTION & TRANSLATION

Left-hand column						Right-hand column					
		Δ						HP			
	ε̄	Kε̄		5	25			M̄ᵃ		90	8100
ς̄	ς̄	ΛϚ̄	6	6	36	P	P	M̄	100	100	10,000
Z̄	Z̄	MΘ̄	7	7	49	Σ	Σ	M̄	200	200	40,000
H̄	H̄	ΞΔ̄	8	8	64	T	T	M̄ᶿ	300	300	90,000
Θ̄	Θ̄	ΠᾹ	9	9	81	Y	Y	M̄	400	400	160,000
Ī	Ī	P̄	10	10	100	Φ	Φ	M̄	500	500	250,000
K̄	K̄	Ȳ	20	20	400	X	X	M̄	600	600	360,000
Λ̄	Λ̄	Π̄	30	30	900	Ψ	Ψ	M̄	700	700	490,000
M̄	M̄	ᾹX	40	40	1600	Ω	Ω	M̄	800	800	640,000

Fig. 17–4. Table of squares in a Greek papyrus from the third century B.C. The left-hand column gives the squares of numbers 1 to 10, then by tens up to 40; the right-hand column gives the squares from 50 to 800. (The upper part of each column is missing.)

TRANSCRIPTION

A	(times) A	(equals)	A		Γ	(times) A	(equals)	Γ
B	—— A	——	B		Γ	—— B	——	ς
B	—— B	——	Δ		Γ	—— Γ	——	Θ
B	—— Γ	——	ς		Γ	—— Δ	——	IB
B	—— Δ	——	H		Γ	—— E	——	IE
B	—— E	——	I		Γ	—— ς	——	IH
B	—— ς	——	IB		Γ	—— Z	——	KA
B	—— Z	——	IΔ		Γ	—— H	——	KΔ
B	—— H	——	Iς		Γ	—— Θ	——	KZ
B	—— Θ	——	IH		Γ	—— I	——	Λ
B	—— I	——	K					

Fig. 17–5. Greek numeral letters in a table of multiplication by 2 and 3.

which completed those at the beginning of the scroll and enabled him to learn the principles of the Greek numeration from 1 to 640,000; and that was perhaps its main purpose. The few mathematical notions contained in the scroll must have been intended less to teach the pupil arithmetic than to teach him to recognize and read the signs used in the system of numeration" (Guéraud and Jouguet).

But how did the scribe who wrote this papyrus express numbers from 1 to 640,000 when the highest Greek numeral letter has the value of only 900? Up to 9000, he simply used the numeral letters for units and gave each of them a small distinguishing mark at the upper left:

Å	Ḃ	Ì̈	Δ̇	Ė	͗ς	'Z	'H	'Θ
1000	2000	3000	4000	5000	6000	7000	8000	9000

This is a common practice in Greek papyruses. It is found, for example, in a letter dated in the fourteenth year of the reign of Ptolemy XI of Alexandria (103 B.C.), in which the quantity of 5000 reeds is expressed in this form: Ė

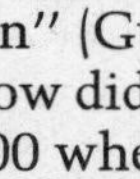

When he reached 10,000 the scribe expressed it by the letter M (mu)—the initial of the Greek word *murios*, "myriad" or "ten thousand"—with an alpha above it. He then wrote multiples of the myriad in this form:

$$\overset{\alpha}{M} \quad \overset{\beta}{M} \quad \overset{\gamma}{M} \quad \overset{\delta}{M} \quad \overset{\epsilon}{M} \dots \overset{\iota\alpha}{M} \quad \overset{\iota\beta}{M} \dots$$

10,000 20,000 30,000 40,000 50,000 110,000 120,000

(1 myriad, 2 myriads, 3 myriads, etc.), and was able to reach 640,000 without difficulty. He could, of course, have extended the process to the 9999th myriad, which he would have expressed in this way:

$$\overset{'\theta \ni \varphi \theta}{M} \qquad \left(\overset{9999}{M} = 99{,}990{,}000 \right)$$

Such notations for large numbers were often used, with variants, by Greek mathematicians.

In the writings of Aristarchus of Samos (310?–230? B.C.) we find this notation for the number 71,755,875 (example cited by Dédron and Itard):

$$'\zeta \rho o \epsilon\ \mathbf{M}\ '\epsilon \omega o \epsilon$$

7175 × 10,000 + 5875 ⟶

Diophantus of Alexandria (third century A.D.) altered the system slightly; in his version, myriads are generally separated from thousands by a dot. For him, the notation:

$$\delta \tau o \beta . '\eta \varphi \zeta$$

4372 • 8097 ⟶

meant 4372 myriads and 8097 units, or 43,728,097 (example cited by Daremberg and Saglio).

Another method of noting large numbers was proposed by the mathematician and astronomer Apollonius of Perga (262?–180? B.C.) and transmitted to us by Pappus of Alexandria (late third and early fourth centuries A.D.). It uses powers of the myriad and is based on a division of the whole numbers into several consecutive "orders." First there is the "elementary order," composed of numbers 1 to 9999—that is, all numbers lower than the myriad. Then comes the order of "first myriads," which includes all multiples of the myriad by numbers 1 to 9999—that is, numbers 10,000, 20,000, 30,000, and so on to 9999 × 10,000 = 99,990,000. To note a number of this order, the indication of the number of myriads it contains is preceded by the sign $\overset{\alpha}{M}$. For example, $\overset{\alpha}{M} \underset{\underset{664}{--\rightarrow}}{\chi\xi\delta}$ means 664 × 10,000 = 6,640,000 (and is read as "first myriads 664").

Next comes the order of "second myriads," including all multiples of a myriad of myriads by the numbers from 1 to 9999—that is, it contains numbers 100,000,000, 200,000,000, 300,000,000, and so on to 9999 × 100,000,000 = 999,900,000,000. Each of these numbers is expressed by an indication of the number of hundreds of millions it contains, preceded by the sign $\overset{\beta}{M}$. For example, $\overset{\beta}{M} \underset{\underset{5863}{----\rightarrow}}{\acute{\epsilon}\omega\xi\gamma}$ means 5863 × 100,000,000

= 586,300,000,000 ("second myriads 5863").

Then comes the "third myriads" (sign: $\overset{\gamma}{M}$) which begin with 100,000,000 × 10,000 = 1,000,000,000,000; then the "fourth myriads" (sign: $\overset{\delta}{M}$), and so on. (Note this difference between Apollonius' method and the one illustrated by the papyrus in fig. 17–4: in the papyrus the symbols $\overset{\alpha}{M}\,\overset{\beta}{M}\,\overset{\gamma}{M}\,\overset{\delta}{M}\,\overset{\epsilon}{M}$ are associated with consecutive multiples of the myriad [10,000, 10,000 × 2, 10,000 × 3, etc.], while for Apollonius these same symbols are associated with successive powers of the myriad [10,000, $10,000^2$, $10,000^3$, etc.].)

Other numbers can be expressed by breaking them down into the orders defined above. This is illustrated by the example in figure 17–6, given by Pappus of Alexandria himself and cited by Dédron and Itard, concerning the notation of 5,462,360,064,000,000 ("third myriads 5462, second myriads 3600, and first myriads 6400"):

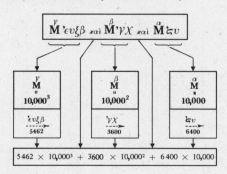

Fig. 17–6.

(The Greek καί can be translated as "plus.")

To write still larger numbers, Archimedes (287?–212 B.C.) proposed a more sophisticated system of notation in his *Psammites* (ψαμμίτης in Greek), or *Sand-reckoner*, concerning the question of how many grains of sand could be contained in a sphere with a radius equal to the distance from the earth to the fixed stars. "Having to deal with numbers greater than a myriad myriads, he used *octads,* double orders containing eight signs instead of four. The first octad included numbers 1 to 99,999,999; the second, numbers beginning with a hundred million, and so on. Numbers were first, second, etc., according to whether they belonged to the first octad, the second, etc. He considers the case of a hundred million octads: here begins the second *period;* then comes the third, the fourth, and so on up to the *myriad-myriadth* period, whose first unit can be expressed by the figure 1 followed by a number of zeros of which I will try to give an idea by pointing out that the unit of numbers of the first period is equal to 10^8 raised to the myriad-myriadth power $[(10^8)^{10^8}]$, that is, *1 followed by 800,000,000 zeros.* This is enough in itself to show how far Greek mathematicians carried the study and application of arithmetic" (Ruelle). Archimedes concluded that the number of grains of sand that could be contained in the celestial sphere was smaller than the eighth term of the eighth octad, which we would express as 1 followed by sixty-four zeros.

His system had only a theoretical purpose and did not win favor with Greek mathematicians; they evidently preferred Apollonius' system.

Notations for dates	Transcriptions & translations	
Κ	Κ	20
ΚΑ	ΚΑ	21
ΚΑ	ΚΑ	21
ΚΒ	ΚΒ	22
ΚΓ	ΚΓ	23
ΚΔ	ΚΔ	24
ΚΕ	ΚΕ	25
ΚΖ	ΚΖ	27
ΚΗ	ΚΗ	28
Λ	Λ	30
ΛΑ	ΛΑ	31
ΛΒ	ΛΒ	32
ΛΓ	ΛΓ	33
ΛΔ	ΛΔ	34
ΛΕ	ΛΕ	35
ΛϹ	ΛϚ	36
ΛΖ	ΛΖ	37
ΛΗ	ΛΗ	38

Fig. 17–7. Numerical notations on the oldest Greek coins dated by means of numeral letters. Their dates refer to the reign of Ptolemy Philadelphus (286–246 B.C.). The earliest one is from 266–265 B.C., the latest from 248–247 B.C. These coins, now in the British Museum, are described in Reginald Poole's catalog (see bibliography).

With the exception of the various notations devised for numbers greater than a myriad, the Greek alphabetic numeration is closely similar to the Hebrew system, so much so that we may ask which of the two inspired the other.

In an attempt to answer that question, I will succinctly review the present state of our knowledge on the subject, beginning with the Greeks.

Some of the oldest known examples of Greek numeral letters are on coins dating from the reign of Ptolemy Philadelphus (286–246 B.C.), the second sovereign of the Macedonian dynasty that began ruling Egypt soon after the death of Alexander the Great (fig. 17–7).

The oldest of these coins dates from 266–265 B.C., but this does not justify concluding, as some authors have done, that the Greek alphabetic numeration goes back no farther than the first half of the third century B.C., because a document showing use of that numeration nearly half a century earlier is now known: a papyrus from Elephantine dated 311–310 B.C., the oldest Greek papyrus discovered so far. It is a marriage contract from the seventh year of the reign of Alexander, son of Alexander the Great, and it gives the amount of the dowry as "alpha drachmas" in this form:

transcription: ⊢ A ;
translation: drachma A

The numeral letter used here probably has the value of 1000—or else the provider of the bride's dowry was so stingy that he parted with only one drachma!

The evidence thus indicates that use of the Greek alphabetic numeration goes back at least to the end of the fourth century B.C.

But the oldest examples of the Hebrew system go back only to the beginning of the first century B.C. or, at most, to the last few years of the second century.

"YEAR 2 OF THE LIBERATION OF ISRAEL"

Fig. 17–8. Coin from the time of the second Jewish rebellion, A.D. 132–134. Kadman Numismatic Museum, Israel.

The coin shown in figure 17–8 is a shekel struck by Simon Bar Kokhba, who seized Jerusalem during the second Jewish rebellion, A.D. 132–134. It bears an inscription in Early Hebrew letters and is dated, using the numeral letter beth, in "year 2 of the liberation of Israel," that is, A.D. 133.

A	B	C
שׁל בשׂראל שׁ שׁקל ישׂראל ש ב 2	שׁל בשׂראל שׁ שׁקל ישׂראל ש ג 3	שׁל בשׂראל שׁ שׁקל ישׂראל שׁח 5
"SHEKEL OF ISRAEL YEAR 2"	"SHEKEL OF ISRAEL YEAR 3"	"SHEKEL OF ISRAEL YEAR 5"

Fig. 17–9. Coins struck during the first Jewish rebellion, A.D. 66–73. They are shekels dated respectively Year 2 (A.D. 67), Year 3 (A.D. 68) and Year 5 (A.D. 70), by means of Early Hebrew numeral letters. Kadman Numismatic Museum, Israel.

.... של מלך אלה מת ₪ו₪..............
[א]כ מל הכ תנש סור[אלכסנדר]
25

"KING ALEXANDER YEAR 25"

Fig. 17–10. Three coins struck in 78 B.C., during the reign of Alexander Jannaeus. Kadman Numismatic Museum, Israel.

Other examples are coins from the time of the first Jewish rebellion, A.D. 66–73 (fig. 17–9), and Hasmonean coins, the oldest of which go back to the end of the first century B.C. The coins shown in figure 17–10 are among the oldest known Jewish coins dated by means of Hebrew numeral letters. They were struck in 78 B.C. during the reign of Alexander Jannaeus, fourth ethnarch of the Hasmonean dynasty, who took the title of king. The language of their inscriptions is Aramaic and the letters are Early Hebrew.

Also worth mentioning is a fragment of a parchment scroll found at Khirbat Qumran, near the northwestern shore of the Dead Sea, about eight miles from Jericho (fig. 17–11). Probably dating from the first century A.D., it is a copy of the rules of the Essenian community at Qumran. The letter gimel (= 3) appears at about the same distance from the top as from the right-hand edge, and "was written in Herodian style, no doubt by a young apprentice who had adopted the new Judean mode of writing, whereas the main scribe of the manuscript used the mode he had learned in his youth" (Milik). This letter served to number the sheet, because the fragment is from the first column of the third sheet of the scroll.

Fig. 17–11. Fragment of a parchment scroll found at Khirbat Qumran. Reproduced by permission of J. T. Milik.

Finally, a bulla (seal) of Jonathan the High Priest, probably dating from the Hasmonean period, has the letter mem at the end of its inscription (fig. 17–12). Although the meaning of this letter is still not clear, it may have been a numeral letter whose value (40) referred to the time of Simon Maccabee, who in 142 B.C. was recognized by Demetrius II as "high priest, military leader and civil ruler of the Jews." If so, it dates from the fortieth year after the beginning of Simon's reign, that is, from 103 B.C., and is the oldest known example of the use of Hebrew numeral letters.

Length: 13 mm
Width: 12 mm
Thickness: 2–3 mm

"JONATHAN HIGH PRIEST
JERUSALEM
M"

Fig. 17–12. Clay seal of Jonathan the High Priest, probably dating from the Hasmonean period. It served to seal a papyrus scroll, to which it was attached by a string. Israel Museum, Jerusalem.

These are obviously examples of a usage that was still in its infancy. Moreover, Israelite inscriptions reveal two kinds of numerical notation that have nothing to do with alphabetic numeration: one, used mainly during the monarchic period, is a borrowing of the Egyptian hieratic numerals, and the other, used from the Persian period to the beginning of the Christian era, is a borrowing of the Aramaic numerals.

All this would seem to indicate that, in common use at least, the Jews did not adopt numeral letters until a relatively late time in their history—and probably under the Greek influence of the Hellenistic period.

That influence is not surprising, for while the Jewish people unquestionably played a major part in the history of religion, all through antiquity they were influenced in various ways by their neighbors, whether allies or conquerors. We know the importance of the cultural relations that Palestine had with Egypt, Phoenicia, and Mesopotamia during the monarchic period, and we also know that from the eighth century B.C. onward the land of Israel successively fell under the domination of the Assyrians, the Babylonians, the Persians, the Ptolemies (who Hellenized the whole region), the Seleucids, and finally the Romans. Reflective of these influences on the Israelites are: the borrowing of the Phoenician alphabet and the Egyptian hieratic numerals during the monarchic period; the borrowing of the Assyro-Babylonian sexagesimal system for weights and measures; the probable borrowing of the ancient Canaanites' lunar calendar, in which the appearance of the new moon determined the beginning of a month; the adoption of the names of the months in the ancient calendar of Nippur, whose use, beginning in the time of Hammurabi, became general in Mesopotamia (the names were Nisan, Aiar, Siwan, Tammuz, Ab, Elul, Teshret, Arashamna, Kisilimmu, Tebet, Shebat, and Adar, nearly the same as in modern Hebrew); and the borrowing of the Aramaic language and writing (in the time of Jesus, Aramaic was still the only language commonly used in Judea, and the Gemara, one of the two divisions of the Talmud, was written in it).

These considerations support the view that the Hebrew alphabetic numeration was derived from the Greek, and this view is certainly not contradicted by Jewish inscriptions made later, during the Diaspora. Between the first century B.C. and the seventh century A.D., when use of the Hebrew alphabetic numeration was becoming increasingly common in the Jewish world, several Jewish scribes in the Mediterranean basin (from Italy to northern Syria and from Phrygia to Egypt), who could write as well in Hebrew as in Greek or Latin, often continued to use the Greek numeral letters.

In the Near East and the eastern Mediterranean in general, from antiquity till the end of the Middle Ages, the Greek alphabetic numeration played a part almost as important as that of the Roman system in western Europe (figs. 17–13 to 17–19).

EGYPT	Coptic inscription concerning Luke and his two works in the New Testament.
	$$\overline{KH} \quad \overline{KΔ} \quad \overline{KZ}$$ 28 24 27
	Jewish gravestones at Tell el-Yahudiyeh (about six miles north of Cairo), 1st century A.D.
	$$\overline{IB} \quad \overline{IΓ} \quad \overline{KΓ} \quad \overline{ΛЄ} \quad N \quad \overline{PB}$$ 12 13 23 35 50 102
PHRYGIA	Jewish inscription dated A.D. 253–254.
	$$TΛH$$ 338
ETHIOPIA	Inscription at Aksum, 3rd century A.D.
	$$KΔ \qquad ΓPIB \qquad ?C KΔ$$ 24 3112 6224
LATIUM	Jewish catacombs of the Via Nomentana, Labicana, and Appia Pignatelli.
	$$ΛΓ \qquad KΔ \qquad ΞΘ$$ 33 21 69
NORTHERN SYRIA	Mosaic in a synagogue, A.D. 392
	$$ΨΓ$$ 703
SOUTH OF THE DEAD SEA	Jewish gravestone inscription, A.D. 389–390
	$$\overline{ΣΠ} \qquad \overline{CΠΓ}$$ 86 283

Fig. 17–13. A few indications of the expansion of the Greek alphabetic numeration.

TRANSJORDAN PALESTINE	Mosaic compass card found in Peraea, east of the Jordan.
	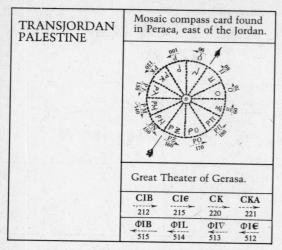
	Great Theater of Gerasa.

CIB	CIЄ	CK	CKA
212	215	220	221

ΦIB	ΦIL	ΦIV	ΦIЄ
515	514	513	512

Fig. 17–13 (continued).

1	2	3	4	5	6	7	8	9
a̅	B̅	Г̅	a̅	e̅	e̅	ꙃ̅	H̅	ϑ̅

10	20	30	40	50	60	70	80	90
·ı̅	к̅	λ̅	u̅	N̅	ξ̅	o̅	п̅	ϥ̅

100	200	300	400	500	600	700	800	900
P̅	C̅	T̅	Ƴ̅	ф̅	x̅	ѱ̅	ш̅	ⳤ̅

1000	2000	3000	4000	5000	6000	7000	8000	9000
a̿	B̿	Г̿	a̿	e̿	e̿	ꙃ̿	H̿	ϑ̿

10,000	20,000	30,000	40,000	50,000	60,000	70,000	80,000	90,000
ı̿	к̿	λ̿	u̿	N̿	ξ̿	o̿	п̿	ϥ̿

100,000	200,000	300,000	400,000	500,000	600,000	700,000	800,000	900,000
P̿	C̿	T̿	Ƴ̿	ф̿	x̿	ѱ̿	ш̿	ⳤ̿

Fig. 17–14. The Coptic numeration. The writing of the Egyptian Christians has an alphabet of thirty-one letters, of which twenty-four are directly derived from the Greek alphabet and the rest from Egyptian demotic writing. It uses the twenty-four letters of Greek origin, plus the archaic Greek letters digamma, koppa, and san, as numerical signs, with the same values they have in the Greek system. In the Coptic system, numbers are represented by letters with a single line above them up to 999 and with two lines from 1000 on.

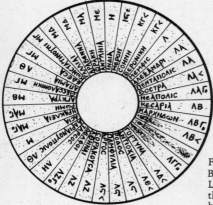

Fig. 17–15. Part of a portable sundial dating from the Byzantine period, now in the Hermitage Museum, Leningrad. This disk gives the names of places where the sundial can be used, with their latitudes expressed by means of Greek numeral letters, arranged in order of increasing magnitude, going clockwise.

TRANSCRIPTION		TRANSLATION	
ΙΝΔΙΑ	Η	India	8
ΜΕΡΟΗ	ΙϚΕ (error; read: ΙϚ <)	Merowe	16 ½
ϹΟΗΝΗ	ΚΓ < (*)	Syene	23 ½
ΒΕΡΟΝΙΚΗ	ΚΓ <	Beronice	23 ½
ΜΕΜΦΙϹ	Λ	Memphis	30
ΑΛΕΞΑΝΔΡΙ	ΛΑ	Alexandria	31
ΠΕΝΤΑΠΟΛΙϹ	ΛΑ	Pentapolis	31
ΒΟϹΤΡΑ	ΛΑ <	Busra	31 ½
ΝΕΑΠΟΛΙϹ	ΛΑ Γο (**)	Neapolis	31 ⅔
ΚΕϹΑΡΙΑ	ΛΒ	Caesarea	32
ΚΑΡΧΗΔωΝ	ΛΒ Γο	Carthage	32 ⅔
....	ΛΒ <	....	32 ½
....		....	33 ⅔
....	ΛΓ Γο		
ΓΟΡΤΥΝΑ	ΛΔ <	Gortyna	34 ½
ΑΝΤΙΟΧΙΑ	ΛΕ <	Antioch	35 ½
ΡΟΔΟϹ	ΛϚ	Rhodes	36
ΠΑΜΦΥΛΙΑ	ΛϚ	Pamphylia	36
ΑΡΓΟϹ	ΛϚ <	Argos	36 ½
ϹΟΡΑΚΟΥϹΑ	ΛΖ	Syracuse	37
ΑΘΕΝΑΙ	ΛΖ	Athens	37
ΔΕΛΦΟΙ	ΛΖ Γο	Delphi	37 ⅔
ΤΑΡϹΟϹ	ΛΗ	Tarsus	38
ΑΔΡΙΑΝΟΥΠΟΛΙϹ	ΛΘ	Adrianopolis	39
ΑϹΙΑ	Μ	Province of Asia	40
ΗΡΑΚΛΕΙΑ	ΜΑ Γο	Heraclea	41 ⅔
ΡωΜΗ	ΜΑ Γο	Rome	41 ⅔
ΑΓΚΥΡΑ	ΜΒ	Ankara	42
ΘΕϹϹΑΛΟΝΙΚΗ	ΜΓ	Thessalonica	43
ΑΠΑΜΙΑ	ΛΘ	Apamea	39
ΕΔΕϹΑ	ΜΓ	Edessa	43
ΚωΝϹΤΑΤΙΝΟΥΠΙΙ	ΜΓ	Constantinople	43
ΓΑΛΛΙΑΙ	ΜΔ	Gaul	44
ΑΡΑΒΕΝΝΑ	ΜΔ	Aravenna	44
ΘΡΑΚΗ	ΜΑ	Thrace	41(***)
ΑΚΥΛΗΙΑ	ΜΕ	Aquileia	45

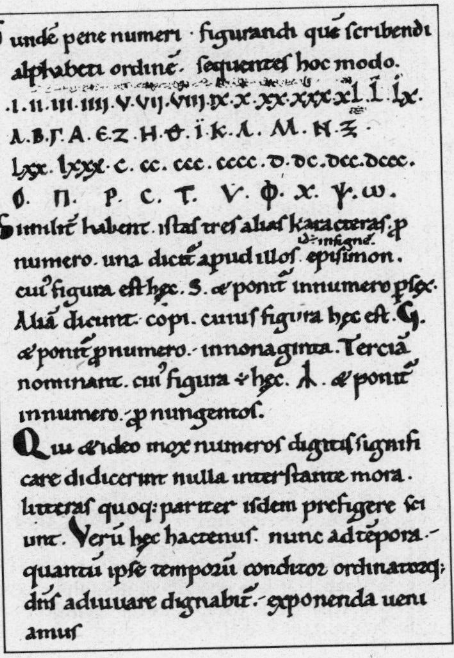

Fig. 17–15 (continued).
*The sign < indicates the fraction ½.
**Γο· = ⅔
***Perhaps 44.

Fig. 17–16. Portion of a Spanish manuscript on the subject of the Venerable Bede's finger counting (see figs. 3–6 to 3–9, 3–24), copied in about 1130, probably at Santa Maria de Ripoll in Catalonia. To explain the finger numerals drawn on the folios following this one, the scribe used both the Latin system and the Greek alphabetic system. National Library, Madrid.

Letter	Value	Letter	Value
a	1	t	300
b	2	u	400
g	3	vi	500
d	4	p'	600
e	5	k'	700
v	6	γ	800
z	7	q	900
h	8	š	1000
t'	9	tš	2000
i	10	ts	3000
k	20	dz	4000
l	30	ts'	5000
m	40	tš'	6000
n	50	ḫ	7000
ï	60	h	8000
o	70	dž	9000
p	80	h	10,000
ž	90		
r	100		
s	200		

Fig. 17–17. The Georgian alphabetic numeral system: a numeration and form of writing influenced by Greek in the early Christian era. The Georgian language is predominant between Armenia and the Caucasus in the U.S.S.R. There are two distinct forms of writing in Georgia: one called *khutsuri*, "priestly," whose letters are shown here, and the other called *mkhedruli*, "military." According to tradition, in the fifth century a scholar named Mesrob formed the Georgian alphabet, along with the Armenian writing to which it is related, on the basis of the Greek system.

#							
1	a	10	i	100	r		
2	b	20	k	200	s		
3	g	30	l	300	t		
4	d	40	m	400	w		
5	e	50	n	500	f		
6	q	60	y	600	ch		
7	z	70	u	700	hw		
8	h	80	p	800	o		
9	th	90		900			

Fig. 17–18. Another alphabetic numeral system influenced by Greek in the early Christian era: the Gothic system. "Near the northeastern borders of the Empire, Greek-language Christianity reached the eastern Germans, the Goths, on the Danube. They later lost their language and merged with various populations from Crimea to North Africa, leaving among other vestiges, the term 'Gothic.' . . . Ulfilas (311–384), a Christian Goth who had become a bishop, translated the Bible, or at least most of it, into his native language. To record his translation, he devised a form of writing (Gothic) which consists of Greek uncial letters and a few supplementary signs" (Cohen).

A	1	K	10	T	100
B	2	L	20	V	200
C	3	M	30	X	300
D	4	N	40	Y	400
E	5	O	50	Z	500
F	6	P	60		
G	7	Q	70		
H	8	R	80		
I	9	S	90		

Fig. 17–19. Numerical alphabet used by certain mystics during the Middle Ages and the Renaissance: an adaptation of the Greek numeration to the Latin alphabet. It was described, notably, by Athanasius Kircher in his *Oedipi Aegyptiaci* (1653).

18

The Phoenician Origin of Numeral Letters: A Legend

Of the various northwestern Semitic peoples, the Phoenicians and the Aramaeans were undoubtedly those who, long before the Hebrews, played the most important part in the history of the ancient Near East. The Phoenicians, great traders and bold seafarers who had settled in the land of Canaan, on the coast of Syria-Palestine, in the middle of the third millennium B.C., were probably the first to achieve the extreme simplification of writing that we call the alphabet. The Aramaeans were also skilled traders but they traveled by land. Beginning in the late second millennium B.C., they made their presence felt in the whole region. From Anatolia to Egypt and from Mesopotamia to the Mediterranean coast, their language and culture gradually established a dominant position and maintained it till the advent of Islam. Through the Aramaeans, Phoenician alphabetic writing was adopted by all the peoples of the Near East, from the Arabian peninsula to Mesopotamia and Egypt, and from the coast of Syria-Palestine to the edge of the Indian subcontinent.

But while the alphabet was in all probability invented by the Phoe-

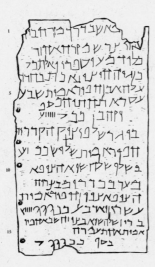

Fig. 18–1. Fragment of a copper scroll, first century B.C., that belonged to members of the Jewish sect of Khirbat Qumran, on the shore of the Dead Sea. Found in the third Qumran cave. Reproduced by permission of J. T. Milik.

275

nicians (or at least the Western Semites), can we also attribute the first use of numeral letters to them? In other words, did they always use the letters of their alphabet as true numerical signs, each having a fixed numerical value, as in the later Greek and Hebrew systems?

However plausible it may seem at first sight, this conjecture (treated as a dogma by a series of authors who have confidently repeated it for more than a century) is still unsupported by any evidence.

All of the many Phoenician and Aramaic inscriptions discovered so far, even the latest of them, have revealed only one kind of numerical notation, and it has nothing to do with the use of numeral letters. There are very few Hebrew documents earlier than the beginning of the Christian era that show the use of letters to express numbers. As we have already seen, the numerical signs that the Hebrews used during the monarchic period (or, at least, between the ninth and sixth centuries B.C.) were the Egyptian hieratic numerals (fig. 15–11).

Later, having adopted the language and writing of the Aramaeans, the Jews also borrowed their written numeration. One of the many papyruses left to us by the members of the Jewish military colony established in the fifth century B.C. on the island of Elephantine, in the Nile, has the notations for 80 and 90:

20 + 20 + 20 + 20
80

10 + 20 + 20 + 20 + 20
90

These signs are completely different from the Hebrew numeral letters pe (= 80) and tsade (= 90).

The scroll shown in figure 18–1 has numerical signs similar to those above (fig 18–2):

	NUMERICAL NOTATIONS IN THE DOCUMENT	VALUES	IF THE SCRIBE HAD USED NUMERAL LETTERS, HE WOULD HAVE WRITTEN:
line 9	ܝܥܝ 3 + 1	4	ד (ד) 4
line 7	ܝ׀׀׀׀ܝ 2 + 5 + 10	17	ܝ׀ (יז) 7 + 10
line 16	ܝ׀׀׀׀ܪܪ 2 + 4 + 20 + 20 + 20	66	ܟ׀ (סו) 6 + 60
line 13	ܝ ܪܪܪ 10 + 20 + 20 + 20	70	ܥ (ע) 70

Fig. 18–2.

At Khirbat el-Kom, a site in Israel between Lachish and Hebron, archaeologists have discovered a document that disproves the assertion that the Aramaeans used their letters as numerals. It is a bilingual ostracon (fig. 18–3) written in Aramaic and Greek, probably dating from the first half of the third century B.C. It served as a receipt of the sum of thirty-two drachmas (*zuz* in the Aramaic version) that a Semitic moneylender named Qos-Yada lent to a Greek named Nikeratos. Following a practice that was current all over the Hellenistic world at the time, Nikeratos noted the sum of thirty-two drachmas (and the date of the transaction) with Greek alphabetic numerals, while Qos-Yada, the Semite, indicated the sum of thirty-two *zuz* in this form: 20 + 10 + 1 + 1, using numerals of the kind we have seen earlier. If he had known how to use numeral letters, he would no doubt have written the sum in this simpler form:

$$\text{or}$$

2 + 30

It is therefore highly unlikely that the Western Semites used an alphabetic numeration before the Greek system was developed.

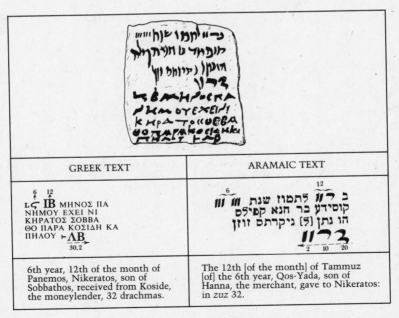

GREEK TEXT	ARAMAIC TEXT

6th year, 12th of the month of Panemos, Nikeratos, son of Sobbathos, received from Koside, the moneylender, 32 drachmas.

The 12th [of the month] of Tammuz [of] the 6th year, Qos-Yada, son of Hanna, the merchant, gave to Nikeratos: in *zuz* 32.

Fig. 18–3. Bilingual ostracon found at Khirbat el-Kom in Israel, probably dating from the sixth year of Ptolemy Philadelphus, i.e., 277 B.C.

HEBREW LETTERS	ARCHAIC PHOENICIAN	PALMYRAN	ESTRANGELO	NESTORIAN	SERTA		NAMES, TRANSCRIPTIONS, AND NUMERICAL VALUES OF SYRIAC LETTERS		
aleph							ŌLAP	'	1
beth							BĒTH	b bh	2
gimel							GŌMAL	g gh	3
daleth							DŌLATH	d dh	4
he							HĒ	h	5
vav							WAW	w	6
zayin							ZAIN	z	7
heth							ḤĒT	ḥ	8
teth							ṬĒT	ṭ	9
yod							YUD	y	10
kaph							KŌP	k kh	20
lamed							LŌMAD	l	30
mem							MIM	m	40
nun							NUN	n	50
samekh							SEMKAT	s	60
ayin							ʿE	ʿ	70
pe							PE	p ph	80
tsade							ṢŌDE	ṣ	90
qoph							QUF	q	100
resh							RISH	r	200
shin							SHIN	sh	300
tav							TAW	t	400

Fig. 19–1. Syriac alphabets compared with the Phoenician, Aramaic (Palmyran), and Hebrew alphabets. The use of Syriac letters as numerical signs is shown, notably, by a manuscript now in the British Museum, in which the order given above appears.

19

Syriac Numerals

The Maronite Christians of Syria, who generally speak Arabic, have preserved a relatively ancient form of writing known as serta or peshitta, mainly for liturgical use. The Nestorian Christians, who live mostly in the region of Lake Urmia, near the borders that Iran shares with Iraq, Turkey, and the Soviet Union, still speak an Aramaic dialect that they record in what is known as Nestorian writing.

These two systems have a twenty-two-letter alphabet derived from the much older estrangelo alphabet, which was used for a Semitic language belonging to the Aramaic group: Syriac.

The Nestorian form is more rounded than estrangelo and, graphically speaking, is intermediate between it and serta, which is more cursive (fig. 19–1). The letters are written from right to left, are connected to each other, and, as in Arabic writing, have different forms depending on their position within a word: initial, medial, final, or isolated. (In fig. 19–1 the Syriac letters are shown only in their isolated forms.)

The oldest known Syriac inscriptions evidently date from the first century A.D., and estrangelo writing does not seem to have been used until the sixth or seventh century. The Nestorian Christians, who were rather numerous in Persia during the Sassanid dynasty (226–651), gradually altered it until it took on the form known as Nestorian in the ninth or tenth century. The Jacobites (members of the Syrian Monophysite Church), most of whom lived in the Byzantine Empire, altered it more rapidly, turning it into serta by the seventh or eighth century.

Estrangelo is only a variant of the Aramaic alphabet and is therefore derived from the Phoenician alphabet. Like all Northwest Semitic alphabets, it preserves the order of the twenty-two original Phoenician letters (figs. 16–3, 19–1).

In serta and Nestorian, letters were used as numerical signs (and are sometimes still used as such today). The system is the same as that of the Hebrew alphabetic numeration: the first nine letters are for units, the next nine for tens, and the last four for the first four hundreds. And, as in Hebrew, hundreds from 500 to 900 are expressed by additive combinations as follows:

$$500 = 400 + 100$$
$$600 = 400 + 200$$
$$700 = 400 + 300$$
$$800 = 400 + 400$$
$$900 = 400 + 400 + 100$$

Thousands are indicated by a kind of accent below the letters representing units, and ten thousands by a horizontal line below them (fig. 19–2). Similar conventions enabled the Maronites to note numbers up to and beyond 10,000,000,000.

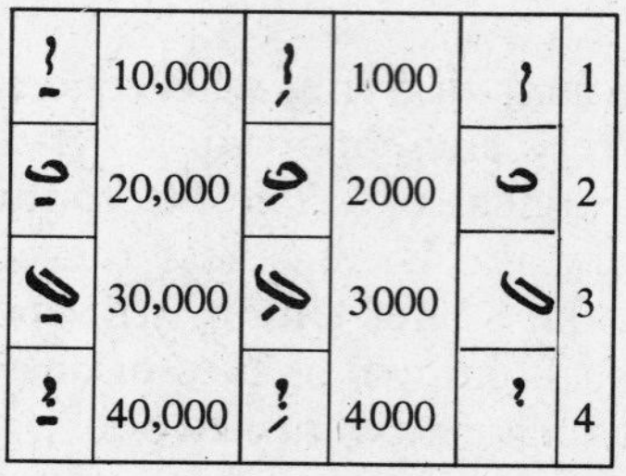

?	10,000	?	1.000	?	1
ب	20,000	؟	2000	ب	2
؟	30,000	؟	3000	؟	3
?	40,000	?	4000	?	4

Fig. 19–2.

With a few exceptions, this system is closely similar to the Hebrew alphabetic numeration. It was developed rather late in the history of Syriac writing, however, since the oldest documents showing its use go back no farther than the sixth or seventh century. The earliest Syriac documents reveal only one kind of numerical notation, related to the "classic" Aramaic system and having no connection with any alphabetic numeration.

Two Syriac inscriptions dated in year 476 of the Seleucid era (A.D. 165–166) have these numerical signs:

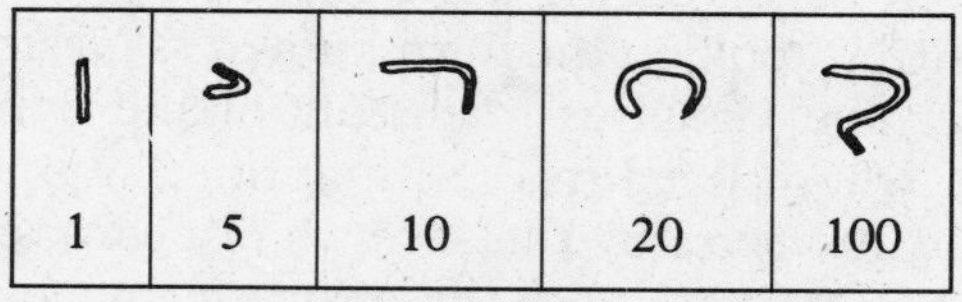

1	ے	٦	∩	?
1	5	10	20	100

And the function of the signs can easily be seen in the two following examples, from the inscriptions themselves:

6 + 10 + 20 + 20 + 20 + 100 × 4   1 + 5 + 10 + 20 + 20 + 20 + 100 × 4

476   476

We do not have enough evidence to say exactly when numeral letters were introduced into Syriac writing, but there are good reasons for believing that they were introduced as a result of Jewish influence on Christian and Gnostic groups in Syria-Palestine.

The folded sections of a Syriac manuscript that probably dates from the seventh or eighth century (now in the British Museum) are numbered with Syriac numeral letters, but the old signs for the same numbers are written near the letters. Does this mean that at the time when the manuscript was written the system of numeral letters was not yet accepted by everyone? Or does it mean that, by then, use of the old signs was already an archaism maintained by tradition, and that the alphabetic numeration had become so common that most Syrians regarded it as the only normal notation? With the evidence now available to us, we cannot answer those questions.

20

Arabic Numeral Letters

T he twenty-eight letters of the Arabic alphabet (fig. 20–1) are not arranged in the same order as the letters of the Phoenician, Aramaic, and Hebrew alphabets. Here, for example, are the first ten letters of the Arabic and Hebrew alphabets:

Arabic	Hebrew
alif	aleph
ba	beth
ta	gimel
tha	daleth
jim	he
ha	vav
kha	zayin
dal	heth
dhal	teth
ra	yod

For several reasons, however, we might expect to find the order of the twenty-two Northwest Semitic letters preserved in the Arabic alphabet.

First, it should be recalled that most alphabets now in use are derived from the Phoenician alphabet. The Roman alphabet, for example, came from the Etruscan, which came from the Greek, which came directly from the Phoenician. The Arabic and Hebrew alphabets are derived from two variants of the ancient Aramaic alphabet, which came from the Phoenician. The order of the twenty-two Phoenician letters was preserved almost intact by most western and eastern traditions: we find it in Early Hebrew, Aramaic (fig. 16–3), modern Hebrew (fig. 16–4), Syriac (fig. 19–1), Greek (figs. 17–1, 17–2), Georgian (fig. 17–17), Gothic (fig. 17–18), and Etruscan (fig. 20–2). "We know the order of the Phoenician letters from the strict concordance between the old Etruscan 'spelling books' . . . and the many Hebrew poems in the Old Testament that present the alphabet in acrostics. The earliest Etruscan 'spelling books' kept

LETTERS						NUMERICAL VALUES	
Form in isolation	Name	Phonetic value	Form in initial position	Form in medial position	Form in final position	In the East	In north-west Africa
ا	alif	ʾ				1	1
ب	ba	b				2	2
ت	ta	t				400	400
ث	tha	th				500	500
ج	jim	j				3	3
ح	ha	ḥ				8	8
خ	kha	kh				600	600
د	dal	d				4	4
ذ	dhal	dh				700	700
ر	ra	r				200	200
ز	zay	z				7	7
س	sin	s				60	300
ش	shin	sh				300	1000
ص	sad	ṣ				90	60

LETTERS						NUMERICAL VALUES	
Form in isolation	Name	Phonetic value	Form in initial position	Form in medial position	Form in final position	In the East	In north-west Africa
ض	dad	ḍ				800	90
ط	ta	ṭ				9	9
ظ	za	ẓ				900	800
ع	ayn	ʿ				70	70
غ	ghayn	gh				1000	900
ف	fa	f				80	80
ق	qaf	q				100	100
ك	kaf	k				20	20
ل	lam	l				30	30
م	mim	m				40	40
ن	nun	n				50	50
ه	ha	h				5	5
و	waw	w				6	6
ي	ya	y				10	10

Fig. 20–1. The Arabic alphabet in its modern form.

PHOENICIAN, ARAMAIC, & HEBREW LETTERS		SYRIAC LETTERS		GREEK LETTERS				ITALIC ALPHABETS			
				NAME	ARCHAIC	CLASSIC	NUMERICAL VALUE	OSCAN	UMBRIAN	ETRUSCAN	
ALEPH	ʾ	ŌLAP	ʾ	ALPHA	a	a	1				a
BETH	b	BĒTH	b	BETA	b	b	2				b
GIMEL	g	GŌMAL	g	GAMMA	g	g	3				g
DALETH	d	DŌLATH	d	DELTA	d	d	4				d
HE	h	HĒ	h	EPSILON	e	e	5				e
VAV	w	WAW	w	DIGAMMA	w	→	6				v
ZAYIN	z	ZAIN	z	ZETA	dz	dz	7				z
HETH	ḥ	HĒT	ḥ	ETA	ē	ē	8				th
TETH	ṭ	TET	ṭ	THETA	th	th	9				i
YOD	y	YUD	y	IOTA	i	i	10				k
KAPH	k	KŌP	k	KAPPA	k	k	20				l
LAMED	l	LŌMAD	l	LAMBDA	l	l	30				m
MEM	m	MIM	m	MU	m	m	40				n
NUN	n	NUN	n	NU	n	n	50				s?
SAMEKH	s	SEMKAT	s	XI	z	ks	60				o
AYIN	ʿ	ʿE	ʿ	OMICRON	o	o	70				p
PE	p;f	PE	p;f	PI	p	p	80				q
TSADE	ṣ	ṢŌDE	ṣ	SAN	s	→	900				r
QOPH	q	QUF	q	KOPPA	k	→	90				s
RESH	r	RISH	r	RHO	r	r	100				t
SHIN	s;sh	SHIN	sh	SIGMA	s	s	200				u
TAV	t	TAW	t	TAU	t	t	300				f
				upsilon		u	400				kh
				phi		ph	500				dh
				khi		kh	600				c
				psi		ps	700				
				omega		ō	800				

Fig. 20–2. In most cases the order of the twenty-two Phoenician letters was kept without alteration. The names shown here for the Phoenician letters are not attested before the sixth century B.C., but their order and phonetic values are very old: they go back at least to the fourteenth century B.C.

the twenty-two letters of the Phoenician alphabet, which makes their testimony even more valuable" (Février).

The order of the Phoenician letters is very old. Excavations at Ras Shamra, near Latakia, Syria, revealed that the people of Ugarit, an ancient city that evidently flourished between the fifteenth and thirteenth centuries B.C., recorded their language (a Semitic language related to Phoenician, Hebrew, and Aramaic) with an alphabet of thirty cuneiform signs. This was a very important discovery: it gave us the oldest known complete Northwest Semitic alphabet, for the oldest known examples of the Phoenician "linear" alphabet go back no farther than the twelfth century B.C. It is nearly certain, however, that this cuneiform alphabet was not the first alphabet in history. It seems highly probable that the Northwest Semitic "linear" alphabet, the ancestor of all modern alphabets, was completely constituted by the fifteenth century B.C. and that the scribes of Ugarit were acquainted with it and merely adapted its letters to cuneiform writing. The discovery of several Ugaritic "spelling books" (fig. 20–3) has shown that the thirty cuneiform signs of the Ugaritic alphabet were generally arranged in accordance with the traditional order of the Northwest Semitic letters and that the eight Ugaritic letters that had no equivalent in the Phoenician alphabet were either interspersed among the twenty-two basic letters or placed at the end of the alphabet.

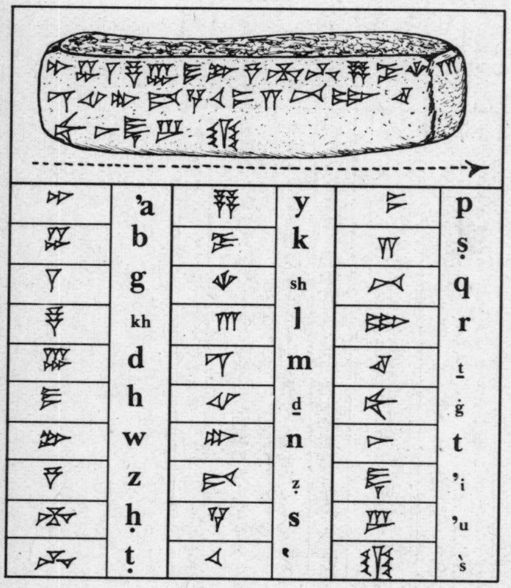

ᵈ	'a	ᵈ	y	ᵈ	p
ᵈ	b	ᵈ	k	ᵈ	ṣ
ᵈ	g	ᵈ	sh	ᵈ	q
ᵈ	kh	ᵈ	l	ᵈ	r
ᵈ	d	ᵈ	m	ᵈ	t
ᵈ	h	ᵈ	ḏ	ᵈ	ġ
ᵈ	w	ᵈ	n	ᵈ	t
ᵈ	z	ᵈ	ẓ	ᵈ	'i
ᵈ	ḥ	ᵈ	s	ᵈ	'u
ᵈ	ṭ	ᵈ	ʿ	ᵈ	ṡ

Fig. 20–3. Ugaritic "spelling book" from the fourteenth century B.C., found intact at Ras Shamra in 1948. Damascus Museum.

We may now raise the question of why the Arabs changed the traditional order of the Semitic letters. The answer is found in the history of their numeration.

A system of numerical notation that the Arabs have often used, and that they "seem to have regarded as belonging to them essentially and by preference" (Woepcke), employs the letters of the Arabic alphabet with a numerical value assigned to each of them, as shown in figure 20–1. This is the system that they call *ḥuruf al-jumal*, which means something like "reckoninig a *sum* by means of *letters*."

حروف الجمل

Ḥuruf Al-Jumal,

But the values assigned to six of the letters by the Eastern Arabs differ from those assigned to the same letters by the Western Arabs of northwestern Africa:

	West	*East*
س sin	300	60
ص sad	60	90
ش shin	1000	300
ض dad	90	800
ظ za	800	900
غ ghayn	900	1000

The numerical values of the Arabic letters can be ordered as follows:

1, 2, 3, 4, . . . 10, 20, 30, 40, . . . 100, 200, 300, 400, . . . 1000

If, referring to figure 20–1, we arrange the numeral letters of the Eastern Arabic system (the older of the two) in accordance with this sequence, the order we obtain is that of the Northwest Semitic letters (fig. 20–4). And if we compare the Arabic numeral letters, as shown in figure 20–4, with the Hebrew (fig. 16–4) and Syriac (fig. 19–1) alphabetic numerations, it is easy to see that the three systems correspond exactly to each other for values equal to or smaller than 400. This proves that "in an early system of numeration, the North Semitic alphabet was preserved, with the additional letters of the Arabic alphabet placed after it in order to reach 1000" (Cohen).

ا	ALIF	'	1		SIN	s	60
	BA	b	2		AYN	ʿ	70
	JIM	j	3		FA	f	80
	DAL	d	4		SAD	ṣ	90
	HA	h	5		QAF	q	100
	WAW	w	6		RA	r	200
	ZAY	z	7		SHIN	sh	300
	ḤA	ḥ	8		TA	t	400
	ṬA	ṭ	9		tha	th	500
	YA	y	10		kha	kh	600
	KAF	k	20		dhal	dh	700
	LAM	l	30		dad	ḍ	800
	MIM	m	40		za	ẓ	900
	NUN	n	50		ghayn	gh	1000

Fig. 20–4. The Arabic letters arranged in accordance with the sequence of numerical values in the alphabetic numeration of the Eastern Arabs.

We may conclude that the Arabs adopted the numeral alphabet in imitation of the Jews and Syriac Christians for the first twenty-two letters (that is, for numbers equal to or smaller than 400) and in imitation of the Greco-Byzantines for the six remaining letters (numbers from 500 to 1000). "After the conquest of Egypt, Syria, and Mesopotamia, Arab scribes began noting numbers either by writing the words for them or by using characters borrowed from the Greek alphabet" (Youshkevitch). Thus in an Arabic manuscript giving a translation of the Gospels (fig. 20–5), the verses are numbered by means of Greek letters.

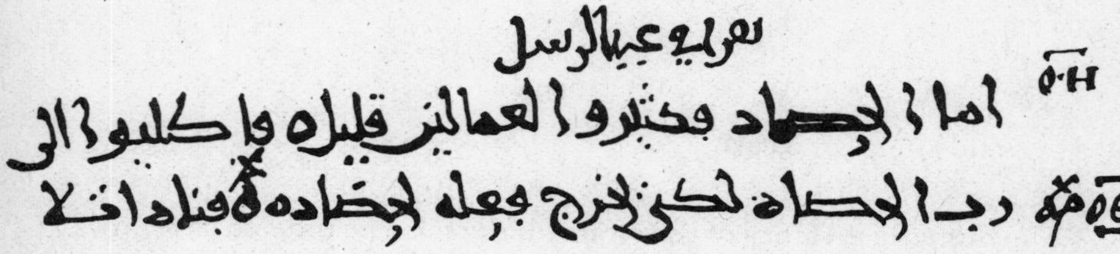

Fig. 20–5. Portion of a ninth-century Arabic manuscript giving a translation of the Gospels, with the verses numbered by means of Greek letters. To the right of the first line: oH = 78; to the right of the second line: oθ = 79. Vatican Library.

And in an Arabic economics text written on papyrus and dated year 248 of the Hegira (A.D. 862–863), now in the Egyptian Library, numbers are expressed exclusively in the Greek system. This practice continued in Arabic documents for several centuries, but had completely disappeared by the twelfth century.

This does not mean, however, that the Arabic alphabetic numeration was not introduced until that time: it was in use before the ninth century.

In a mathematical manuscript copied at Shiraz between 358 and 361 of the Hegira (between A.D. 969 and 971), all the Arabic numeral letters are used in accordance with the eastern system. On an astrolabe (an instrument for determining the altitude of celestial bodies) dated 315 of the Hegira (A.D. 927–928), that date is expressed in Arabic numeral letters written in the style known as Kufic (figs. 20–6, 20–7). Older documents show that the adoption of this system goes back to the eighth century or, at the earliest, the end of the seventh.

Fig. 20–6. Detail of an Arab astrolabe of the archaic eastern type, signed Bastulus and dated 315 of the Hegira (A.D. 927–928), with the date expressed in numeral letters of the eastern system. The inscription is in Kufic characters with a few diacritical marks. The astrolabe is said to have once belonged to King Faruk of Egypt. Reproduced by permission of A. Brieux.

صنعة بـطولس ٮٮة شٮه	"WORK OF BASTULUS YEAR 315"

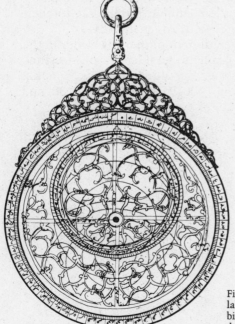

Fig. 20–7. Eighteenth-century Persian astrolabe signed Mohammed Muqim, with Arabic numeral letters used for marking 360 degrees at intervals of five.

After adding six letters to the twenty-two of the Northwest Semitic alphabet they had inherited, and developing their alphabetic numeration without changing the traditional order of the letters, Arab grammarians of the seventh or eighth century—who "worked mainly in Mesopotamia, where Jewish and Christian studies had flourished and continued, with Greek influences" (Cohen)—disrupted the original order, probably for pedagogical reasons, by bringing together letters whose forms closely resembled each other. Thus the letters ba, ta, and tha were grouped in the alphabet, as were jim, ḥa, and kha:

خ	ح	ج	ث	ت	ب
KHA	ḤA	JIM	THA	TA	BA
600	8	3	500	400	2

And finally, the Eastern Arabs devised eight meaningless combinations of syllables as aids to remembering the order of the numeral letters (fig. 20–8).

All this shows that the order known as *Abjad* (or *Abajad, Aboujad, Aboujed,* etc., according to the vocalizations given by different authors), which sometimes governs the letters of the Arabic alphabet, corresponds to neither their phonetic values nor their graphic forms, but to their numerical values in the system of the Eastern Arabs (fig. 20–9).

Six of the twenty-eight letters have different numerical values in the system of the Western Arabs (fig. 20–1), and numeral letters are classified differently. They are divided into nine memory-aid words corresponding to these groups of values: 1, 10, 100, and 1000; 2, 20, and 200; 3, 30, and 300, etc. (fig. 20–10).

	MEMORY AIDS		ASSOCIATED VALUES	
ABAJAD	ابجد	د ج ب ا d j b a	4.3.2.1
HAWAZIN	هوز	ز و ه z w h	7.6.5
ḤUṬIYA	حطي	ي ط ح y ṭ h	10.9.8
KALAMUNA	كلمن	ن م ل ك n m l k	50.40.30.20
SA'AFAṢ	سعفص	ص ف ع س ṣ f ʿ s	90.80.70.60
QURASHAT	قرشت	ت ش ر ق t sh r q	400.300.200.100
THAKHUDH	ثخذ	ذ خ ث dh kh th	700.600.500
ḌAḌHUGH	ضظغ	غ ظ ض gh dh ḍ	1000.900.800

Fig. 20–8. Memory aids for the eastern system of Arabic numeral letters (fig. 20–4).

604	خ د 4 600	خد	12	ي ب 2 10	يب
472	ت ع ب 2 70 400	تعب	58	ن ح 8 10	نح
1283	غ ر ف ج 3 80 200 1000	غرفج	96	ص و 6 90	صو
1631	ا ل خ غ 1 30 600 1000	غخلا	169	ق س ط 9 60 100	قسط
1629	غ خ ك ط 9 20 600 1000	غخكط	315	ش ي ه 5 10 300	شيه

Fig. 20–9. Numbers expressed in Arabic numeral letters of the eastern system. They are written from right to left, in decreasing order of value. Like ordinary Arabic letters, the numeral letters are usually connected to each other, and their forms change slightly according to their position in a numerical combination. These examples are from an Arabic manuscript copied at Shiraz in about A.D. 970, now in the National Library, Paris.

1	ا Alif	10	ي Ya	100	ق Qaf			
2	ب Ba	20	ك Kaf	200	ر Ra			
3	ج Jim	30	ل Lam	300	س Sin			
4	د Dal	40	م Mim	400	ت Ta			
5	ه Ha	50	ن Nun	500	ث Tha			
6	و Waw	60	ص Sad	600	خ Kha			
7	ز Zay	70	ع Ayin	700	ذ Dhal			
8	ح Ha	80	ف Fa	800	ظ Za			
9	ط Ta	90	ض Dad	900	غ Ghayn			
				1000	ش Shin			

AYQASH	ايقش	ا ي ق ش — sh q y a — 1000 100 10 1
BAKAR	بكر	ب ك ر — r k b — 200 20 2
JALAS	جلس	ج ل س — s l j — 300 30 3
DAMAT	دمت	د م ت — t m d — 400 40 4
HANAF	هنث	ه ن ث — th n h — 500 50 5
WASAKH	وصخ	و ص خ — kh ṣ w — 600 60 6
ZA'ADH	زعذ	ز ع ذ — dh ʿ z — 700 70 7
HAFADH	حفظ	ح ف ظ — ẓ f ḥ — 800 80 8
TADUGH	طضغ	ط ض غ — gh ḍ ṭ — 900 90 9

Fig. 20–10. The alphabetic numeration used by the Arabs of northwestern Africa, and the memory aids associated with it.

21

Numerals, Letters, Magic, Mysticism, and Divination

The art and composition of chronograms

In Jewish and Moslem writings from the Middle Ages on, there are abundant examples of the chronogram, a means of expressing dates that is a genuine art, with elements of calligraphy and poetry. It "consists in grouping, within a meaningful word or phrase, letters whose combined numerical values give the date of a past or future event" (Colin).

Here, for example, is an inscription on a Jewish gravestone in Toledo, Spain:

שנת אגלי טל על חמשת אלפים

THOUSAND FIVE   ON   DEWDROP   YEAR

<---

YEAR: "DEWDROP ON FIVE THOUSAND"

Taken literally, this means practically nothing, but if we take the sum of the numerical values of the letters composing the Hebrew word for "dewdrop":

א ג ל י   ט   ל   "DEWDROP"

30 9 10 30 3 1

<---------------------

83

we find that the phrase expresses a date: the person buried in the grave died in the year "eighty-three (dewdrop) on five thousand," that is, year 5083 of the Jewish calendar (A.D. 1322–23).

For the year 5144, another gravestone in Toledo gives this expression numerically equivalent to 144:

שנת היינה אין אב

2 1 50 10 1   5 50 10 10 5   YEAR

<-------------------

YEAR: "WE ARE LEFT FATHERLESS"

144

(In Hebrew chronograms, thousands are commonly omitted, just as we sometimes shorten 1980 to '80, without danger of confusion when it is clear that we are referring to our own century.)

As in the example above, Hebrew chronograms are often distinguished from ordinary words by three dots placed above each of their elements.

On other Jewish gravestones, the year 5109 (A.D. 1348–49) is expressed in one of these two forms:

לֹּ הֹחֹיֹיֹם
40 10 10 8 5 6 30
"FOR LIFE"
109

מֹנֹוֹחֹה
5 8 6 50 40
"REST"
109

Chronograms are also found in Islamic countries, notably Turkey, Iraq, Iran, and the state of Bihar in India, but they evidently go back no farther than the eleventh century A.D.

To commemorate the death of King Sher of Bihar (in northeastern India), who was killed by an explosion in year 952 of the Hegira (A.D. 1545), this chronogram was composed.

زأتش مرد
"DIED OF BURNS"

ز أ ت ش   م ر د
4 200 40   300 400 1 7
952

Another example was given by al-Biruni (973–1048) in his *Chronology of Ancient Nations*. This erudite mathematician and astronomer, who accused the Jews of having deliberately changed their calendar by diminishing the number of years since the Creation, so that the date of Jesus's birth was no longer in accord with prophecies concerning the coming of the Messiah, boldly stated that they expected the Messiah to come in year 1335 of the Seleucid era (A.D. 1024), and he expressed that date with the following chronogram:

نجاة الخلق من الكفر بمحمد

"MOHAMMED SAVES THE WORLD FROM UNBELIEF."

ن ج ا ة   ا ل خ ل ق   م ن   ا ل ك ف ر   ب م ح م د
4 40 8 40 2   200 80 20 30 1   50 40   100 30 600 30 1   5 1 3 50
1335

Chronograms are also common in Morocco, but began there no earlier than the sixteenth century, according to the present state of documentary evidence. They were often used not only in versified inscriptions commemorating one event or another, but also by poets, historians, and biographers, including the court secretary and poet Mohammed Ben Ahmad al-Maklato (died 1630) and the poets Mohammed al-Mudara (died 1734) and Abd-al-Wahab Adaraq (died 1746), who each composed a didactic historical summary, concerning the notables of Fez and the saints of Meknes, respectively.

In inscriptions and manuscripts, chronograms were sometimes written in a color that made them stand out from the rest of the text, and in manuscripts thicker letters were also used occasionally. Arabic chronograms were always announced by *fi* ("in") or *sanat ama* . . . ("in the year . . .").

The following chronogram is from an Arabic inscription found by G. S. Colin more than fifty years ago in the south chamber of a building known as the Qubbat-al-Bukhari, in Tangiers:

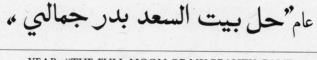

عام "حل بيت السعد بدر جمالي "

YEAR: "THE FULL MOON OF MY BEAUTY CAME
INTO THE CHAMBER OF HAPPINESS."

ح ل ب ي ت ا ل س ع د ب د ر ج م ا ل ي

10 30 1 40 3 200 4 2 4 70 300 30 1 400 10 2 30 8

1145

It indicates that the building was constructed in year 1145 of the Hegira, which began on June 24, A.D. 1732. It is a good example of an art in which great imagination was required to compose a sentence whose numerical value would be the exact expression of a certain date.

Interpretations and speculations of Gnostics, Cabalists, magicians, and soothsayers

The system of assigning numerical values to the letters of an alphabet gave rise to curious doctrines based on the procedure of interpreting a word, group of words, or group of letters on the basis of its numerical value, sometimes relating it to another word or expression that may or may not have the same value. The Jewish form of this procedure is called gematria, the Greek form isopsephia, and the Moslem form *hisab al-jumal*.

Gematria has led to all sorts of homiletic interpretations as well as various speculative and divinatory calculations. It was often used in rabbinical literature, particularly in the Talmud (a collection of Jewish laws, traditions, and opinions, compiled by rabbis) and the Midrash (compilations traditionally regarded as commentaries on the Hebrew Scriptures). But it is found chiefly in esoteric literature: the "magic" applications of Cabalism made extensive use of it.

Though I am not an expert on the subject, I would like to give a brief description and a few examples of the main religious and literary practices, and the strange predictive calculations, that have arisen from gematria, in order to show how far its practitioners have extended their speculations and reasoning.

Some rabbis link the Hebrew words *yayin* ("wine") and *sod* ("secret") because of the saying "From wine will come the secret" (*Nikhnas yayin yatsa sod*, which has essentially the same meaning as the Latin *In vino veritas*), and because the two words have the same numerical value in the common Hebrew system (fig. 16–4).

סוד   יין

4   6   60   50   10   10

SOD   YAYIN

70   70

In his *Pardes Rimonim*, Moses Cordovero gives another example by linking the words *gevurah*, "strength," and *aryieh*, "lion":

אריה   גבורה

5   10 200   1   5   200 6   2   3

ARYIEH   GEVURAH

216   216

These two words have the same numerical value; the lion is traditionally regarded as a symbol of the divine majesty, bravery, strength, and power of Yahweh, and *gevurah* is one of the attributes of God.

Another example: the Messiah is often referred to as *Semah*, "the Germ," or *Menahem*, "the Consoler," because those two words have the same numerical value:

צֶמַח   מנחם
8 40 90   40 8 50 40
ṢEMAḤ   MENAHEM
138   138

Since the words *Mashiyaḥ*, "Messiah," and *naḥash*, "serpent," have the same value:

נָחָשׁ   משיח
300 8 50   8 10 300 40
NAHASH   MASHIYAH

it has been concluded that when the Messiah comes he will struggle against Satan and defeat him.

Some believers in gematria have said that the world was created at the beginning of the Jewish civil year because the first two words of the Torah, *Berishit bara*, "In the beginning [God] created," have the same numerical value as *Berosh hashanah nibṛa*, "It was created at the beginning of the year":

בָּרָא בְּרֵאשִׁית
1 200 2   400 10 300 1 200 2

1116   BERESHIT BARA

נברא הַשָׁנָה בְּרֹאשׁ
1 200 2 50   5 50 300 5   300 1 200 2

1116   BEROSH HASHANAH NIBRAH

In Genesis 32:4 Jacob says, "I have been living with Laban"; in Hebrew: *Im Laban garti*. According to Rashi's commentary on that sentence, it means that Jacob did not follow the impious Laban's bad examples while he was living with him, but continued to observe the 613 commandments of the Jewish religion, because the numerical value of the Hebrew word *garti* is 613:

גרתי
10 400 200 3
GARTI
613

When Abram learned that his nephew Lot had been defeated in battle and captured by his enemies, "he mustered his retainers, men born in his household, three hundred and eighteen of them, and pursued as far as Dan" (Genesis 14:14). Then, after he had rescued his nephew, "the word of the Lord came to Abram in a vision. He said, 'Do not be afraid, Abram, I am giving you a very great reward.' Abram replied, 'Lord God, what canst thou give me? I have no standing among men, for the heir to my household is Eliezer of Damascus' " (Genesis 15:1–2).

One of the Haggadic rules for interpreting the Torah explains that the 318 men mentioned in the passage are none other than Eliezer—that is, Abram defeated his enemies with the help only of Eliezer, his steward, whose Hebrew name means "My God is help" and has the numerical value of 318:

אליעזר

200 7 70 10 30 1
ELI'EZER
318

The words *ahavah*, "love," and *eḥad*, "one," are equivalent in numerical value:

אחד   אהבה

4 8 1   5 2 5 1
EḤAD   AHAVAH
13   13

These two terms, it is said, correspond to the central point of the biblical ethic, the concept of God as love, since "one" represents the One God of Israel and love is basic to the conception of the world. Furthermore the sum of their numerical values is 26, the number assigned to the Divine Name of Yahweh:

יהוה

5 6 5 10
YHWH
26

The common Semitic word for "god" is *el*, but the Old Testament seems to use it only in compound names (Israel, Ishmael, Eliezer, etc.). To designate God, the Torah mainly uses the form *Elohim* (which is plural), this name being regarded as the one that expresses all supernatural forces and powers. It also uses the attributes of God, such as *Hai*, "Living," and *Shadai*, "Almighty." But Yahweh, or YHWH, the Divine Tetragrammaton, is the only true Name of God. It embodies the eternal nature of God because it is composed of the three Hebrew forms of the verb "to be":

היה	HaYaH	"HE WAS"
הוה	HoWeH	"HE IS"
יהיה	YiHYeH	"HE WILL BE"

In invoking God by this name, one stresses his intervention and solicitude in all things. It must therefore not be written or spoken in ordinary circumstances, and to avoid "taking the name of the Lord in vain" it is read as *Adonai*, "My Lord."

There have been all sorts of speculations about the numerical value of 26 attributed to the Divine Tetragrammaton in the classical system.

Some authors versed in gematria have pointed out that it is in verse 26 of the first chapter of Genesis that God says, "Let us make man in our image"; that Adam and Moses are separated by twenty-six generations; that twenty-six descendants are mentioned in Shem's genealogy, etc. According to them, the fact that God made Eve from one of Adam's ribs is reflected in the difference of 26 between the Hebrew names of Adam (= 45) and Eve (= 19):

חוה	אדם
5 6 8	40 4 1
<----------	<----------
HAWAH	ADAM
19	45

The method provided by the ordinary Hebrew alphabetic numeration (fig. 16–4) is by no means the only one that has been used by rabbis and Cabalists for interpretations of this kind. A manuscript preserved in the Bodleian Library at Oxford enumerates more than seventy different types of gematria.

One of these methods consists in giving the letters the number of units they have in the usual system, but disregarding tens and hundreds. The letter mem, מ , for example, whose usual value is 40, is given the value of 4, and shin, ש , whose usual value is 300, is given the value of 3 (fig. 21–1, col. A). Another way of determining these same values is to add the digits of the number indicating each letter's position in the alphabet: since mem occupies the thirteenth position, its value is 1 + 3 = 4.

Using this method, some exegetes have linked the name of Yahweh and the divine attribute *tov*, "good":

יהוה	טוב
5 6 5 1	2 6 9
<---------	<---------
YHWH	TOV
17	"Good"
	17

Another method consists in giving each letter the square of its usual value. The letter gimel, for example, whose usual value is 3, is given the value of 9 (fig. 21–1, col. B).

Another method assigns the value of 1 to the first letter, the sum of the first two numbers to the second letter, the sum of the first three numbers to the third letter, and so on (fig. 21–1, col. C). Thus the letter yod, tenth in the alphabet, has a value equal to the sum of the first ten numbers:

$$1 + 2 + 3 + \ldots + 9 + 10 = 55$$

Still another method attributes to each letter the numerical value of its name, calculated according to the usual system (fig. 21–1, col. D). The letter aleph, א , for example, has the value of 111:

ALEPH   אלף

80   30   1

<---------

111

Comparisons can be made between two words evaluated either by a single method or by two different ones. For example, *Maqom*, "Place," another appellation of God, has been linked with the name Yahweh, the former being evaluated by the usual method and the latter by taking the squares of the usual values:

מקום	יהוה
40 6 100 40	5² 6² 5² 10²
<---------	<---------
MAQOM	YHWH
186	186

These few examples (out of a vast multitude that could be cited) should be enough to give some idea of the complexity of Cabalistic calculation and the extent of the research that has led exegetes not only to interpretation of certain passages in the Torah but also to all sorts of speculations.

The same procedures were known to the Greeks, at least in late antiquity.

Among Greek poets, such as Leonidas of Alexandria, who lived in

Position in the alphabet	Usual value	A	B	C	D	
1	א 1	1	1^2	1	111 ——— אלף	ALEPH
2	ב 2	2	2^2	1+2	412 ——— בית	BETH
3	ג 3	3	3^2	1+2+3	73 ——— גמל	GIMEL
4	ד 4	4	4^2	1+2+3+4	434 ——— דלת	DALETH
5	ה 5	5	5^2	1+2+3+4+5	6 ——— הא	HE
6	ו 6	6	6^2	1+2+3+4 ...+6	12 ——— וו	VAV
7	ז 7	7	7^2	1+2+3+4 ...+7	67 ——— זין	ZAYIN
8	ח 8	8	8^2	1+2+3+4 ...+8	4ı8 ——— חית	HETH
9	ט 9	9	9^2	1+2+3+4 ...+9	419 ——— טית	TETH
10	י 10	1	10^2	1+2+3+4 ...+10	20 ——— יוד	YOD
11	כ 20	2	20^2	1+2+3+4 ... +11	100 ——— כף	KAPH
12	ל 30	3	30^2	1+2+3+4 ...+12	74 ——— למד	LAMED
13	מ 40	4	40^2	1+2+3+4 ... +13	90 ——— מים	MEM
14	נ 50	5	50^2	1+2+3+4 ...+14	110 ——— נון	NUN
15	ס 60	6	60^2	1+2+3+4 ...+ 15	120 ——— סמך	SAMEKH
16	ע 70	7	70^2	1+2+3+4 ...+16	130 ——— עין	AYIN
17	פ 80	8	80^2	1+2+3+4 ...+17	85 ——— פה	PE
18	צ 90	9	90^2	1+2+3+4 ...+ 18	104 ——— צדי	TSADE
19	ק 100	1	100^2	1+2+3+4 ...+19	186 ——— קוף	QOPH
20	ר 200	2	200^2	1+2+3+4 ...+20	510 ——— ריש	RESH
21	ש 300	3	300^2	1+2+3+4 ...+21	360 ——— שין	SHIN
22	ת 400	4	400^2	1+2+3+4 ...+22	406 ——— תו	TAV

Fig. 21–1. A few of the many methods of numerically evaluating Hebrew letters that have been used in interpretations by rabbis and cabalists.

the time of Nero, they gave rise to literary compositions of a special kind: isopsephic distichs and epigrams. A distich (couplet) is isopsephic if the sum of the numerical value of the letters in the first line is equal to the sum of those of the letters in the second line. An epigram (short poem expressing a single thought) is isopsephic if all the distichs it contains are isopsephic and the corresponding value is constant.

More generally, isopsephia consists in using the numerical values of letters, as in Hebrew gematria, for the purpose of establishing connections between words or groups of words.

Some isopsephic inscriptions found in Pergamum were supposedly composed by the father of the great physician Galen, a mathematician whose son said of him that he had "mastered everything it was possible to know about geometry and the science of numbers."

An inscription found at Pompeii, in which a man named Amerimnus pays homage to the lady "whose honorable name is 45," also contains this: "I love her whose number is 545."

The author of the *Pseudo-Callisthenes*, a Greek work written in about the second century A.D., says that the Egyptian god Serapis revealed his name to Alexander the Great in this way: "Take two hundred, one, one hundred, one, four times twenty, and ten. Then place the first of those numbers at the end and you will know which god I am." Following the god's instructions to the letter, if I may put it that way, we obtain:

$$200 \quad 1 \quad 100 \quad 1 \quad 80 \quad 10 \quad 200$$

which, by the Greek system shown in figure 17–2, gives the name:

$$\Sigma \quad A \quad P \quad \Pi \quad I \quad \Sigma$$
$$200 \quad 1 \quad 100 \quad 1 \quad 80 \quad 10 \quad 200$$
-->
"SARAPIS"

In connection with the murder of Agrippina, Suetonius compares the name of her son Nero, written in Greek, with the sentence *Idian metera apekteine*, "He killed his own mother," and shows that they have the same numerical value in the Greek system:

N E P Ω N	I Δ I A N M H T E P A A Π E K T E I N E
50 5 100 800 50	10 4 10 1 50 40 8 300 5 100 1 1 80 5 20 300 5 10 50 5
-------- >	--- >
"NERO"	"HE KILLED HIS OWN MOTHER"
1005	1005

"The Greeks do not seem to have begun speculating about the numerical values of letters until rather late. This practice must have entered Greek thought after it came in contact with Jewish thought. The famous passage on the 'number of the beast' in the Book of Revelation shows that the Jews were familiar with such mystic calculations long before they developed gematria and the Cabala. The Greeks and the Jews were remarkably gifted for both arithmetical calculation and transcendental speculation. The liked all kinds of subtleties, including those of number mysticism, which brought both aptitudes into play. The Pythagorean school, the most superstitious of philosophical sects, and the one most deeply impregnated with Oriental influences, had already indulged in number mysticism. In late antiquity this form of mysticism spread at an amazing rate. It gave rise to divination by numbers; it inspired sibyls, soothsayers, and pagan *theologoi;* it disquieted the Fathers of the Church, who were not always able to resist its fascination. Isopsephia is one of its methods" (Perdrizet).

Father Theophanes Kerameus stressed the numerical equivalence of *Theos,* "God," *agios,* "holy," and *agathos,* "good":

"GOD"	"HOLY"	"GOOD"
Θ Ε Ο Σ	Α Γ Ι Ο Σ	Α Γ Α Θ Ο Σ
9 5 70 200	1 3 10 70 200	1 3 1 9 70 200
------------>	------------>	------------>
284	284	284

He also saw a representation of the Universal Church in the name of Rebecca (wife of Isaac and mother of the twins Jacob and Esau) because, according to him, there were 153 species of fish in the sea, all of them taken by Simon and his fellow fishermen in the abundant catch described in Luke 5:4–7, and 153 was the numerical value of the Greek form of Rebecca's name:

$$\text{P E B E K K A}$$
$$100 \quad 5 \quad 2 \quad 5 \quad 20 \quad 20 \quad 1$$
$$\text{------------------>}$$
$$153$$

In Revelation 22:13 Jesus says, "I am the Alpha and the Omega." This expression employs the first and last letters of the Greek alphabet. In Gnostic and Christian thought it represents the totality of space and time, as well as the key to knowledge. In saying that he is the Alpha and the Omega, Jesus therefore declares himself to be the beginning and end of all things. He identifies himself with the Holy Spirit and, according to Christian doctrine, with God himself. After his baptism, Jesus "saw the spirit of God descending like a dove to alight upon him" (Matthew 3:16). The Greek word for "dove," *peristera,* has the numerical value of 801, which is the same as that of the letters alpha and omega taken together. The expression "the Alpha and the Omega" is therefore a mystical way of affirming the Christian dogma of the Trinity:

A & Ω	Π Ε Ρ Ι Σ Τ Ε Ρ Α
1 800	80 5 100 10 200 300 5 100 1
------>	--------------------------->
801	801

Another procedure that was often used in the Middle Ages consisted in attributing supernatural virtues to numbers according to the forms of the signs that represented them.

There is an example of this in the Epistle of Barnabas, which was included in the fourth-century Codex Sinaiticus of the New Testament. In Abram's victory over his enemies with the aid of 318 men, the author sees a reference to the cross (in the shape of the Greek letter tau, T) and the first two letters of the Greek name of Jesus (Ιησους):

$$T \; + \; IH \; = \; 318$$
$$\;\;\;300 \quad\;\; 10+8$$

And he states that 318 was the number of people saved by the crucifixion of Jesus.

In his *De Pascha Computus*, Cyprian writes that 365 is a sacred number because it is the sum of 300 (T, symbol of the cross), 18 (IH, first two letters of the Greek name of Jesus), 31 (the number of years that Jesus lived, according to Cyprian) and 16 (the year of Tiberius's reign in which Jesus was crucified). This may explain why some heretics believed that the world would end in A.D. 365.

To support the doctrine that Jesus is the son of God, some Christian mystics linked the Hebrew *ab qal*, which refers to the cloud in Isaiah 19:1: "See how the Lord comes riding swiftly upon a cloud," and the word *bar*, "son":

עב קל   בר

   30 100   2   70   200 2
‹- - - - - - - - - - -   ‹- - - - - -
       202   202

Gnosticism (from the Greek *gnosis*, "knowledge") is a religious doctrine that appeared in the early centuries of the Christian era among Judeo-Christians but was sternly opposed by rabbis and the apostles of the New Testament. It was based essentially on the hope of obtaining salvation through esoteric knowledge of the divine realm, transmitted by initiation. The Gnostics made extensive use of isopsephia.

"From a text whose probable source is Hippolytus, it appears that in certain Gnostic sects isopsephia was, so to speak, the normal form of symbolism and catechesis. It did not serve only . . . to envelop a revelation in mystery; though it hid in some cases, it revealed in others, shedding light on things that would not have been understood without it.

"Gnosticism appears to us laden with an enormous burden of Egyptian superstitions. It claimed to seek knowledge of the Universal Principle; actually, it chiefly sought the means of learning God's name and then, with the help of magic, the old magic of Isis, forcing God to let man raise himself to his level. A person's name was part of him, like his

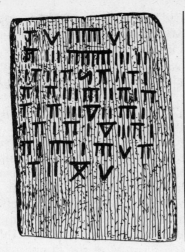

TRANSCRIPTION

I V $\overline{\text{IIIIII}}$ V I

$\overline{\text{II}}$ III $\overline{\text{IIIIIIII}}$ III II

I $\overline{\text{I}}$ II $\overline{\text{II}}$ $\hat{\text{V}}\overline{\text{I}}$ $\overline{\text{II}}$ II$\overline{\text{I}}$I

$\overline{\text{II}}$ I $\overline{\text{II}}$ III $\overline{\text{III}}$ II $\overline{\text{II}}$ I $\overline{\text{II}}$

$\overline{\text{I}}$ I $\overline{\text{II}}$ III $\overline{\text{V}}$ II $\overline{\text{III}}$ I

I $\overline{\text{II}}$ I $\overline{\text{II}}$ I $\overline{\text{V}}$ III $\overline{\text{II}}$ I

$\overline{\text{II}}$ I $\overline{\text{IIII}}$ I $\overline{\text{III}}$ v $\overline{\text{II}}$

$\overline{\text{I}}$ II $\overline{\text{X}}$ v

Fig. 21–2. Wooden tablet found in North Africa, dating from the late fifth century A.D. The sum of each line is 18. (The lines drawn above some of the Roman numerals indicate partial totals.) We do not know if it is an arithmetical document or a "magic" tablet related to speculations on the numerical values of Greek or Hebrew letters.

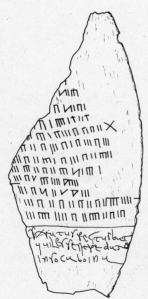

Fig. 21–3. One of many slates found in the Salamanca region of Spain; this one was discovered at Santibáñez de la Sierra and dates from about the sixth century. It is evidently the same kind of document as the one shown in figure 21–2: the sum of each intact line is 26.

shadow or his breath; it was even more than that: it was identical with the person, it was the person himself.

"Knowing God's name was therefore the problem that Gnosticism set for itself. At first sight it seemed insoluble—how could one know the name of the Ineffable? So the Gnostics did not claim that they could know the actual name of God; but they believed it was possible to determine its formula and that was enough, because the formula of the divine name contained all its magic power. That formula was the number of the divine name.

"According to Basilides, the supreme God of Gnosticism combined within himself the 365 secondary gods who presided over the days of the year. . . . The Gnostics referred to him by circumlocutions such as 'He whose number is 365 (ου εστιν η ψηαος ΤΞΕ).' From this came the magic power of the seven vowels, the seven notes of the scale, the seven planets, the seven metals (gold, silver, tin, copper, iron, lead, mercury), and the four weeks of the lunar month. Whatever the name of the Ineffable might be, the Gnostics were sure it partook of the two magic numbers, 7 and 365. Lacking the unknowable name of God, one should be able to find a designation that would be like the formula of the divine name, and to find it one had to unite and combine the mystic numbers 7 and 365. It was for this reason that Basilides coined the word *Abrasax*, which has seven letters with numerical values whose sum is 365:

$$
\begin{array}{ccccccc}
\text{A} & \text{B} & \text{P} & \text{A} & \Sigma & \text{A} & \Xi \\
1 & 2 & 100 & 1 & 200 & 1 & 60
\end{array}
$$

$$
\text{-----------------------------} \longrightarrow
$$

$$365$$

"Holiness was the first characteristic of God (or of God's name, since the two were equivalent). Aγιος ο Θεος (*Agios o Theos*), says the seraphic hymn. 'Hallowed be thy name,' says the Lord's Prayer—that is, 'Let God's holiness be proclaimed.'

"God's name remained unknown, but the Gnostics knew that it had the characteristic of being the supremely holy name. Therefore nothing was more suitable for designating the Ineffable than the expression *Agion Onoma* ('Holy Name'), and the Gnostics often used it. But it was not only for this somewhat metaphysical or theological reason, or because they had borrowed that appellation from the Jews; it was also for a mystical reason that was peculiar to them: by a coincidence in which they saw a revelation, the biblical expression *Agion Onoma* ('Holy Name') had the same number as *Abrasax*, 365:

$$
\begin{array}{ccccccccc}
\text{A} & \Gamma & \text{I} & \text{O} & \text{N} & \text{O} & \text{N} & \text{O} & \text{M} & \text{A} \\
1 & 3 & 10 & 70 & 50 & & 70 & 50 & 70 & 40 & 1
\end{array}
$$

$$
\text{--} \longrightarrow
$$

$$365$$

"Once they had set out on that path, the Gnostics made other discoveries that were equally striking to them.

"Being already involved with magic, they were naturally led to syncretism. Through isopsephia they found a way to identify their supreme God with the national god of Egypt: the Nile—which for the Egyptians was none other than Osiris—was the god of the year, because the regularity of its floods corresponded to the regular course of the years, and the number of *Neilos*, the Greek name of the Nile, was 365:

$$
\begin{array}{cccccc}
\text{N} & \text{E} & \text{I} & \Lambda & \text{O} & \Sigma \\
50 & 5 & 10 & 30 & 70 & 200
\end{array}
$$
----------------------->
365

"Also by isopsephia, the Gnostics performed another curious feat of syncretism. The prodigious diffusion of the Mithras cult in the second and third centuries of the Christian era is well known. The Gnostics noticed that ΜΕΙΘΡΑΣ, the Greek form of Mithras, had the value of 365:

$$
\begin{array}{cccccc}
\text{M} & \text{E} & \text{I} & \Theta & \text{P} & \text{A} & \Sigma \\
40 & 5 & 10 & 9 & 100 & 1 & 200
\end{array}
$$
------------------------->
365

"And so the sun god of Iran was the same as the 'Ruler of the 365 Days'" (Perdrizet). (Figs. 21–2, 21–3.)

In Greek and Coptic inscriptions the letters koppa and theta (ϙ Θ) sometimes appear after a benediction, an imprecation, or an exhortation to praise. This combination of letters remained enigmatic until the nineteenth century, when Wessely showed that they are a mystic way of expressing "Amen," since they have the same numerical value as the Greek form of that word (Ἀμήν):

$$
\begin{array}{cccc}
\text{A} & \text{M} & \text{H} & \text{N} \\
1 & 40 & 8 & 50
\end{array}
\qquad
\begin{array}{cc}
\text{ϙ} & \Theta \\
90 & 9
\end{array}
$$
- - - - - - - >　　　- - >
99　　　　　　　99

The dedication of a mosaic in the monastery of Khoziba, near Jericho, begins:

Φ Λ Ε　　ΜΝΗΣΦΗΤΙ ΤΟΥ ΔΟΥΛΟΥΣΟΥ
"Φ Λ Ε　　Remember thy servant"

What does the group phi-lambda-epsilon mean? The answer was found by Smirnoff. These letters correspond to the Greek word for "Lord" (Κυριε), whose numerical value is 535:

$$
\begin{array}{ccc}
\Phi & \Lambda & E \\
500 & 30 & 5
\end{array}
\quad\longrightarrow\quad
\begin{array}{ccccc}
K & Y & P & I & E \\
20 & 400 & 100 & 10 & 5
\end{array}\longrightarrow
$$

$$535 \qquad\qquad 535$$

Christian mystics have speculated at great length on the number 666, attributed by John the Apostle to the Beast of the Apocalypse. The beast has been identifed with the Antichrist, who, according to Christian doctrine, will appear in the last days before the Second Coming, commit countless crimes, spread terror, turn nations against each other, and finally be overcome by Christ when he returns to this world. In Revelation 13:16–18 John writes, "Moreover, it [the beast] caused everyone, great and small, rich and poor, slave and free, to be branded with a mark on his right hand or forehead, and no one was allowed to buy or sell unless he bore this beast's mark, either name or number. (Here is the key: and anyone who has intelligence may work out the number of the beast. The number represents a man's name, and the numerical value of its letters is six hundred and sixty-six.)"

This is a clear reference to a system of numerically evaluating letters; but since the system is not specified, the name of the beast has tried the ingenuity of interpreters for centuries, and many different solutions have been proposed.

The number 666 is said to represent "a man's name"; some interpreters have taken this to mean that it indicates a prominent historical figure whose name has the value of 666 in the Hebrew, Greek, or Latin system. Nero, the first Roman emperor to persecute the Christians, was sometimes identified as the Beast of the Apocalypse because his name, accompanied by the title of Caesar, has the value of 666 in the Hebrew system:

קסר נרון

50 6 200 50   200 60 100

QSAR NERON
666

Other interpreters, believing that 666 represented a *type* of man, attributed it to Latins in general because the Greek word *Lateinos* had that value:

$$\Lambda \quad A \quad T \quad E \quad I \quad N \quad O \quad \Sigma$$

| 30 | 1 | 300 | 5 | 10 | 50 | 70 | 200 |

-->

666

During the Wars of Religion in France a Catholic mystic named Petrus Bungus wrote a book (published at Bergamo in 1584–85) in which he demonstrated, to his own satisfaction, that the German reformer Martin Luther was the Antichrist because his name had the value of 666 in the Latin system:

$$L \quad V \quad T \quad H \quad E \quad R \quad N \quad V \quad C$$

| 30 | 200 | 100 | 8 | 5 | 80 | 40 | 200 | 3 |

-->

666

Luther's disciples were quick to reply; they considered the words that were supposedly on the papal tiara, *Vicarius Filii Dei* ("Vicar of the Son of God"), added up the values of the Roman numerals contained in them:

$$V \quad I \quad C_{A} R \quad I \quad V_{s} \quad F \quad I \quad L \quad I \quad I \quad D \quad E \quad I$$

| 5 | 1 | 100 | | 1 | 5 | | 1 | 50 | 1 | 1 | 500 | | 1 |

--->

666

and drew the conclusion that can easily be imagined.

Numerical evaluation of words was the basis of a practice called *hisab al-nim*, used by Moslem soothsayers in time of war for predicting which side would win. The Arab historian ibn-Khaldun (1332–1406) described it as follows:

"Here is how to carry out the operation. First, add the numerical values of the letters composing the name of each king; these are conventional values assigned to the letters of the alphabet, going from one to a thousand, classified by units, tens, hundreds, and thousands. When the addition has been done, subtract nine from each sum as many times as necessary in order to have two remainders smaller than nine. Compare the two remainders: if one is larger than the other and they are both odd or both even, the king whose name gave the smaller remainder will be victorious. If one remainder is even and the other odd, the king whose name gave the larger remainder will be victorious. If the remainders are equal and even, the king who is attacked will win the victory; if they are equal and odd, the king who attacks will triumph."

Each Arabic letter is the initial of an attribute of Allah: alif is the initial of Allah's name, ba is the initial of *Bāqī*, "He who remains," and

LETTERS	VALUES	ASSOCIATED DIVINE ATTRIBUTES — Names	Meanings	VALUES	LETTERS	VALUES	ASSOCIATED DIVINE ATTRIBUTES — Names	Meanings	VALUES
alif ١	1	الله ALLAH	Allah	66	sin س	60	سمیع SAMI'	Hearer	180
ba ب	2	باقی BĀQĪ	He who remains	113	ayn ع	70	علی 'ALĪ	Lofty	110
jim ج	3	جامع JĀMI'	He who assembles	114	fa ف	80	فتاح FATĀH	Who opens	489
dal د	4	دیان DAYĀN	Judge	65	sad ص	90	صمد SAMAD	Eternal	134
ha ه	5	هادی HĀDĪ	Guide	20	qaf ق	100	قادر QĀDIR	Powerful	305
waw و	6	ولی WALĪ	Master	46	ra ر	200	رب RAB	Lord	202
zay ز	7	زکی ZAKĪ	Purifier	37	shin ش	300	شفیع CHAFI'	Who accepts	460
ha ح	8	حق HAQ	Truth	108	ta ت	400	توّاب TAWAB	Who brings back to righteousness	408
ta ط	9	طاهر TĀHIR	Holy	215	tha ث	500	ثابت THĀBIT	Stable	903
ya ی	10	يسین YASSĪN	Chief	130	kha خ	600	خالق KHĀLIQ	Creator	731
kaf ک	20	کافی KĀFĪ	Sufficient	111	dhal ذ	700	ذاکر DHĀKIR	Who remembers	921
lam ل	30	لطیف LATĪF	Benevolent	129	dad ض	800	ضار DĀR	Punisher	1001
mim م	40	ملک MALIK	King	90	za ظ	900	ظاهر DHĀHIR	Apparent	1106
num ن	50	نور NŪR	Light	256	ghayn غ	1000	غفور GHAFŪR	Indulgent	1285

Fig. 21–4. The system of numerical evaluation used in *da'wa* theology.

so on. This gave rise to a system in which the usual value of each letter is replaced with the numerical value of the divine attribute it represents. The value of alif, for example, which is usually 1, becomes 66, the numerical value of Allah's name calculated by the ordinary *Abjad* method (fig. 20–4). This was the system used in the symbolic theology known as *da'wa* ("invocation") by which mystics and soothsayers made predictive calculations and speculated on the past, present, and future (fig. 21–4).

Here is a North African talisman intended to bring wealth and ward off evil:

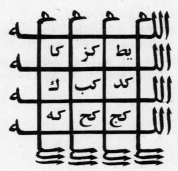

It is a magic square, which means that the same sum is obtained by adding the numbers written in any row, column, or diagonal:

21	26	19
20	22	24
25	18	23

And in this case the sum is 66, the numerical value of Allah's name:

5 30 30 1

ALLAH

A practice that was quite common in antiquity, and persists to this day, consists in using letters to interpret dreams involving numbers. The following example, borrowed from Bouché-Leclercq, concerns the ancient Greeks:

"What was perhaps the most awkward case arose when a dream promised an old man a number of years too large to be added to his present age and too small to represent his whole life. But the problem was not insoluble. If a seventy-year-old man heard someone say to him in a dream, 'You will live fifty years,' it meant that he would live another thirteen years. Since he had already lived seventy years, the 'fifty' could not refer to his entire lifetime, nor could it refer to the time he still had left to live, since that would be impossibly long for a man of his age. He would live another thirteen years because the letter N, which represented fifty, was the thirteenth letter of the alphabet.

"One is almost tempted to admire the ruses of an imperturbable faith which encounters seemingly insurmountable difficulties and turns them into evidence in its favor. Nothing sheds more light on psychological history than this irresistible prestige of preconceived ideas."

PART V

Hybrid Numeration Systems

22

The Drawbacks of the Additive Principle

Roman notations for large numbers

The largest numeral letter of the Roman system, as we know it and sometimes still use it, corresponds only to 1000. Simply applying the additive principle to the seven basic numerals of this system therefore enables us to represent only numbers below 5000. It is impossible, practically speaking, to use those numerals for indicating large numbers: 87,000, for example, would require writing the symbol M eighty-seven times (fig. 22–1)!

In an effort to overcome this difficulty, certain graphic conventions were adopted, but their number was increased so rashly that the Roman

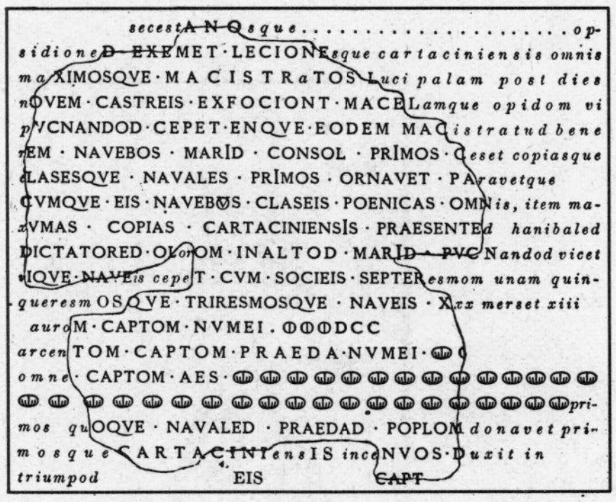
*secest*ANQ*sque* . . . . *op-*
*sidione*D EXEMET · LECIONE*sque cartaciniensis omnis*
*ma*XIMOSQVE · MACISTRaTOS *Luci palam post dies*
*n*OVEM · CASTREIS · EXFOCIONT · MACEL*amque opidom vi*
*p*VCNANDOD · CEPET · ENQVE · EODEM MAC*istratud bene*
*r*EM · NAVEBOS · MARID · CONSOL · PRIMOS · C*eset copiasque*
*c*LASESQVE · NAVALES · PRIMOS · ORNAVET · PA*ravetque*
CVMQVE · EIS · NAVEBVS · CLASEIS · POENICAS · OMN*is, item ma-*
*x*VMAS · COPIAS · CARTACINIENSIS · PRAESENTE*d hanibaled*
DICTATORED · O*lor*OM · INALTOD · MARID · PVCN*andod vicet*
*v*IQVE · NAVE*is cepe*T · CVM · SOCIEIS · SEPTER*esmom unam quin-*
*queresm*OSQVE · TRIRESMOSQVE · NAVEIS · X*xx merset xiii*
*auro*M · CAPTOM · NVMEI . ↀↀↀDCC
*arcen*TOM · CAPTOM · PRAEDA · NVMEI · ↈ
omne · CAPTOM · AES · ↈↈↈↈↈↈↈↈↈↈↈↈↈↈ
ↈↈↈↈↈↈↈↈↈↈↈↈↈↈↈↈↈↈↈↈ*pri-*
*mos qu*OQVE · NAVALED · PRAEDAD · POPLOM *donavet pri-*
mosque CARTACINI*ens*IS *ince*NVOS · D*uxit in*
triumpod EIS CAPT

Fig. 22–1. *Elogium* of Duilius, who defeated the Carthaginians in the Battle of Mylae, 260 B.C. The inscription was reengraved during the reign of Claudius I (A.D. 41–54) in the style of the third century B.C. It was found at the site of the Rostral Column in the Roman Forum and is now in the Palazzo dei Conservatori, Rome. In this transcription, the capital letters correspond to the part of the inscription that is still intact, the italics to the restoration of the missing parts given in *Corpus Inscriptionum Latinarum* (Berlin, 1861).

In lines 15 and 16, the sign for 100,000 occurs twenty-three times in the intact portion, and a total of thirty-four times according to the restoration.

In line 13, the number 3700 is written in this form: ↀↀↀ DCC system lost its cohesion and represented a step backward in relation to other systems. The problems encountered by the Romans—and later the peoples of the medieval West—are worth considering in some detail.

In the time of the republic, the Romans had a relatively simple graphic procedure that enabled them to assign special notations to the numbers 5000, 10,000, 50,000, and 100,000 (fig. 9–2). The main signs, which were used sporadically until the Renaissance (figs. 22–2, 22–3, 22–4, 22–5), are shown in figure 22–6.

TRANSCRIPTION

HS n. CCIↃↃ CCIↃↃ CCIↃↃ CCI IↃↃ
ↃↃ ↃↃ LXXVIIII*

quæ pecunia in sti-
pulatum L. Caecili
Iucundi venit
ob auctione (m) M. Lucre-
ti Leri [mer] cede
quinquagesima minu [s]

Fig. 22–2. Second panel of a triptych found at Pompeii (and therefore necessarily dating from no later than A.D. 79, the year when the city was destroyed). It is an acknowledgment of debt, in common Roman classical writing, drawn up after a public sale. The sum of 38,079 sesterces (lines 1 and 2) is, according to the *Corpus Inscriptionum Latinarum*, the highest one mentioned. It seems to be composed of 37,332 sesterces from the sale itself, plus interest of $1/_{50}$ (*mercede quinquagesima*), or 2 percent, amounting to 746.64 sesterces, rounded off to 747.

Fig. 22–3. Detail from a Portuguese manuscript of 1200, concerning the Venerable Bede's system of finger counting. Public Library, Lisbon.

Roman numerals associated with the finger signs:

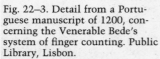

X̄L̄ XXX = 40,030

L̄ XC = 50,090

L̄X̄ = 60,000

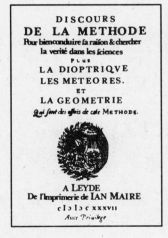

Fig. 22–4. Archaic Roman numerals as used in *Mysticae Numerorum Significationes Opus*, by Petrus Bungus, published at Bergamo in 1584–85.

Fig. 22–5. Title pages of works by Descartes and Spinoza, published in 1637 and 1677 respectively, with the dates in archaic Roman numerals.

If these numerals are compared with each other and with the various archaic forms of the sign for 1000 (fig. 9–2), their common origin becomes apparent; they are all stylizations, recognizable in varying degrees, of the five original signs.

Five of these original signs were formed by a simple geometric procedure. The first sign for 1000 was a circle with a vertical line through its center; one more circle was added to make the sign for 10,000, and

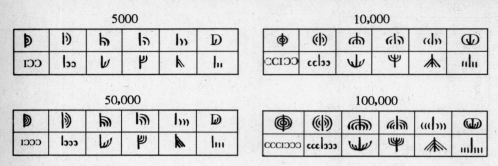

Fig. 22–6.

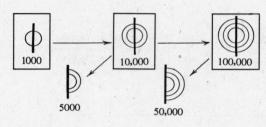

Fig. 22–7.

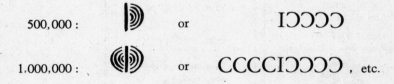

500,000 :   or   IↃↃↃ

1,000,000 :   or   CCCCIↃↃↃↃ , etc.

Fig. 22–8.

two more for 100,000; half of the sign for 10,000 then became the sign for 5000, and half of the sign for 100,000 became the sign for 50,000 (figs. 22–7, 22–9).

Following this principle, the Romans could have noted the numbers 500,000, 1,000,000, 5,000,000, etc. in analogous forms (fig. 22–8).

But probably because such forms make it difficult for the eye to identify numerals above 100,000, and perhaps also because they had no special names for units above 100,000, the Romans evidently did not extend the procedure. (In his *Historia Naturalis*, Pliny wrote that for the

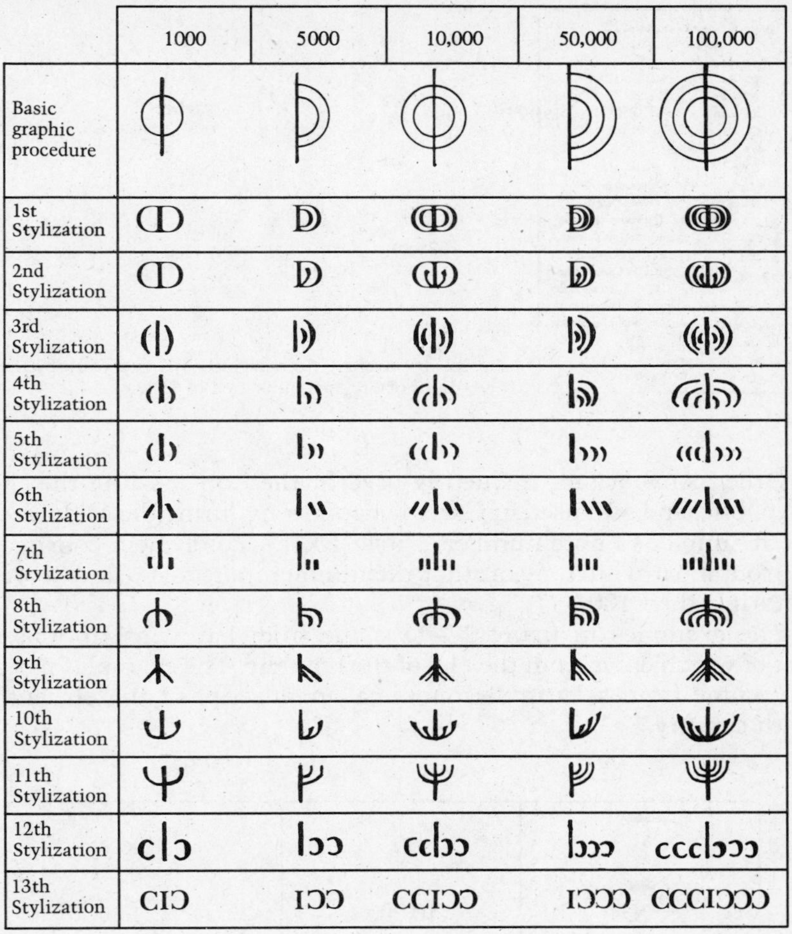

	1000	5000	10,000	50,000	100,000
Basic graphic procedure	⏀	⟩	⊕	⟩⟩	⊚
1st Stylization	ⅭⅮ	Ⅾ	ⅭⅮ	Ⅾ	ⅭⅭⅮ
2nd Stylization	ⅭⅮ	Ⅾ	ⅭⅮ	Ⅾ	ⅭⅭⅮ
3rd Stylization	(Ɩ)	Ɩ⟩	(Ɩ⟩)	Ɩ⟩	((Ɩ⟩)
4th Stylization	⟨Ɩ⟩	Ɩ⟩	⟨Ɩ⟩	Ɩ⟩	⟨⟨Ɩ⟩
5th Stylization	⟨Ɩ⟩	Ɩ⟩⟩	⟨⟨Ɩ⟩⟩	Ɩ⟩⟩⟩	⟨⟨Ɩ⟩⟩⟩
6th Stylization	⟋Ɩ⟍	ƖⱵ	⟋⟋ƖⱵ	ƖⱵⱵ	⟋⟋⟋ƖⱵⱵ
7th Stylization	ɩƖɩ	Ɩɩɩ	ɩɩƖɩɩ	Ɩɩɩɩ	ɩɩɩƖɩɩɩ
8th Stylization	⟨Ɩ⟩	Ɩ⟩	⟨Ɩ⟩	Ɩ⟩	⟨⟨Ɩ⟩
9th Stylization	木	⟩	乑	⟩	乑
10th Stylization	⍋	⍳	⍲	⍳⟩	⍲⍲
11th Stylization	Ψ	Ψ	Ψ	Ψ	Ψ
12th Stylization	cƖɔ	Ɩɔɔ	ccƖɔɔ	Ɩɔɔɔ	cccƖɔɔɔ
13th Stylization	CIƆ	IƆƆ	CCIƆƆ	IƆƆƆ	CCCIƆƆƆ

Fig. 22–9. Classification of archaic Roman numerals for 1000, 5000, 10,000, 50,000, and 100,000.

number we call a million, the Romans said *decies centena milia,* "ten hundred thousand.")

Though it is worth noting only as a curiosity, in 1582 a Swiss author named Fregius published a work (fig. 22–10) in which he used the thirteenth stylization in figure 22–9 for numbers up to a million as follows:

I	V	X	L	C	IƆ	CIƆ	IƆƆ	CCIƆƆ	IƆƆƆ	CCCIƆƆƆ	IƆƆƆƆ	CCCCIƆƆƆƆ
1		10		10^2		10^3		10^4		10^5		10^6
	5		$5 \cdot 10$		$5 \cdot 10^2$		$5 \cdot 10^3$		$5 \cdot 10^4$		$5 \cdot 10^5$	

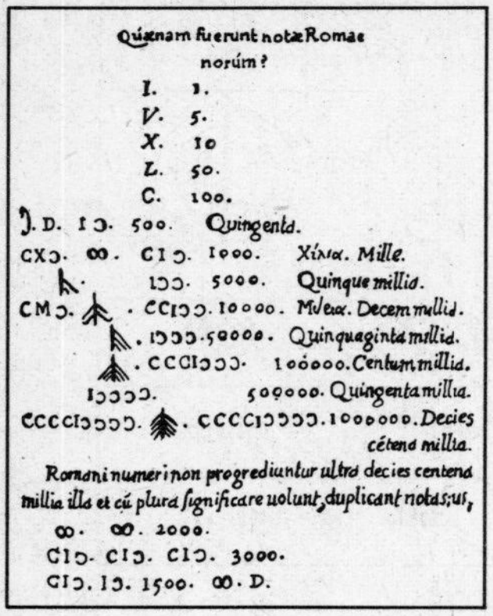

Fig. 22–10. Archaic Roman numerals in a work by the Swiss author Fregius, published in 1582.

Other conventions, frequently used by the Romans after the end of the republic, and still used in Christian countries during the Middle Ages, made it simpler to note numbers above 1000. One of them consisted of a horizontal bar which meant that the number indicated below it was to be multiplied by 1000.

The examples in figure 22–11 come from Latin inscriptions, the oldest of which dates from the end of the republic. The examples in figure 22–12 come from a Latin astronomical manuscript of the eleventh or twelfth century.

$\overline{V}$	5000	5 × 1000
$\overline{X}$	10,000	10 × 1000
$\overline{\text{LXXXIII}}$	83,000	83 × 1000

Fig. 22–11.

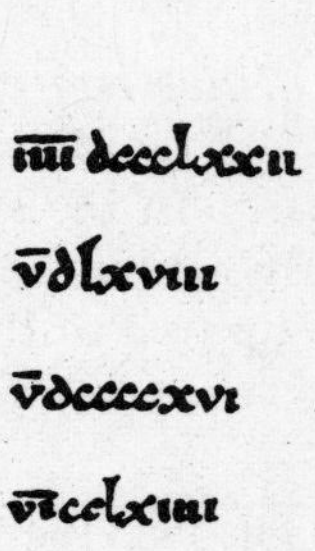

$\overline{\text{IIII}}$ DCCCLXX	4870
$\overline{V}$ DLXVIII	5568
$\overline{V}$ DCCCCXVI	5916
$\overline{VI}$ CCLXIIII	6264

Fig. 22–12.

This writing convention sometimes caused confusion with an older practice: to distinguish letters indicating numbers from those used phonetically, the Romans drew a horizontal bar above their numeral letters, notably in abbreviations such as $\overline{\text{II}}$VIR = *duumvir* and $\overline{\text{III}}$VIR = *triumvir*. It was probably for this reason that during the reign of Hadrian (second century A.D.) multiplication by 1000 was sometimes indicated as follows:

35,000	⌐XXXV¬ 35 × 1000
557,274	⌐DLVII¬ CCLXXIV 557 · 1000　　274

But this notation usually had a different function: that of indicating that the number placed within the frame was to be multiplied by 100,000. Like the horizontal bar, the frame made it possible to represent numbers from 1000 to 500,000,000:

$$\lceil M\ D\ C\ L\ I \rceil \qquad \overline{L\ X\ X\ V\ I\ I\ I} \qquad C\ C\ C\ X\ V\ I$$

$$\underset{\times}{1651} \quad + \quad \underset{\times}{78} \quad + \quad 316$$
$$100{,}000 \qquad\qquad 1000$$

$$-------------------->$$

$$165{,}178{,}316$$

The examples in figure 22–13 come from inscriptions made during the imperial period.

⌐XII¬	1,200,000	12 × 100,000
⌐XIII¬	1,300,000	13 × 100,000
⌐∞ ∞¬	200,000,000	2000 × 100,000

Fig. 22–13.

But this system could cause mistakes: $\overline{\text{IL}}$VII (or $\lceil \text{LVII}$), for example, which represents the number 56 × 100,000 = 5,600,000, could easily be confused with $\overline{\text{LVII}}$, indicating 57 × 1000 = 57,000.

Suetonius tells the following story. After the death of his mother, Livia, Tiberius was required to make payments to some of her other heirs, one of whom was Galba, the future emperor. The will stipulated that Galba was to receive 50,000,000 sesterces: HS $\overline{\text{CCCCC}}$

$$500 \times 100,000$$

Tiberius reduced this to half a million: **HS** $\overline{\text{CCCCC}}$

$$500 \times 1000$$

on the pretext that the sum represented as HS $\overline{\text{CCCCC}}$ could only be 500 × 1000 because the vertical lines of the frame were too short to indicate multiplication by 100,000.

According to some authors, the Romans used a logical extension of the horizontal-bar notation that, in conjunction with the frame, enabled them to represent all numbers between 1000 and 5,000,000,000; examples:

1,000,000,000	$\overline{\overline{\text{M}}}$	= 1000 × 1,000,000
2,300,000,000	$\overline{\overline{\text{MMCCC}}}$	= 2300 × 1,000,000

I have found no instances of this, however, in any of the Roman inscriptions known to me.

Appearance and extension of the multiplicative principle

Let us return to the Greek acrophonic numeration. The version of it used in Athens, for example, at first had only these signs:

I	Γ	Δ	H	X	M
1	5	10	100	1000	10,000

As we have seen, the users of this system simplified it by introducing special signs for 50, 500, 5000, 50,000, etc. But instead of indicating those numbers with signs totally independent of the others, they had the idea of using the following combinations, which are obviously based on the multiplicative principle:

⌐Δ⌐	⌐H⌐	⌐X⌐	⌐M⌐
5 × 10	5 × 100	5 × 1000	5 × 10,000

The same idea was exploited by the Cretans in the second half of the second millennium B.C. To these numerical signs:

I	–	O	◇
1	10	100	1000

they added a sign for 10,000 by combining the horizontal bar for 10 with the sign for 1000 (fig. 14–5):　　⊖

1000 × 10

A similar idea was used in the third millennium B.C. by the Sumerians, who indicated 600 and 36,000 by combining the signs for 60 and 3600, respectively, with the sign for 10 (fig. 14–32).

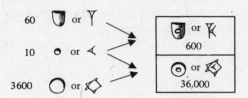

And the Elamites—who about 3300 B.C., as we have seen, began representing numbers with objects of conventional shapes and sizes (fig. 10–14)—used this principle when they symbolized 10, 300, and 3000 by these tokens:

They thus had the idea (which was highly abstract for their time) of indicating multiplication by 10 by placing a small circular mark (the "written" symbol for the sphere representing 10) on the large cone representing 300.

In Egyptian stone inscriptions, beginning in the time of the New Kingdom (second half of the second millennium B.C.), we find a remarkable irregularity in relation to the classical hieroglyphic system: a numerical notation based not on the additive principle but on the multiplicative principle. The rule is as follows: when a tadpole (the hieroglyphic numeral for 100,000) is placed above a numerical expression with a lower value, the latter becomes a multiplier. For example, the sign for 18 with a tadpole above it no longer stands for 100,018 (100,000 + 18) but for 1,800,000 (100,000 × 18), a number that would be expressed in the classical system by juxtaposing eight tadpoles and the sign for 1,000,000. In a Ptolemaic inscription (first to third century B.C.) we find the number 27,000,000 expressed as 100,000 × 27:

100,000

270

27,000,000

We know that this does not stand for 100,000 + 270 because that number would have been noted by juxtaposing the tadpole and the representation of 270:

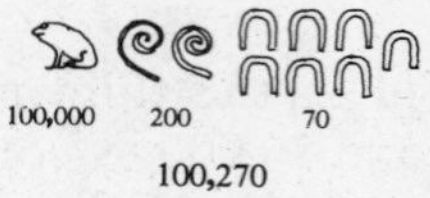

This irregularity in Egyptian stone inscriptions resulted from the influence of hieratic writing on the hieroglyphic system: during the Middle Kingdom, to note large numbers and write more rapidly, scribes had begun abandoning the additive principle in favor of the multiplicative principle (fig. 22–14).

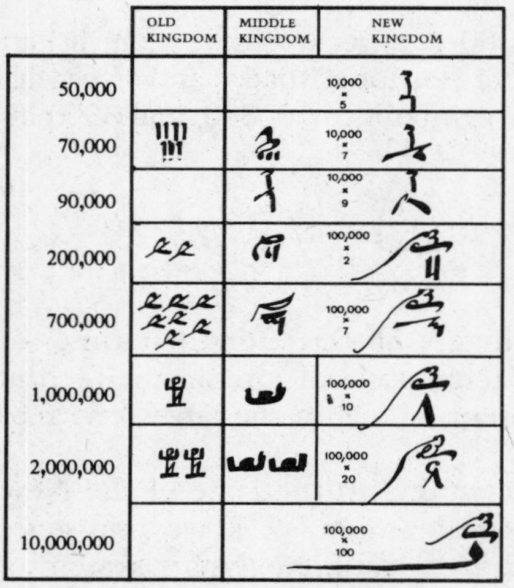

Fig. 22–14. Egyptian hieratic numerals for numbers above 10,000.

Let us now consider a written numeration used in India shortly before and after the beginning of the Christian era, as revealed by Buddhist inscriptions on the walls of caves at Nana Ghat (about seventy-five miles from Poona) and Nasik (about 120 miles from Bombay). The system in the Nana Ghat inscriptions (fig. 22–15) dates from the second century B.C., while the Nasik system dates from the first or second century A.D. These notations are among the oldest known in India (except, of course, for those of the Indus culture, which goes back to about the twenty-fifth century B.C.). They are associated with strictly Indian forms of writing and are therefore derived from the ancient script known as Brahmi.

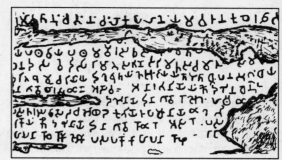

Fig. 22–15. Portion of an inscription on a wall of the Nana Ghat cave.

—	=		¥		9	?		?
1	2		4		6	7		9
α α	O				⊣		∞	
10	20				60		80	
Ӈ Ӿ	Ӿ		ӾД			Ӭ		
100	200		400			700		
T			Ty		Ŧψ			
1000			4000		6000			
Тα	Тo							
10,000	20,000							

—	=	≡	¥	¥	Ӷ	Ꜣ	Ꜥ	1	Ꜩ	Ꝡ	?
1	2	3	4		5	6		7	8	9	
α ⍺	θ		'		Ꜣ			Ӽ			
10	20				40			70			
η	ӷ				Ꜩӻ						
100	200				400						
9	9	Ꝑ	9Ꜩ							95	
1000	2000	3000	4000							8000	
								Ꜩӻ			
								70,000			

(left) Fig. 22–16. Numerals in the inscriptions of the Nana Ghat cave (second century B.C.).

(right) Fig. 22–17. Numerals in the inscriptions of the Nasik caves (first or second century A.D.).

Despite gaps in the notations as we know them (figs. 22–16, 22–17), it can be seen that they belong to a decimal numeration essentially based on the additive principle, with a special sign for each of these numbers.

1	2	3	4	5	6	7	8	9
10	20	30	40	50	60	70	80	90
100	200	300	400	500	600	700	800	900
1000	2000	3000	4000	5000	6000	7000	8000	9000, etc.

We know the values of the signs because the words for the corresponding numbers are written next to them. Some numbers are represented according to the multiplicative principle: 400, 700, 4000, 6000, 10,000, and 20,000, for example, which in the Nana Ghat inscriptions are expressed in these forms:

𝐻𝐻	𝐻𝐻	𝐻𝐻	𝐻𝐻	𝐻𝐻	𝐻𝐻
400	700	4000	6000	10,000	20,000

𝐻𝐻	𝐻𝐻	𝐻𝐻	𝐻𝐻	𝐻𝐻	𝐻𝐻
100 × 4	100 × 7	1000 × 4	1000 × 6	1000 × 10	1000 × 20

Thus in some cases, to reduce the effort of memory required, the additive principle was discarded and numerals were formed with multiplicative combinations of previously established signs.

When the multiplicative principle was first introduced into a numeration initially based on the additive principle, it was used only to form a limited number of new symbols corresponding either to some relatively high power of the base that till then had had no written representation, or to intermediate values obtained by means of an auxiliary base. But in the second stage of this evolution various peoples conceived the idea of extending the use of the multiplicative principle to avoid tedious repetitions of signs for certain powers of the base, or to avoid excessive efforts of memory such as those required, for example, by the use of special symbols for the successive units of a single order. And in this way numerical notations of a new type were developed. We will now consider the main examples of them.

23

The Common
Mesopotamian Numeration
System

At the time when writing was first being developed in Mesopotamia, the Sumerians were preponderant culturally as well as numerically. But near them, particularly in northern Babylonia, lived the Akkadians, a Semitic people whose name comes from that of Akkad, the large Mesopotamian city that for nearly two centuries was the capital of the powerful empire founded by Sargon I after his victory over the Sumerians about 2350 B.C.

The Akkadians assimilated most of the Sumerian cultural heritage and developed it to a considerable extent. They adapted their predecessors' cuneiform characters to their own language and traditions and altered the system of cuneiform writing in successive stages until it lost the essentially mnemonic aspect it originally had and became the graphic expression of an independent literary tradition.

The adaptation of Sumerian cuneiform characters to the Akkadian language took place in the second half of the third millennium B.C., when for the first time the Semites gained control of Mesopotamia. Their political power was accompanied by a cultural ascendancy that led them to promote their language and therefore write and use it systematically.

Along with the Sumerians' cuneiform characters, the Akkadians adopted their numeration (fig. 23–1). But because they, like most Semites, were in the habit of counting by hundreds and thousands, they introduced strictly decimal notations into the Sumerian sexagesimal system. Finding no signs for 100 and 1000 in that system, they devised a phonetic means of writing those two numbers. Using the principle of acrophony, they designated 100 by this cuneiform group:

initial of   "hundred"

ME   **ME - AT**

The Akkadian word for "thousand" is LIM. At first it was written as:

LI - IM   or   **LI - IM**

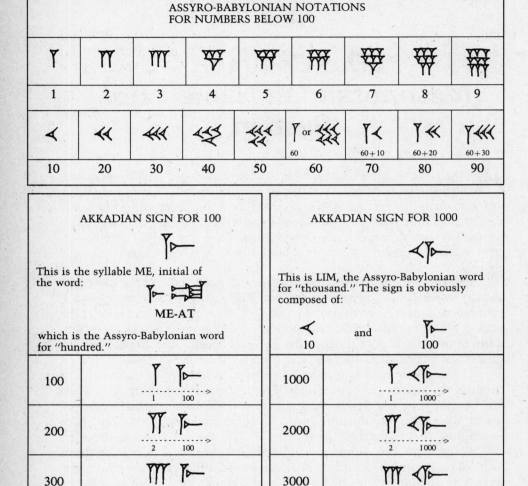

Fig. 23–1. The common Assyro-Babylonian numeration, an adaptation of the Sumerian cuneiform system to the numerical traditions of the Semites.

But the form that prevailed was: ⟨⊦—

LIM

This cuneiform group has the phonetic value of the syllable LIM, and also an ideographic value, since it is composed of a crescent (= 10) and the group ME (= 100), so that it corresponds to: LIM = 10 ME = 10 × 100.

The numeration that the Akkadians acquired from the Sumerians became a kind of mixed system in which special signs were attributed to each of the numbers:

$$1, 10, 60, 10^2, 60 \times 10, 60^2, 60^2 \times 10 \ldots$$

in these forms:

⟆	⟨	⟆ or SHU-SHI	⊦ ME	⟅	⟨⊦— LIM	◇
1	10	60	100	600	1000	3600

The following examples are from economic tablets, going back to the nineteenth century B.C., discovered in clandestine excavations in Mesopotamia.

60 40	2 ME	1 ME 3	1 ME 50 4
----->	----->	----->	----->
100	200	103	154

Other examples, from a northern Babylonian account tablet concerning sheep, seventeenth century B.C.:

1 SHU-SHI 3	60 10 3	60 20 5	1 ME 1 SHU-SHI 8
----->	----->	----->	----->
63	73	85	168

Thus in the first half of the second millennium B.C., the inhabitants of Mesopotamia sometimes used the Sumerian sexagesimal system, sometimes a Semitic decimal system, and sometimes a system that included both bases, 60 and 10. But when the Akkadian language had decisively supplanted Sumerian in Mesopotamia, it was the cuneiform decimal notaton that prevailed in common use. From then on, the written numeration used only the signs:

ME LIM

1 10 100 1000

In this system, units and tens were indicated by repeating the vertical wedge or the crescent as often as necessary, with the signs grouped in arrangements that differed considerably from their Sumerian counterparts. Multiples of 100 and 1000 were represented according to the multiplicative principle (fig. 23–1).

In the examples below, from Assyrian cuneiform tablets describing Sargon II's eighth campaign in Armenia (714 B.C.), the numbers concern the booty taken and are expressed in a strictly decimal system:

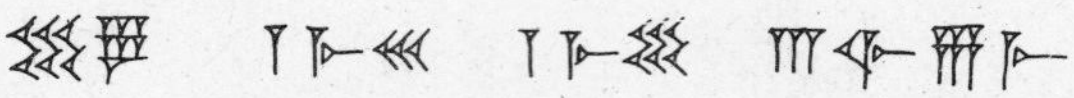

60 7 → 67 | 1 ME 30 → 130 | 1 ME 60 → 160 | 3 LIM 6 ME → 3600

And so the Semites eventually brought about a radical transformation of the original Sumerian system. The number 60, however, still had a special place in the common Assyro-Babylonian system. Although the decimal combination  was usually assigned to it (at least in the first millennium B.C.), the Akkadians sometimes noted it in the Sumerian form ; or by writing the word for it in this form: 

1 SHU-SHI

("one sixty") to avoid confusion with the vertical wedge for 1; or in the abridged form: 

1 SHU

Furthermore they continued to write numbers 70, 80, and 90 in the old Sumerian way:

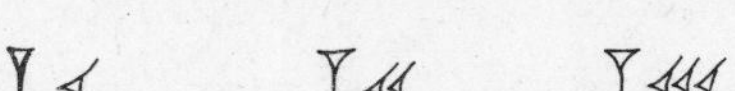

60 . 10 → 70 | 60 . 20 → 80 | 60 . 30 → 90

As Thureau-Dangin has pointed out, these forms in the Akkadian numeration were relics of the old Sumerian sexagesimal system, just as, in the French oral decimal numeration, the terms *quatre-vingts* ("eighty," literally "four-twenties") and *quatre-vingt-dix* ("ninety," literally "four-twenty-ten") are relics of a vanished vigesimal system.

This peculiarity of the common Akkadian cuneiform numeration is illustrated by a passage from *The Black Stone of Esarhaddon.* The city of Babylon was destroyed in 689 B.C. by Sennacherib, but his son Esarhaddon began rebuilding it immediately after his accession to the throne in 680. The passage in question says that after writing in the Book of

Destiny that Babylon was to remain deserted for seventy years, the god Marduk changed his mind and reversed the signs, which reduced the time of desertion to eleven years. This becomes understandable when we know that in the Assyro-Babylonian cuneiform system the number 70 was written as: Ⓨ ⨞

60 10

and that by reversing those two signs we obtain the notation for 11:

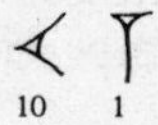

⨞ Ⓨ

10 1

Although the strictly decimal Akkadian system became the one commonly used in Mesopotamia, the old sexagesimal units 600 and 3600 were never completely abandoned. The signs for them appear in some contracts and economic documents, as well as in religious, historical, and commemorative texts. The original Sumerian sign for 3600 was gradually altered as shown in figure 23–2.

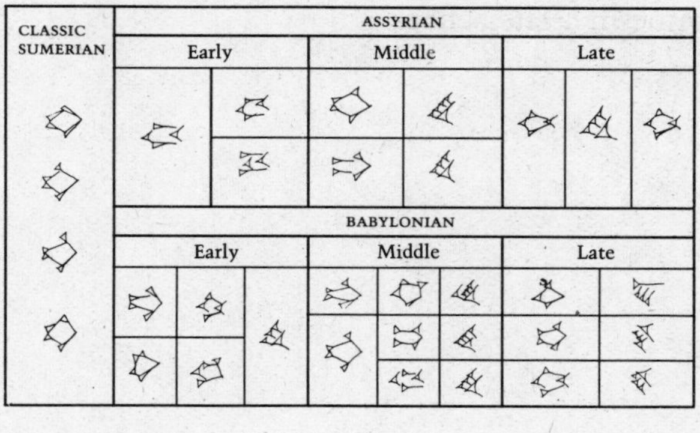

CLASSIC SUMERIAN	ASSYRIAN		
	Early	Middle	Late

CLASSIC SUMERIAN	BABYLONIAN		
	Early	Middle	Late

Fig. 23–2.

The following example is significant. In one of the inscriptions of Sargon II, the wall of Dur Sharrukin ("Fortress of Sargon"), an ancient city at the site of the modern village of Khorsabad, Iraq, is said to be 16,280 cubits long. (The cubit, KUSH, had a length of about eighteen inches; six cubits made a QANUM, and 60 an USH.) The measurement is expressed not as in this form:

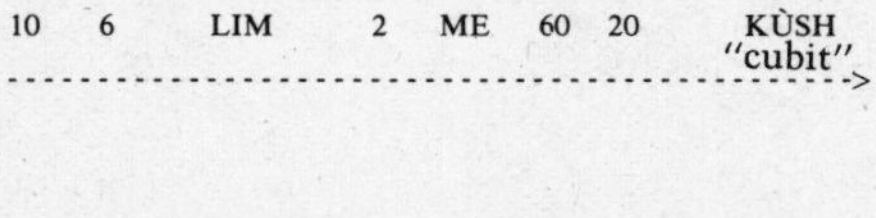

10   6   LIM   2   ME   60   20   KUSH
"cubit"

but in this one:

| 14,400 cubits | | | | 1 800 cubits | 1×60 cubits | 3×6 cubits | 2 cubits |

(3600 . 3600 . 3600 . 3600 . 600 . 600 . 600 . 1 USH . 3 QA-NI . 2 KÙSH)

Three main stages can be distinguished in the history of the common Assyro-Babylonian cuneiform numeration (fig. 23–3):

1. Use of the Sumerian sexagesimal system, during the period when the Sumerian cultural heritage was still being assimilated.
2. Use of a mixed system combining sexagesimal and decimal units.
3. Use of a strictly decimal system adapted to Semitic numerical traditions.

But while the Sumerian sexagesimal system finally disappeared from common use in Mesopotamia under Semitic domination, it continued to exist in mathematical and astronomical texts, in a remarkable form that we will examine in a later chapter.

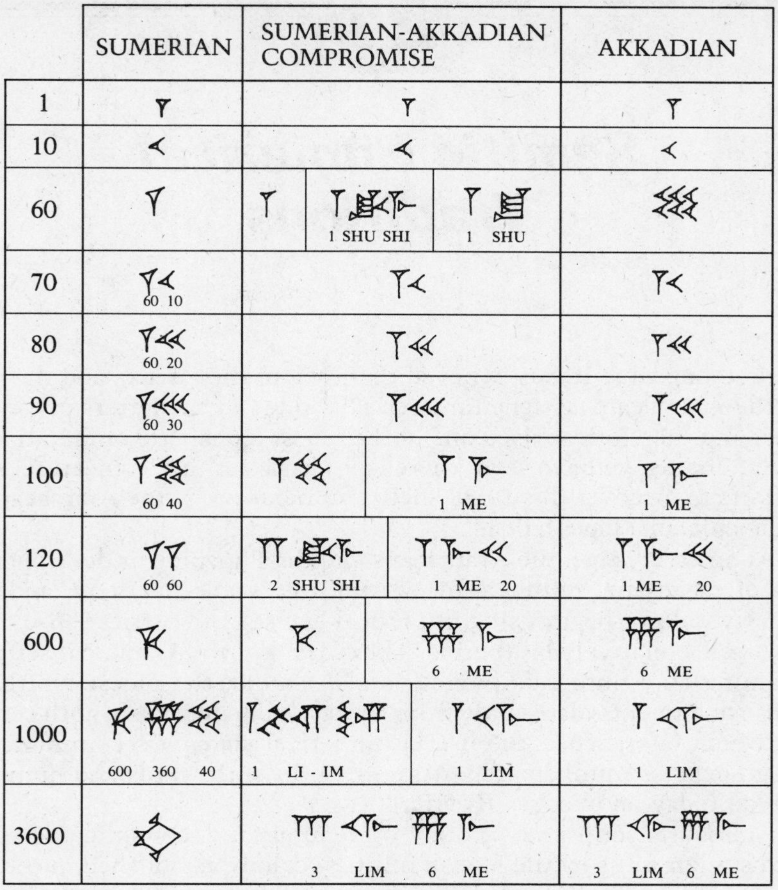

	SUMERIAN	SUMERIAN-AKKADIAN COMPROMISE			AKKADIAN
1	𒁹	𒁹			𒁹
10	𒌋	𒌋			𒌋
60	𒁹	𒁹	𒁹 1 SHU SHI	𒁹 1 SHU	𒐏
70	𒁹𒌋 60 10	𒁹𒌋			𒁹𒌋
80	𒁹𒌋𒌋 60 20	𒁹𒌋𒌋			𒁹𒌋𒌋
90	𒁹𒌋𒌋𒌋 60 30	𒁹𒌋𒌋𒌋			𒁹𒌋𒌋𒌋
100	𒁹𒐏 60 40	𒁹𒐏	𒁹 𒈨 1 ME		𒁹 𒈨 1 ME
120	𒁹𒁹 60 60	𒁹𒁹 2 SHU SHI	𒁹 𒈨 𒌋𒌋 1 ME 20		𒁹 𒈨 𒌋𒌋 1 ME 20
600	𒐕	𒐕	𒐚 𒈨 6 ME		𒐚 𒈨 6 ME
1000	𒐕 𒐚 𒐏 600 360 40	𒇷𒅎 1 LI - IM	𒁹 𒇻 1 LIM		𒁹 𒇻 1 LIM
3600	𒐏	𒐈 𒇻 𒐚 𒈨 3 LIM 6 ME			𒐈 𒇻 𒐚 𒈨 3 LIM 6 ME

Fig. 23–3. Evolution of the cuneiform numeration in Mesopotamia.

24

Semitic Numerical Traditions

For a long time it was believed that, before the Greeks and the Jews, the Aramaeans assigned numerical values to the letters of their alphabet and therefore devised one of the oldest alphabetic numerations in history. But, as we have seen, there is no basis for that conjecture, since all efforts to discover the use of such a numeration by the Aramaeans (or the Phoenicians) have failed.

At an early stage, the Aramaeans and the Phoenicians developed the habit of expressing numbers by writing the words for them, and they evidently did not begin using numerical signs, in the strict sense of the term, until a relatively late period. The oldest known Aramaic inscription containing true numerals (which is also among the oldest Northwest Semitic documents) dates only from the second half of the eighth century B.C. "Much later, when the use of numerical signs was common, they liked to express numbers by writing them in both words and numerals, as we do today on checks" (Février).

Aramaic inscriptions—notably the numerous economic and juridic papyruses left to us by the Jewish military colony established in the fifth century B.C. on the island of Elephantine, in the Nile—reveal a non-alphabetic type of numerical notation. In the oldest varieties of this system, the number 1 was indicated by a single vertical stroke, which was repeated as many times as necessary for the other numbers up to 9. The strokes were often grouped by threes to make it easier to count them. There was a special sign not only for 10 but also for 20. Other numbers below 100 were expressed by repetitions of vertical strokes and the signs for 10 and 20 (fig. 24–1). So for numbers 1 to 99 the system was based on the well-known principle that juxtaposition of two or more signs indicates addition of the corresponding numbers.

But while we might expect the consecutive tens to be represented, according to the additive principle, by as many repetitions of the sign for 10 as necessary, they are in fact represented by combinations apparently stemming from the base 20:

10, 20, 20 + 10, 20 + 20, 20 + 20 + 10, etc.

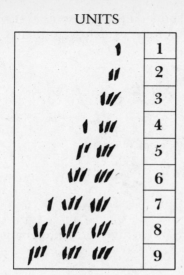

UNITS

1	1
11	2
111	3
1 111	4
11 111	5
111 111	6
1 111 111	7
11 111 111	8
111 111 111	9

SIGN FOR 10

SIGN FOR 20

NOTATIONS FOR OTHER TENS

	30
	40
	50
	60
	70
	80
	90

NUMBERS BELOW 100

	18
	38
	98

Fig. 24–1. Aramaic notations for numbers below 100. All signs are from Elephantine papyruses.

Is this, as has often been said, a trace of a vigesimal system that was once used by the Northwestern Semites and then abandoned?

If we examine the various forms of the sign for 20, and notice that the sign for 10 was first a straight horizontal stroke and then became a horizontal stroke curved downward on the right (fig. 24–1), it becomes clear that the sign for 20 was derived from the sign for 10 in accordance with a normal line of development that is common to all cursive scripts. The Aramaic sign for 20 resulted from the habits of ancient scribes who wrote on papyrus or stone with a brush dipped in some sort of colored liquid: it began as two horizontal strokes which were then joined and gave rise to a wide variety of forms:

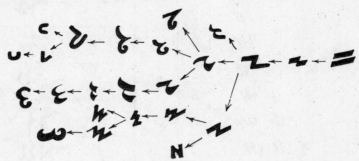

That is, the consecutive tens were originally represented by horizontal strokes grouped in twos:

80	70	60	50	40	30	20	10

This is confirmed by the oldest known Aramaic inscription containing true numerals, since the number 70 was expressed in it first by writing the word for it and then by means of seven horizontal lines grouped in pairs, as shown above. (The inscription dates from the second half of the eighth century B.C. and was engraved on the colossal statue of a king named Panamu; it probably comes from the Gercin hill, about four miles northeast of Zencirli in northern Syria, not far from the border between Turkey and Syria.) Thus, except for the arrangement of its signs, the Aramaic system for noting numbers below 100 was originally the same as the Cretan linear systems (figs. 14–5, 14–6) and the Hittite hieroglyphic numeration (fig. 14–8).

From this, some writers on the history of numerical notations have concluded that the Aramaeans' written numeration was a primitive system using only the additive principle and that it was narrowly limited because it had no signs for going beyond the second power of its base. This conclusion is unjustified because, while it is true that the system is primitive with regard to numbers below 100, it has an interesting feature not shared by most of the other systems we have studied so far: the Elephantine papyruses show that users of Aramaic writing had a special sign not only for 100 but also for 1000 and 10,000, and that for multiples of these units, instead of simply repeating the sign as often as necessary (as was done by the Egyptians, Cretans, and Hittites, for example), they used the multiplicative principle by placing as many strokes as necessary to the right of the sign for the unit (fig. 24–2).

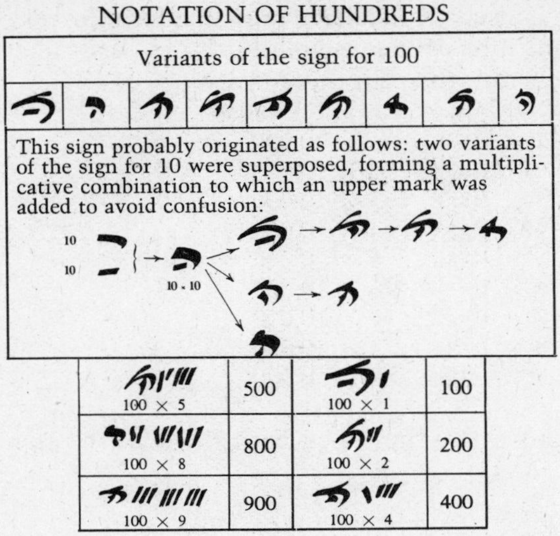
NOTATION OF HUNDREDS
Variants of the sign for 100
This sign probably originated as follows: two variants of the sign for 10 were superposed, forming a multiplicative combination to which an upper mark was added to avoid confusion:
10
10
10 . 10
100 × 5
500
100 × 1
100
100 × 8
800
100 × 2
200
100 × 9
900
100 × 4
400

Fig. 24–2. The Aramaic written numeration for multiples of 100, 1000, and 10,000, as shown by the Elephantine papyruses. Signs drawn with solid strokes are known to have been used; those drawn with double lines are reconstructions based on comparative study.

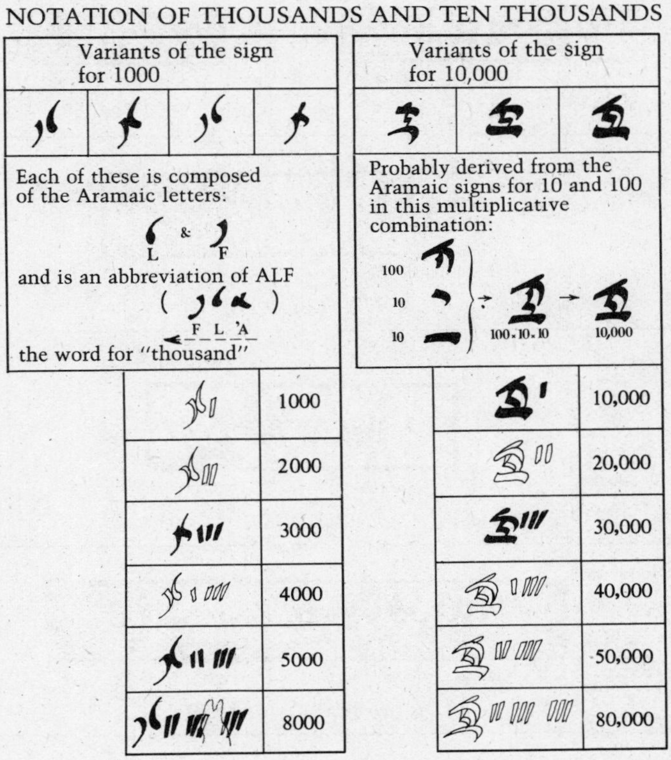
NOTATION OF THOUSANDS AND TEN THOUSANDS
Variants of the sign for 1000
Each of these is composed of the Aramaic letters:
L & F
and is an abbreviation of ALF
F L 'A
the word for "thousand"
1000
2000
3000
4000
5000
8000
Variants of the sign for 10,000
Probably derived from the Aramaic signs for 10 and 100 in this multiplicative combination:
100
10
10
100.10.10
10,000
10,000
20,000
30,000
40,000
50,000
80,000

To sum up, the Aramaeans used the additive principle for numbers 1 to 99, but they used the multiplicative principle for hundreds, thousands, and ten thousands, and wrote them in this form:

$$
\begin{array}{lll}
1 \times 100 & 1 \times 1000 & 1 \times 10{,}000 \\
2 \times 100 & 2 \times 1000 & 2 \times 10{,}000 \\
3 \times 100 & 3 \times 1000 & 3 \times 10{,}000 \\
4 \times 100 & 4 \times 1000 & 4 \times 10{,}000 \\
5 \times 100 & 5 \times 1000 & 5 \times 10{,}000 \\
6 \times 100 & 6 \times 1000 & 6 \times 10{,}000 \\
7 \times 100 & 7 \times 1000 & 7 \times 10{,}000 \\
8 \times 100 & 8 \times 1000 & 8 \times 10{,}000 \\
9 \times 100 & 9 \times 1000 & 9 \times 10{,}000
\end{array}
$$

And for other numbers above 100 they used both the additive and multiplicative principles (fig. 24–3).

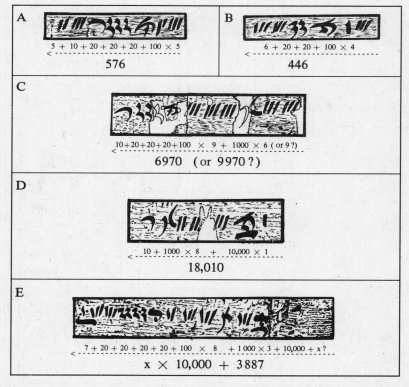

Fig. 24–3. Interpretation and deciphering of five number notations in the Elephantine papyruses.

The use of such a system generally corresponds to the numerical traditions of the Semitic peoples. It was found among nearly all the Northwestern Semites, which is not surprising, since the Phoenicians, Palmyrans, Nabataeans, etc. used notations that belonged to the same category, differing only in details, and probably had the same origin as the Elephantine Aramaic numeration (fig. 24–4).

It was also found among the Eastern Semites. As we have seen, the Assyrians and Babylonians, who inherited the additive sexagesimal numeration of the Sumerians, drastically altered its structure while keeping its cuneiform signs. Because they were in the habit of counting by hundreds and thousands, but found no numerals for 100 and 1000 in their predecessors' system, they had the idea of writing those two numbers by a phonetic device and indicating multiples of them by using the multipli-

HATRAN Early centuries of the Christian era			NABATAEAN Beginning in the 2nd century B.C.			PALMYRAN Early Christian era			PHOENICIAN Beginning in the 6th century B.C.		
Units			Units			Units			Units		
5	4	1	5	4	1	5	4	1	5	4	1
	9			9			9			9	
Ten			Ten			Ten			Ten		
Twenty			Twenty			Twenty			Twenty		
Hundreds			Hundreds			Hundreds			Hundreds		
100 × 1	100 × 1	100 × 1	100 × 1			100 × 1			100 × 1	100 × 1	100 × 1
100 × 2			100 × 2			100 × 2			100 × 2		100 × 2
100 × 3			100 × 3			100 × 3			100 × 3		100 × 3
100 × 4			100 × 4			100 × 4			100 × 4		100 × 4

Fig. 24–4. Numerations of Northwestern Semitic peoples.

cative principle, rather than the additive principle (fig. 23–1). The Southwestern Semites also had this tradition.

In the fourth century A.D., probably as a result of contact with Christian missionaries from Egypt and Syria-Palestine, the Ethiopians borrowed the Greek alphabetic numeration, but radically altered its principle for numbers above 100. After adopting the first nineteen numeral letters of the Greek system for noting the first 100 whole numbers, they indicated hundreds and thousands by placing the letter rho (P), whose value was 100, to the right of the letters associated with the corresponding units and tens. This meant that, instead of representing numbers 200, 300, . . . 900 and 1000, 2000, . . . 9000 in the Greek manner, by these letters:

$$\underset{200}{\Sigma} \quad \underset{300}{T} \quad \underset{400}{Y} \quad \ldots \quad \underset{900}{\lambda} \quad \underset{1000}{'A} \quad \underset{2000}{'B} \quad \ldots \quad \underset{9000}{'\Theta}$$

they expressed them in this form (fig. 24–5):

$$\underset{2 \times 100}{BP} \quad \ldots \quad \underset{8 \times 100}{HP} \quad \ldots \quad \underset{20 \times 100}{KP} \quad \ldots \quad \underset{80 \times 100}{\Pi P}$$

$$\underset{200}{\dashrightarrow} \quad \underset{800}{\dashrightarrow} \quad \underset{2000}{\dashrightarrow} \quad \underset{8000}{\dashrightarrow}$$

They then represented 10,000 by joining two signs for 100 (a combination that can be transcribed as PP) to indicate 100 multiplied by itself, and multiples of 10,000 by placing this double sign to the right of the letters for the corresponding units and tens (fig. 24–5).

$$\underset{2 \times 10,000}{B\,PP} \quad \ldots \quad \underset{8 \times 10,000}{H\,PP} \quad \ldots \quad \underset{20 \times 10,000}{K\,PP} \quad \ldots \quad \underset{80 \times 10,000}{\Pi\,PP}$$

$$\underset{20,000}{\dashrightarrow} \quad \underset{80,000}{\dashrightarrow} \quad \underset{200,000}{\dashrightarrow} \quad \underset{800,000}{\dashrightarrow}$$

The old numerals that the Ethiopians sometimes still use today are variants, generally more rounded, of the numerical signs in inscriptions at Aksum (capital of the ancient Axumite empire, near the modern city of Adwa) and follow the same principle. They are derived from the first nineteen letters of the Greek alphabetic numeration (fig. 17–2) and, to make it clear that they represent numbers, they are always placed between two horizontal lines with a hook at each end, a practice that began in the fifteenth century.

The Eastern Arabs, who (no doubt in imitation of the Greeks, Jews, and Syrian Christians) devised an additive numeration enabling them to note numbers up to 1000 (fig. 20–4), preferred to indicate thousands, ten thousands, and hundred thousands by using the multiplicative principle,

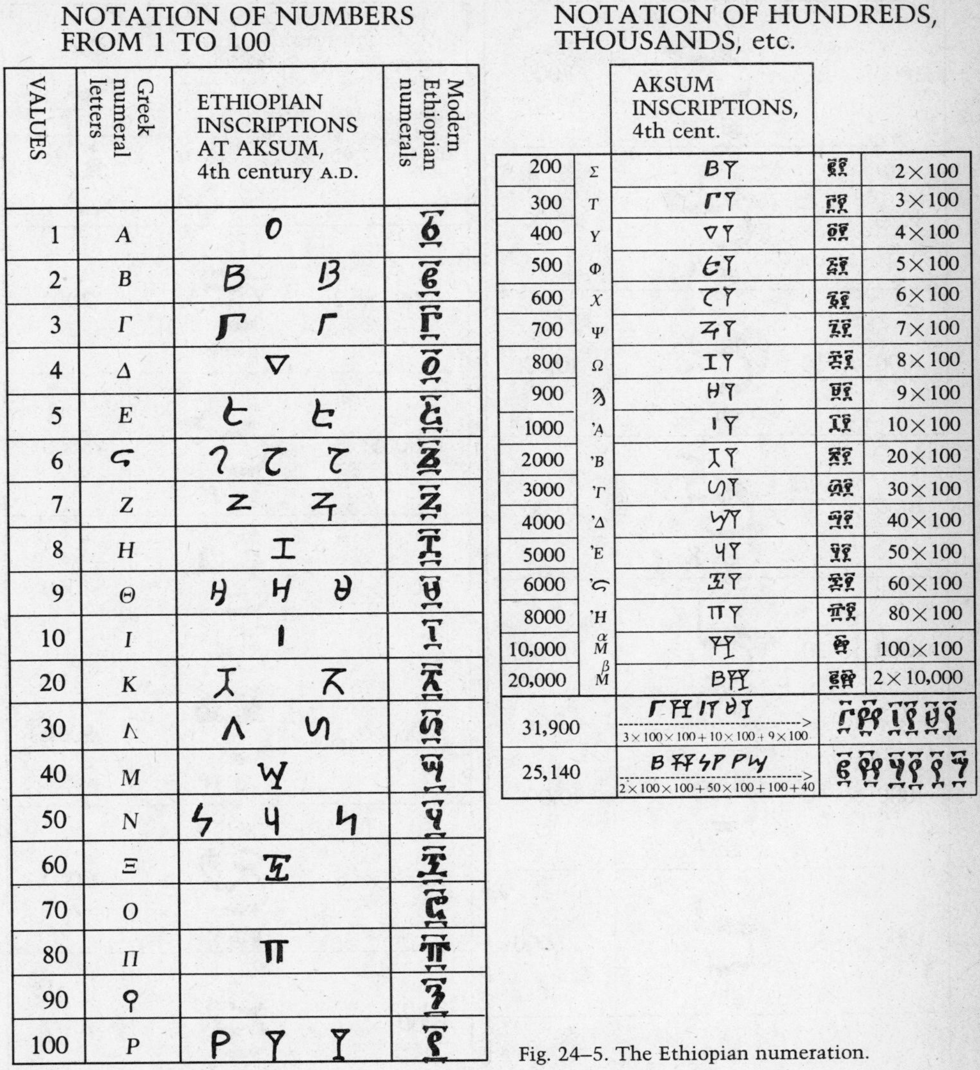
NOTATION OF NUMBERS FROM 1 TO 100
VALUES
Greek numeral letters
ETHIOPIAN INSCRIPTIONS AT AKSUM, 4th century A.D.
Modern Ethiopian numerals
1 Α
2 Β
3 Γ
4 Δ
5 Ε
6 ϛ
7 Ζ
8 Η
9 Θ
10 Ι
20 Κ
30 Λ
40 Μ
50 Ν
60 Ξ
70 Ο
80 Π
90 ϙ
100 Ρ
NOTATION OF HUNDREDS, THOUSANDS, etc.
AKSUM INSCRIPTIONS, 4th cent.
200 Σ 2×100
300 Τ 3×100
400 Υ 4×100
500 Φ 5×100
600 Χ 6×100
700 Ψ 7×100
800 Ω 8×100
900 ϡ 9×100
1000 ʹΑ 10×100
2000 ʹΒ 20×100
3000 ʹΓ 30×100
4000 ʹΔ 40×100
5000 ʹΕ 50×100
6000 ʹϛ 60×100
8000 ʹΗ 80×100
10,000 α M 100×100
20,000 β M 2×10,000
31,900
3×100×100+10×100+9×100
25,140
2×100×100+50×100+100+40

Fig. 24–5. The Ethiopian numeration.

placing the letter ghayn (1000) to the left of the letters associated with the corresponding units, tens, and hundreds (fig. 24–6).

All these systems belong to a new category that, to borrow Guitel's term, can be called hybrid: instead of being based entirely on the additive principle, they are based (usually beginning with hundreds) on a mixed principle that is both additive and multiplicative.

1000 × 8	حغ gh H	8000
1000 × 9	طغ gh Ṭ	9000
1000 × 10	يغ gh Y	10,000
1000 × 20	كغ gh K	20,000
1000 × 30	لغ gh L	30,000
1000 × 40	مغ gh M	40,000
1000 × 50	نغ gh N	50,000
. . . .		

Arabic numeral letter representing 1000: It is the 28th letter in the *Abjad* system. Its final form is:	غ Ghayin ـغ	
1000 × 2	بغ gh B	2000
1000 × 3	جغ gh J	3000
1000 × 4	دغ gh D	4000
1000 × 5	هغ gh H	5000
1000 × 6	وغ gh W	6000
1000 × 7	زغ gh Z	7000

Fig. 24–6. Representation of units above 1000 in the alphabetic numeration of the Eastern Arabs.

25

The Traditional
Chinese Numeration System*

The modern system

The Chinese ordinarily use a decimal system that has thirteen basic signs associated with numbers 1 to 10, 100, 1000, and 10,000. The simplest and most common forms of those signs are shown in figure 25–1.

They are ordinary writing characters that express the Chinese words for the corresponding numbers: *yī* ("one"), *èr* ("two"), *sān* ("three"), *sì* ("four"), *wǔ* ("five"), *liù* ("six"), *qī* ("seven"), *bā* ("eight"), *jiǔ* ("nine"), *shí* ("ten"), *bǎi* ("hundred"), qiān ("thousand"), *wàn* ("ten thousand").

The modern Chinese numeration is a hybrid system. Tens, hundreds, thousands, and ten thousands are expressed by means of the multiplicative principle (fig. 25–2).

For other numbers, the Chinese use both addition and multiplication (fig. 25–3), breaking down the number 79,564, for example, in this way:

$$7 \times 10,000 + 9 \times 1000 + 5 \times 1000 + 6 \times 10 + 4$$

---

*In this chapter, Chinese words and names will be transcribed in letters of the Roman alphabet according to the system officially adopted by the People's Republic of China in 1958. Some of the letters are not pronounced as they are in English. The main differences are as follows:

  b   English *p*
  c   English *ts*
  d   English *t*
  g   English *k*
  h   At the beginning of a word, it has a sound close to the hard German *ch* in Bach.
  i   At the beginning of a word, *ee* as in "feet"; after *c, ch, r, s, sh,* or *z,* a sound close to German *ö* or French *eu;* after *a* or *u,* it is a diphthong composed of *e* as in "bet" and *ee* as in "feet."
  q   A complex sound that can be described as *ts* with an aspiration.
  r   At the beginning of a word, it has the sound of *s* in "pleasure"; otherwise, it has the sound of German *öl* or French *eul.*
  ü   German *ü* or French *u*
  x   At the beginning of a word, it has a sound close to the soft German *ch* in *ich.*
  z   English *dz*
  zh English *j*

Accents placed above vowels indicate the four tones, or variations in pitch, of the standard Peking dialect; with the letter *a,* for example, *ā* indicates the "high-level" tone, *á* the "high-rising" tone, *ǎ* the "falling-and-rising" tone, and *à* the "falling" tone.

1 一	10 十
2 二	100 百
3 三	1000 千
4 四	10,000 萬 or 万
5 五	
6 六	
7 七	
8 八	
9 九	

Fig. 25–1.

TENS		HUNDREDS		THOUSANDS		TEN THOUSANDS	
10	一十 1 × 10	100	一百 1 × 100	1000	一千 1 × 1000	10,000	一萬 1 × 10,000
20	二十 2 × 10	200	二百 2 × 100	2000	二千 2 × 1000	20,000	二萬 2 × 10,000
30	三十 3 × 10	300	三百 3 × 100	3000	三千 3 × 1000	30,000	三萬 3 × 10,000
40	四十 4 × 10	400	四百 4 × 100	4000	四千 4 × 1000	40,000	四萬 4 × 10,000
50	五十 5 × 10	500	五百 5 × 100	5000	五千 5 × 1000	50,000	五萬 5 × 10,000
60	六十 6 × 10	600	六百 6 × 100	6000	六千 6 × 1000	60,000	六萬 6 × 10,000
70	七十 7 × 10	700	七百 7 × 100	7000	七千 7 × 1000	70,000	七萬 7 × 10,000
80	八十 8 × 10	800	八百 8 × 100	8000	八千 8 × 1000	80,000	八萬 8 × 10,000
90	九十 9 × 10	900	九百 9 × 100	9000	九千 9 × 1000	90,000	九萬 9 × 10,000

Fig. 25–2. Present-day Chinese notations for multiples of the first four powers of 10. (The modern horizontal arrangement is used here.)

TRADITIONAL ARRANGEMENT
(Notations taken from the
page shown in fig. 25-5.)

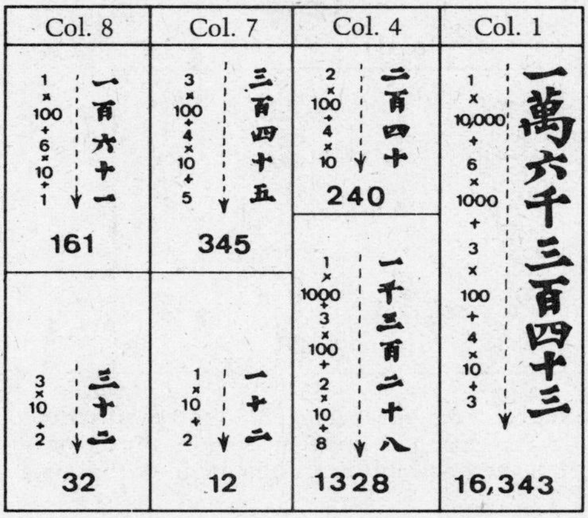

PRESENT ARRANGEMENT

7 × 10,000 + 9 × 1000 + 5 × 100 + 6 × 10 + 4

79,564

Col. 8	Col. 7	Col. 4	Col. 1
1 × 100 + 6 × 10 + 1 → **161**	3 × 100 + 4 × 10 + 5 → **345**	2 × 100 + 4 × 10 → **240**	1 × 10,000 + 6 × 1000 + 3 × 100 + 4 × 10 + 3 → **16,343**
3 × 10 + 2 → **32**	1 × 10 + 2 → **12**	1 × 1000 + 3 × 100 + 2 × 10 + 8 → **1328**	

Fig. 25–3. Examples of Chinese numerical signs representing numbers that are not multiples of the base. (Traditionally, these signs, like other Chinese characters, were arranged vertically in columns that were read from top to bottom and from right to left, but horizontal arrangement going from left to right is now preferred in the People's Republic of China.)

Different forms of Chinese numerals

Each of the thirteen basic characters has several different forms, all of which have the same pronunciation but correspond to various styles of Chinese writing and the uses made of them.

The forms we have considered so far—which will be referred to as "classical" from now on—are those most commonly used today. They are also the simplest, and are used in elementary teaching of Chinese characters. And finally, they are the oldest of the forms still in ordinary use, having remained unchanged since the fourth century A.D. They are derived from the ancient *lìshū* style ("functionary writing") used during the Han period (figs. 25–4, 25–5).

The classical forms belong to the *kǎishū* style of modern Chinese writing, in which the strokes that compose the characters are essentially straight lines of different lengths and orientations, drawn in a certain order according to well-defined rules (fig. 25–6).

Unless another meaning is specified, the term "Chinese writing" nearly always refers to *kǎishū*.

The second way of writing Chinese numerals is known as *guān zí* in Taiwan as well as the People's Republic of China. It is used mainly for such documents as bills of sale, checks, receipts, and invoices. These

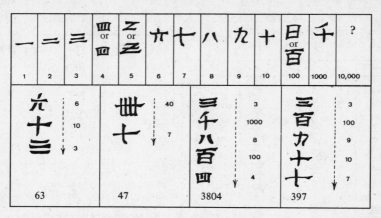

Fig. 25–4. The first of the modern Chinese numerical notations, belonging to the *lìshū* style used during the Han dynasty (206 B.C.–A.D. 220). The examples are reconstructions made on the basis of administrative documents dating from the first century A.D., found in central Asia.

永樂大典卷之一萬六千三百四十三　十翰

筭　筭法十四

異乘同除詳明筭法歌曰。異乘同除法何如。物貫錢來做例兒。先下原錢乘只物。却將原物法除之。將錢買物互乘取。百里千斤以類推。筭者留心能善用。一絲一忽不差池。

九章筭經今有絲一斤。價直二百四十。今有錢一千三百二十八。問得絲幾何。

荅曰。五斤八兩一十二銖五分銖之四。

術曰。以一斤價數爲法。以一斤乘。今有錢數爲實。實如法得絲數。按此術今有之義。以一斤價爲所有率。一斤爲所求率。今有錢爲所有數。而今有之即得。

今有絲一斤。價直三百四十五。今有絲七兩一十二銖。問得錢幾何。

荅曰。一百六十一錢三十二分錢之二十三。

術曰。以一斤銖數爲法。以一斤價數乘七兩一十二銖爲實。實如法得

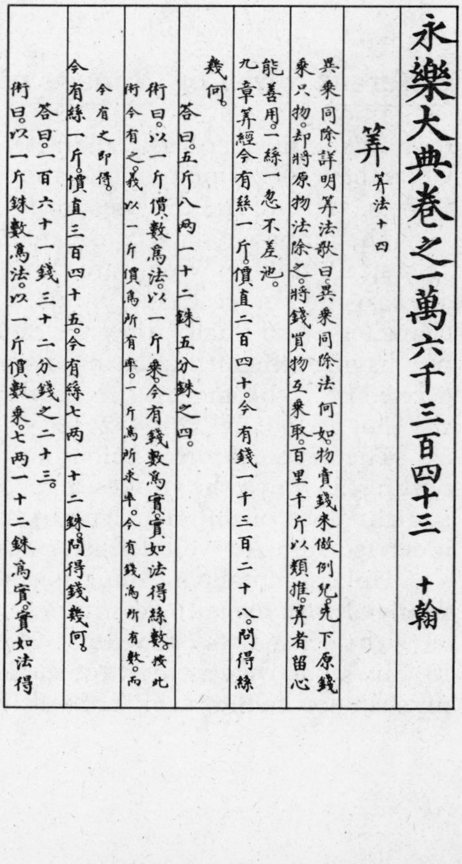

Fig. 25–5. Page of *Yung-Lo Ta Tien,* a fifteenth-century Chinese mathematical document now in the Cambridge University Library.

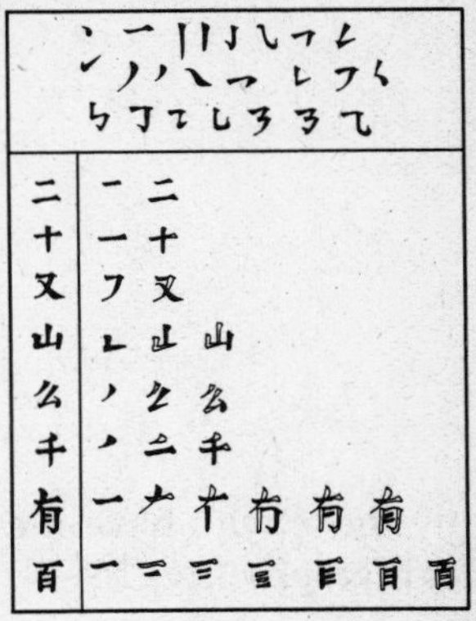

Fig. 25–6. The basic strokes of the regular style of Chinese writing known as *kǎishū*, with examples of their use in several characters, showing the order in which they are drawn.

forms are more complicated than the classical forms, being composed of many more strokes. They were devised for the purpose of avoiding mistakes and fraudulent alterations in financial transactions.

Example: 13,684

Classic notation	一 萬 三 千 六 百 八 十 四
Guān zí notation	壹 萬 參 仟 陸 佰 捌 拾 肆

```
- - - - - - - - - - - - - - - - - - - - - - - - ->
yī    wàn  sān  qiān  liù  bǎi  bā   shí  sì
1 × 10,000 + 3 × 1,000 + 6 × 100 + 8 × 10 + 4
```

The third type of numerals belongs to *xíngshū*, a cursive style commonly used in handwritten letters, rough drafts, personal notes, etc. *Xíngshū* was developed to increase the speed of writing, but without changing the structures of the characters. The only difference is that they are written in a more rapid, flowing way.

Example: 49,265

Classic notation	四 萬 九 千 二 百 六 十 五
Xíngshū notation	*(cursive rendering)*

```
- - - - - - - - - - - - - - - - - - - - - - - - ->
sì    wàn  jiǔ  qiān  èr   bǎi  liù  shí  wǔ
4 × 10,000 + 9 × 1,000 + 2 × 100 + 6 × 10 + 5
```

A desire for still more writing speed, combined with the whimsy and virtuosity of certain artists and calligraphers, soon changed these cursive forms, which resemble the original signs rather closely, into what the Chinese call *cǎoshū* ("grass-shaped style"). It can be understood only

Example: 75,696

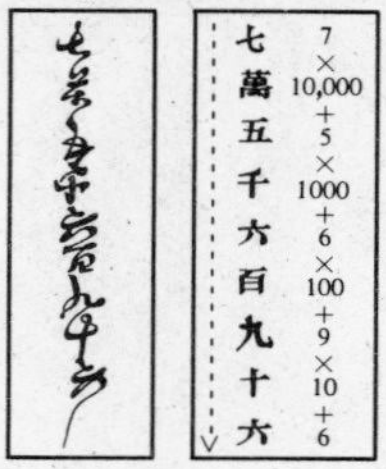

by those who have made a special study of it, so today it is used only in calligraphy (fig. 25–7).

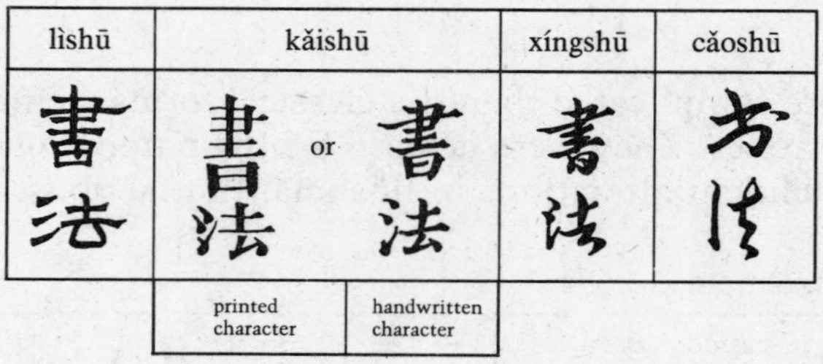

lìshū	kǎishū		xíngshū	cǎoshū
書法	書法 or	書法	書法	方住
	printed character	handwritten character		

Fig. 25–7. Differences among the main styles of Chinese writing. Shown here is the word *shūfǎ* ("calligraphy") in these styles: *lìshū* (used during the Han dynasty), *kǎishū* (used since the fourth century A.D. as a replacement of *lìshū*), *xíngshū* (common cursive style), and *cǎoshū* (a shortened cursive style now used only in calligraphy).

The numerals known as *ngán mà* or *gán zí* ("secret figures or marks") are used in ordinary calculations and in commerce, to indicate prices, for example. These are the numerals that foreigners going to China need to know if they want to understand the amounts written on hotel and restaurant bills (fig. 25–8).

The numerals called *sháng fāng dà zhuàn* have a curiously geometric design (fig. 25–9). With other characters in the same style, they are used in seals. "The characters of Chinese writing are distinctive enough to serve as marks (marks giving power, marks of identity or ownership, trademarks, etc.), and seals commonly bear characters. This is not the case with the civilizations of the West or the Middle East, in which drawings are normally used for seals. Syllabaries and alphabets are composed of signs too uniform to serve as marks, as none of them characterizes a unique reality, since each appears in the compositions of an indefinite number of written words" (Gernet).

I will not carry this list of writing styles any farther because imaginative variations are so numerous in modern China that descriptions of all of them would go beyond the purpose of this book.

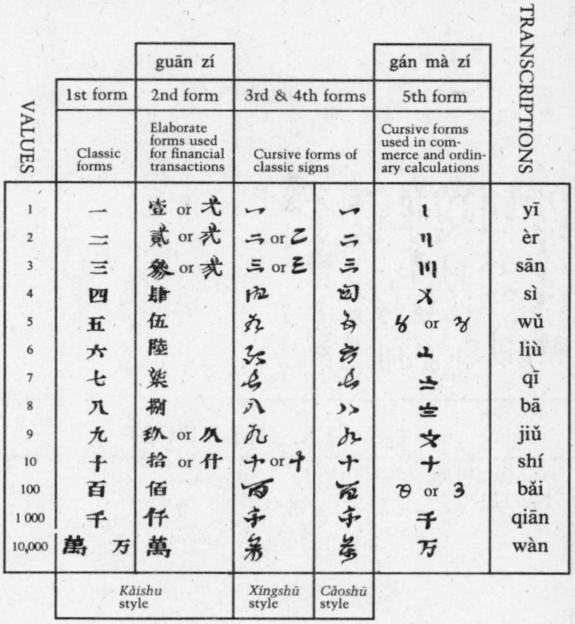

VALUES	guān zí		3rd & 4th forms	gán mà zí	TRANSCRIPTIONS
	1st form	2nd form		5th form	
	Classic forms	Elaborate forms used for financial transactions	Cursive forms of classic signs	Cursive forms used in commerce and ordinary calculations	
1	一	壹 or 弌			yī
2	二	貳 or 弍			èr
3	三	參 or 弎			sān
4	四	肆			sì
5	五	伍			wǔ
6	六	陸			liù
7	七	柒			qī
8	八	捌			bā
9	九	玖 or 久			jiǔ
10	十	拾 or 什			shí
100	百	佰			bǎi
1 000	千	仟			qiān
10,000	萬 万	萬			wàn
	Kǎishu style		Xíngshū style / Cǎoshū style		

Fig. 25–8. Main forms of the thirteen basic signs in the modern Chinese numeration.

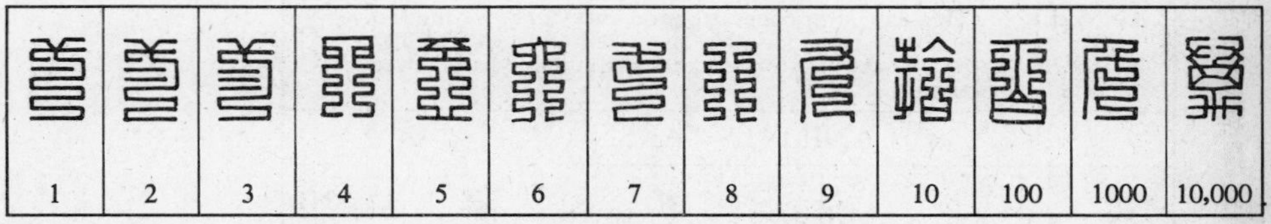

| 1 | 2 | 3 | 4 | 5 | 6 | 7 | 8 | 9 | 10 | 100 | 1000 | 10,000 |

Fig. 25–9. An example of calligraphic whimsy in writing the basic characters of the Chinese numeration: the *sháng fāng dà zhuàn* numerals, still used for seals and signatures.

Origin of the Chinese system

The oldest known examples of Chinese writing and written numeration are on several thousand bones and tortoise shells, most of them discovered since the late nineteenth century at Xiao-dun, a village in northeastern Honan province, near the town of An-yang. Dating from the Yin dynasty (fourteenth to eleventh centuries B.C.), they have inscriptions engraved on one side and, on the other, small cracks caused by heat. They belonged to soothsayer-priests at royal courts and were used in the rite of divination by fire.

According to H. Maspéro in his *La Chine antique* (Paris, 1965), this

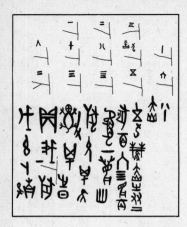

TRANSLATION

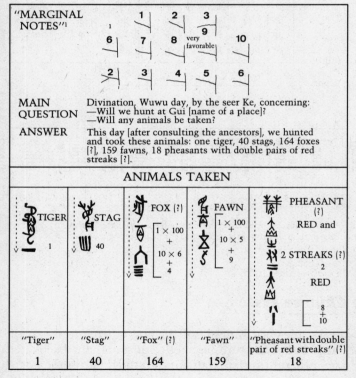

	TIGER	STAG	FOX (?)	FAWN	PHEASANT (?) RED and 2 STREAKS (?) RED
	1	40	1×100 + 10×6 + 4	1×100 + 10×5 + 9	2 $\begin{array}{c} 8 \\ + \\ 10 \end{array}$
	"Tiger"	"Stag"	"Fox" (?)	"Fawn"	"Pheasant with double pair of red streaks" (?)
	1	40	164	159	18

MAIN QUESTION — Divination, Wuwu day, by the seer Ke, concerning:
—Will we hunt at Gui [name of a place]?
—Will any animals be taken?

ANSWER — This day [after consulting the ancestors], we hunted and took these animals: one tiger, 40 stags, 164 foxes [?], 159 fawns, 18 pheasants with double pairs of red streaks [?].

ANIMALS TAKEN

[1] These numerical signs mark various parts of the tortoise shell, probably to indicate the order in which the cracks are to be examined. The writing sign in the part numbered 9 interprets the corresponding crack as a good omen.

Fig. 25–10. Divinatory inscription on the bottom part of a tortoise shell, discovered at Xiao-dun and dating from the Yin dynasty. Translation and interpretation by Léon Vandermeerch.

rite was performed as follows: addressing the royal ancestors (whose worship held a preponderant place in the Chinese religion of the time), the priest inscribed questions on the bottom part of a sanctified tortoise shell, or on the scapula of a stag, bovine, or sheep; he then held the uninscribed side of the shell or bone close to a fire, and the pattern of cracks resulting from the heat of the fire supposedly gave him the answers to his questions (fig. 25–10).

Probably pictographic in its original form, the writing revealed by these divinatory inscriptions seems to have been already considerably developed, since it is neither purely pictographic nor purely ideographic. (Archaic Chinese writing consists essentially of a few hundred elementary signs representing ideas and simple objects, with a number of more complex signs containing two elements, one related to the sound of the corresponding word, the other to its visual or symbolic meaning.) It has reached a rather advanced graphic stage (fig. 25–11). "Stylization and economy of means are carried so far in the oldest known Chinese writing that its signs seem more like letters than drawings" (Gernet).

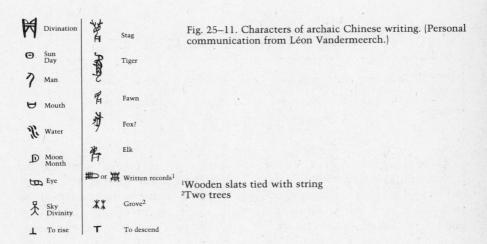

Fig. 25–11. Characters of archaic Chinese writing. (Personal communication from Léon Vandermeerch.)

Divination
Sun Day
Man
Mouth
Water
Moon Month
Eye
Sky Divinity
To rise

Stag
Tiger
Fawn
Fox?
Elk
Written records[1]
Grove[2]
To descend

[1]Wooden slats tied with string
[2]Two trees

"Furthermore, in its very constitution, this writing abounds in abstract formations (opposed or reversed signs, strokes marking a certain part of a sign, representations of human gestures, etc.) and simpler combinations of signs that serve to create new symbols" (Gernet). The system of numeration associated with it appears to be already well on the way toward abstract notation.

Numbers 1 and 10 are represented by a horizontal and a vertical stroke, respectively. (It seems that these two signs naturally occur to the human mind under certain conditions; we know, for example, that the Cretans, Hittites, and Phoenicians used signs of that kind for the same two numbers.) The number 100 is noted by a sign that Needham calls a pine cone, and 1000 by one that is evidently related to the sign for "man" in the corresponding writing.

Numbers 2, 3, and 4 are indicated by repetitions of horizontal strokes. Like everyone else who used this kind of notation, the Chinese made a break at 4 because scarcely anyone is able to recognize a row of more than four elements at a glance, without counting. But instead of continuing the same procedure with more easily recognizable patterns (dividing the signs into two groups, for example, as the Egyptians did, or into two or three groups, as the Babylonians and Phoenicians did), the Chinese preferred to introduce five special signs apparently with no direct visual reference, for the next five numbers (fig. 25–12).

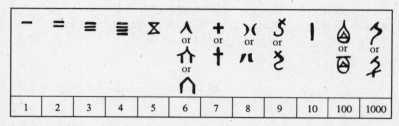

Fig. 25–12. The basic signs of the archaic Chinese numeration. They are found on divinatory bones and tortoise shells from the Yin dynasty (fourteenth to eleventh centuries B.C.) and on bronzes from the Zhu dynasty (tenth to sixth centuries B.C.). Table revised by Léon Vandermeerch.

For certain numbers, this numeration may have used what can be called "phonetic borrowings," signs chosen for their sounds, regardless of their original meanings. (Such signs appear in the writing associated with the numeration.) This may explain, for example, why 1000 was indicated by the sign for "man": the two words probably had the same pronunciation in the Chinese language of the time. Perhaps other signs were chosen for magical or religious reasons and were directly related to the number mysticism of ancient China, each sign supposedly representing, by its form, the "realities" of the number corresponding to it.

Be that as it may, the numerical system that appears in divinatory inscriptions from the second half of the second millennium B.C. has already made considerable progress, mathematically speaking, toward the present Chinese system.

Disregarding numbers 20, 30, and 40 for the moment, tens, hundreds, and thousands are represented according to the multiplicative principle, by combining the corresponding signs with those of the units associated with them. Specifically, tens from 50 to 90 are represented by superposed signs in this form:

$$\begin{array}{ccccc} 10 & 10 & 10 & 10 & 10 \\ \times & \times & \times & \times & \times \\ 5 & 6 & 7 & 8 & 9 \end{array}$$

And this cannot cause confusion with the signs for numbers 15 to 19, which are represented in this form:

$$\begin{array}{ccccc} 5 & 6 & 7 & 8 & 9 \\ + & + & + & + & + \\ 10 & 10 & 10 & 10 & 10 \end{array}$$

Hundreds are noted by placing the signs for the corresponding units above the sign for 100, and thousands are noted by a method analogous to the one used for tens (fig. 25–13). Numbers above 50 that are not multiples of the base are represented by a combination of the additive and multiplicative principles.

The Chinese numeration has therefore been a hybrid system since the earliest known examples of it. And while numbers 20, 30, and 40

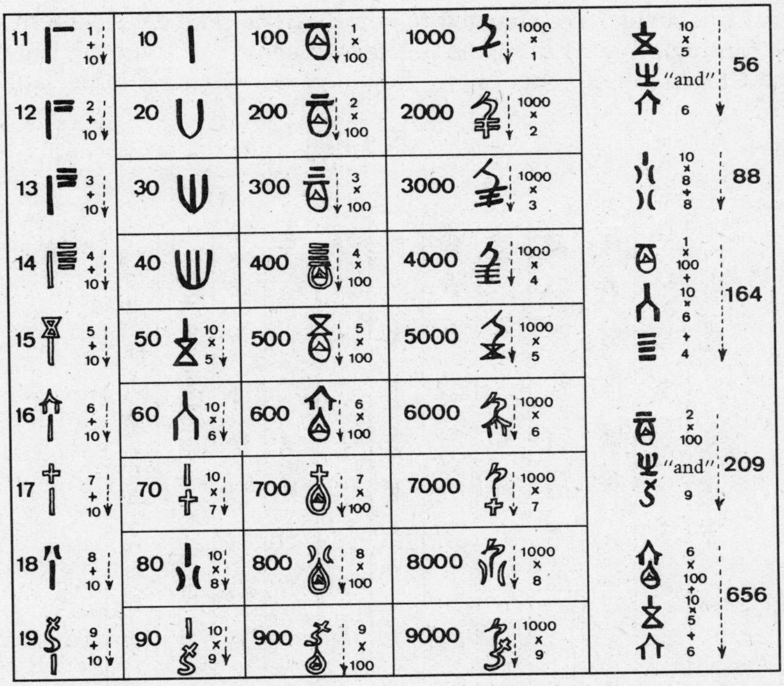

Fig. 25–13. Structure of the archaic Chinese numeration. Signs drawn with solid strokes have been noted in divinatory inscriptions from the Yin dynasty; those drawn with double lines are probable reconstructions. Table revised by Léon Vandermeerch.

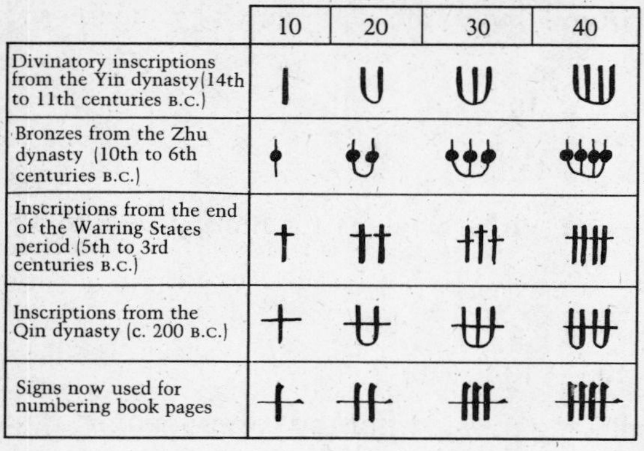

	10	20	30	40
Divinatory inscriptions from the Yin dynasty (14th to 11th centuries B.C.)	I	U	U	W
Bronzes from the Zhu dynasty (10th to 6th centuries B.C.)	♦	♦♦	♦♦♦	♦♦♦♦
Inscriptions from the end of the Warring States period (5th to 3rd centuries B.C.)	†	††	†††	††††
Inscriptions from the Qin dynasty (c. 200 B.C.)	✚	✚✚	✚✚✚	✚✚✚✚
Signs now used for numbering book pages	✚	✚✚	✚✚✚	✚✚✚✚

Fig. 25–14. Stability of the signs for 10, 20, 30, and 40 in the history of the Chinese numeration.

have often been indicated by repetitions of the sign for 10 (fig. 25–14), the reason is that use of the multiplicative principle would not have produced simpler notations in those cases.

The structure of the Chinese written numeration has remained basically the same throughout its long history, though the arrangements and forms of its signs have changed to some extent. The development of its signs can be followed from the fourteenth century B.C. to the beginning of the Christian era; since then, they have not changed appreciably (fig. 25–15).

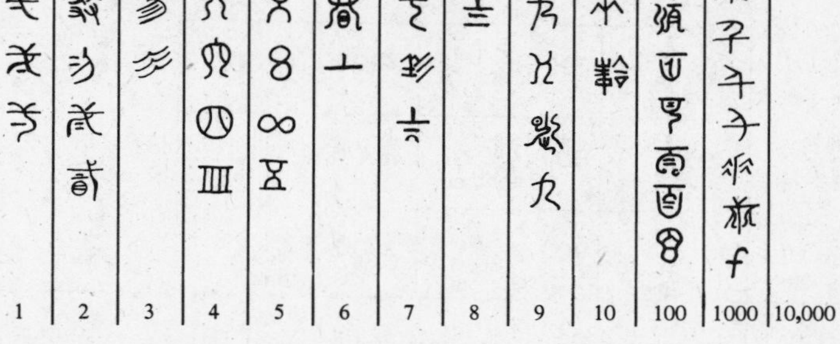

1	2	3	4	5	6	7	8	9	10	100	1000	10,000

Fig. 25–15. Variants of Chinese numerical signs in inscriptions from the end of the Warring States period (fifth to third centuries B.C.). Table revised by Léon Vandermeerch.

Diffusion of the Chinese system in the Far East

Because the Chinese language is divided into many regional dialects, each character of Chinese writing, including the thirteen basic signs of its numeration, may be pronounced differently, depending on whether it is read by an inhabitant of Manchuria, Honan, Peking, Canton, or Singapore. Furthermore, the signs have been pronounced in different ways within each dialect, at various stages of its development. But the characters of Chinese writing, whose internal structures have not changed significantly through the centuries, have kept the same basic meanings, and despite the diversity of their pronunciations they have remained comprehensible to readers in all parts of China.

"For example, the word for 'eat' is pronounced *chī* in Mandarin, and written with a character that we call 'A.' This character is read as *hɛk* in Cantonese. When the Cantonese speak of eating, however, they say *sik*, and if they transcribe their speech they write *sik*, 'eat,' with a character that we will call 'B.' But any educated Chinese, even if in his dialect the word for 'eat' is pronounced neither *chī* nor *sik*, easily understands characters 'A' and 'B,' knowing that they both have the meaning of 'eat' " (Alleton). Chinese writing is therefore an "Esperanto for the eyes," to use B. Karlgren's expression. "People who cannot communicate with each other by speech can easily understand each other when they write their respective languages in Chinese characters; this has always been regarded as one of the most remarkable features of Chinese writing" (Alleton). It is all the more important because it has undoubtedly contributed to the diffusion of Chinese culture in the Far East. Several neighboring countries—such as Korea, Japan, and Vietnam, to name only the best-known cases—adopted Chinese writing and have undergone its cultural influence for centuries, since it has been used not only for noting words of Chinese origin but also for transcribing the local language.

It is for this reason that the Japanese, whose writing has the outward appearance of Chinese writing but has a mixed structure adapted to the Japanese language, have always used the Chinese numerical signs in their various forms (*kǎishū*, cursive, etc.), though today they often use them concurrently with Hindu-Arabic numerals. In Japan, each of these signs has two different pronunciations: one, known as Sino-Japanese, is derived from Chinese pronunciation at the time when the Japanese borrowed the signs and their corresponding words, and the other is specifically Japanese (figs. 25–16, 25–17).

The Chinese numerical signs were also imported into ancient Vietnam, along with Chinese writing in general. They were adopted unchanged into the writing known as *chu nho* ("scholars' writing"), which used Chinese characters with Sino-Vietnamese pronunciations, that is, local pronunciations derived from an old Chinese dialect. Vietnamese

Example: 67,945

NOTATION		PRINCIPLE	PRONUNCIATION		
regular	cursive		Chinese	Sino-Japanese	Japanese
六	六	6 ×10,000 +	liù	roku	mutsu
萬	萬		wàn	man	yorozu
七	七	7 ×1000 +	qī	shichi	nanatsu
千	千		qiān	sen	chi
九	九	9 ×100 +	jiǔ	ku	kokonotsu
百	百		bǎi	hyaku	momo
四	四	4 ×10 +	sì	shi	yotsu
十	十		shí	ju	to
五	五	5	wǔ	go	itsu

Fig. 25–16.

Fig. 25–17. The present Japanese numeration.

	NUMERICAL SIGNS USED IN JAPAN			PRONUNCIATIONS	
	Regular numerals	Cursive numerals	Numerals used in commerce and ordinary calculations	Sino-Japanese	Japanese
1	一	一	╷	ichi	hitotsu
2	二	二	╷╷	ni	futatsu
3	三	三	╷╷╷	san	mitsu
4	四	四	ㄨ	shi	yotsu
5	五	五	ㄨ	go	itsu
6	六	六	亠	roku	mutsu
7	七	七	士	shichi	nanatsu
8	八	八	圭	hachi	yatsu
9	九	九	文	ku	kokonotsu
10	十	十	十	ju	to
100	百	百	ㄗ	hyaku	momo
1000	千	千	千	sen	chi
10,000	萬	萬	万	man	yorozu

scholars, however, remained faithful to the Chinese language, as well as Chinese writing, feeling that it was richer and more complete than their own language. But in common use, evidently beginning in the late thirteenth century A.D., Chinese numerical signs were adapted to the Vietnamese language.

As can be seen from figure 25–18, the signs that resulted were different, at first sight, from their Chinese equivalents. Since the seventeenth century the Vietnamese language has been written with letters of the Roman alphabet, but it was formerly transcribed "in another writing, *chu nom,* composed on the basis of the elements and principles of Chinese writing. Most *chu nom* characters consist of an element of a Chinese

没 仝 它 罪 觧 赳 罷 糁 尬 迹 彔 斦 閧

mot	hay	ba	bon	nam	saou	bai	tam	chin	müoey	tram	ngan	mouon
1	2	3	4	5	6	7	8	9	10	100	1000	10,000

Fig. 25–18. The Vietnamese *chu nom* numerical signs. They were used in the kingdom of Annam, probably beginning in the late thirteenth century A.D., in correspondence, contracts, administrative documents, and popular literary works. In their written numeration, the Vietnamese used the additive and multiplicative principles in the same way as the Chinese:

Example: 518

觧	nam	5
		×
彔	tram	100
		+
没	mot	1
		×
迹	müoey	10
		+
糁	tam	8

character (or the whole character) taken for its meaning, and another element (or whole character) chosen to indicate the pronunciation of the Vietnamese word being transcribed" (Alleton). With this in mind, we can now understand the structure of the complex signs that the Vietnamese devised, in their *chu nom* system, to represent numbers 1 to 9 and the first four powers of 10 (fig. 25–18). Each of them is composed of a Chinese numerical sign and another character, also of Chinese origin. The first of these two elements is the ideographic or symbolic part of the combination and the second indicates the Vietnamese pronunciation of the corresponding number word, using the procedure of phonetic borrowing. This adaptation of the Chinese numerical signs to the Vietnamese language can be seen clearly in the table below.

	2	3	4	6	7	9	10	100	1000	10,000
Chinese signs	二	三	四	六	七	九	十	百	千	万
Chu nom signs	 仝	它	罪	赳	罷	尬	迹	彔	斦	閧

Systems similar to the common Chinese system

We will now consider several systems that are hybrid, like the common Chinese system, but were obviously developed without any Chinese influence. This will enable us to distinguish the similarities and differences among hybrid systems and classify them according to types.

One type of hybrid system is exemplified by the Assyro-Babylonian and Elephantine Aramaic numerations. It uses the base 10 and has special signs for numbers 1, 10, 100, 1000, and 10,000. (The Elephantine Aramaic numeration might seem to have a special sign for 20, but, as we have seen [fig. 24–1], that sign was formed by doubling the sign for 10. In the common Assyro-Babylonian numeration, 1000 is the largest number represented by a special sign.) Multiples of 100, 1000, and 10,000 are expressed according to the multiplicative principle, by placing the signs for those numbers next to the signs for the corresponding units (figs. 23–1, 24–2). Mathematically speaking, however, this system is not identical with the Chinese numeration, since it represents numbers 1 to 99 according to the additive principle, by repeating the signs for 1 and 10 as often as necessary.

Example: 6657

It is thus a decimal system whose basic signs correspond to numbers 1, 10, 10^2, 10^3, and 10^4, and in which the units of these consecutive orders are expressed according to these principles:

Units	Tens	Hundreds	Thousands	Ten thousands
1	10	1×10^2	1×10^3	1×10^4
$1+1$	$10+10$	$(1+1) \times 10^2$	$(1+1) \times 10^3$	$(1+1) \times 10^4$
$1+1+1$	$10+10+10$	$(1+1+1) \times 10^2$	$(1+1+1) \times 10^3$	$(1+1+1) \times 10^4$
...	...	...	...	...

Another type of numeration similar to the Chinese system was once used by the Singhalese-speaking people of the island of Ceylon. It is a decimal numeration, with special signs for numbers 1 to 9, tens, 100, and

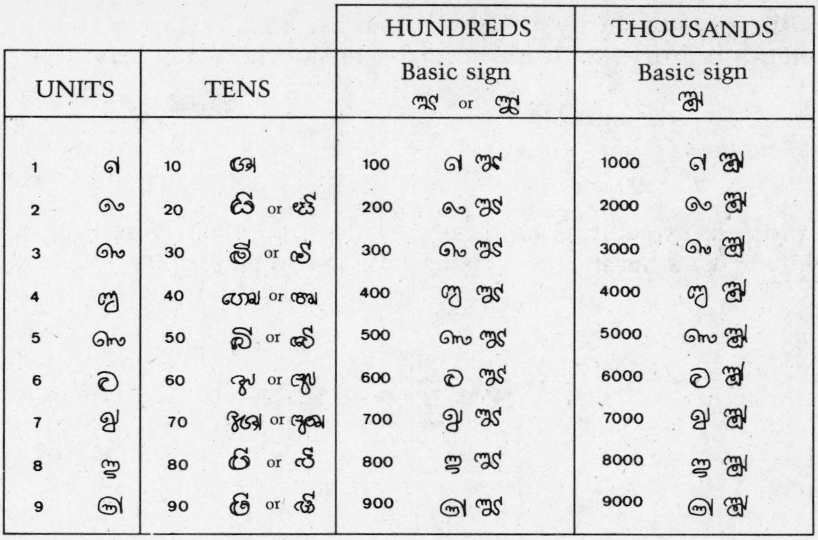

UNITS		TENS		HUNDREDS Basic sign ඏ or ඏ		THOUSANDS Basic sign ඏ	
1	ෙ	10	ෙෂ	100	ෙ ඏ	1000	ෙ ඏ
2	ෙෙ	20	ඬ or ඬ	200	ෙෙ ඏ	2000	ෙෙ ඏ
3	ෙෙ	30	ඬ or ඬ	300	ෙෙ ඏ	3000	ෙෙ ඏ
4	ඏ	40	ෙෙෂ or ෙෙෂ	400	ඏ ඏ	4000	ඏ ඏ
5	ෙෙ	50	ඬ or ඬ	500	ෙෙ ඏ	5000	ෙෙ ඏ
6	ෙ	60	ඬ or ෙඬ	600	ෙ ඏ	6000	ෙ ඏ
7	ඪ	70	ෙෙෂ or ෙෙෂ	700	ඪ ඏ	7000	ඪ ඏ
8	ඏ	80	ඬ or ඬ	800	ඏ ඏ	8000	ඏ ඏ
9	ෙෙ	90	ඬ or ඬ	900	ෙෙ ඏ	9000	ෙෙ ඏ

Fig. 25–19. The Singhalese numeration.

1000, in which multiples of 100 and 1000 are indicated according to the multiplicative principle (fig. 25–19). That is, its basic signs are associated with these numbers:

$$1 \quad 2 \quad 3 \quad 4 \quad 5 \quad 6 \quad 7 \quad 8 \quad 9$$
$$10 \quad 20 \quad 30 \quad 40 \quad 50 \quad 60 \quad 70 \quad 80 \quad 90$$
$$10^2 \text{ and } 10^3$$

and hundreds and thousands are expressed in these forms:

$$1 \times 10^2 \qquad 1 \times 10^3$$
$$2 \times 10^2 \qquad 2 \times 10^3$$
$$3 \times 10^2 \qquad 3 \times 10^3$$
$$4 \times 10^2 \qquad 4 \times 10^4$$
$$\cdots \qquad \cdots$$

Other numbers are expressed according to both the additive and multiplicative principles:

Example: 6657

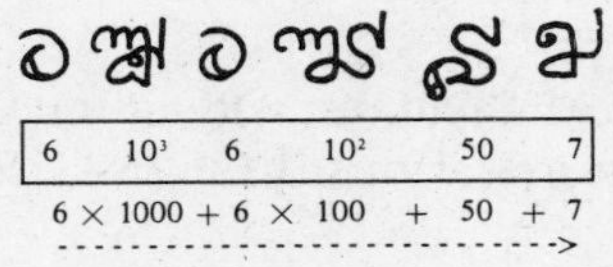

6	10^3	6	10^2	50	7

$$6 \times 1000 + 6 \times 100 \ + \ 50 \ + \ 7$$
- ->

There is also the hybrid system that has been used in Ethiopia since the fourth century A.D. It has special signs for these numbers:

$$1 \quad 2 \quad 3 \quad 4 \quad 5 \quad 6 \quad 7 \quad 8 \quad 9$$

$$10 \quad 20 \quad 30 \quad 40 \quad 50 \quad 60 \quad 70 \quad 80 \quad 90$$

$$100 \text{ and } 10{,}000$$

and proceeds by addition for numbers below 100, then by multiplication for hundreds, thousands, ten thousands, and so on (fig. 24–5):

Example: 6657

60	10²	6	10²	50	7

$$60 \times 100 + 6 \times 100 + 50 + 7$$
-------------------->

At first sight the Ethiopian system seems essentially the same as the Singhalese system, but there is one difference: the Ethiopian system is a numeration of base 100, rather than base 10, since hundreds, thousands, ten thousands, and hundred thousands are expressed in these forms:

$$1 \times 100 \quad 10 \times 100 \quad 1 \times 100^2 \quad 10 \times 100^2$$
$$2 \times 100 \quad 20 \times 100 \quad 2 \times 100^2 \quad 20 \times 100^2$$
$$3 \times 100 \quad 30 \times 100 \quad 3 \times 100^2 \quad 30 \times 100^2$$
$$4 \times 100 \quad 40 \times 100 \quad 4 \times 100^2 \quad 40 \times 100^2$$
$$\cdots \qquad\quad \cdots \qquad\quad \cdots \qquad\quad \cdots$$
$$9 \times 100 \quad 90 \times 100 \quad 9 \times 100^2 \quad 90 \times 100^2$$

Finally, we will consider the system still used by the Tamils of southern India and the system used till the nineteenth century by the Malayalam-speaking people of the Malabar Coast region in southwestern India. Tamil and Malayalam are both Dravidian languages.

Both these systems are of base 10 and have special signs for numbers 1 to 9 and the first three powers of 10. Multiples of 10, 100, and 1000 are expressed according to the multiplicative principle, by placing signs for the corresponding units to the left of the signs for 10, 100, and 1000 (fig. 25–20).

The Tamil, Malayalam, and Chinese systems all belong to the same type of hybrid numerations (figs. 25–2, 25–16).

Example: 6657

TAMIL SYSTEM | MALAYALAM SYSTEM

Example: 6657

CHINESE SYSTEM

6	10³	6	10²	5	10	7		6	10³	6	10²	5	10	7

$6 \times 1000 + 6 \times 100 + 5 \times 10 + 7$ $6 \times 1000 + 6 \times 100 + 5 \times 10 + 7$

ARCHAIC		MODERN	
	10³	1000 ×	六	6 ×
	6	6 +	千	10³	1000 +
	6	6 ×	六	6 ×
	10²	100 +	百	10²	100 +
	10	10 ×	五	5 ×
	5	5 +	十	10 +
	7	7	七	7	7

	Tamil system	Malayalam system			Tamil system	Malayalam system
UNITS 1	声	⏀	**HUNDREDS** 100	m	ᬭ
2	உ	ᬓ	200	உm	ᬓᬭ
3	ᬫ	ᬬ	300	ᬫm	ᬫᬭ
4	ᬆ	ᬯ	400	ᬆm	ᬆᬭ
5	ᬎ	ᬖ	500	ᬎm	ᬖᬭ
6	ᬝ	ᬭ	600	ᬝm	ᬭᬭ
7	ᬏ	ᬖ	700	ᬏm	ᬖᬭ
8	ᬛ	ᬗ	800	ᬛm	ᬗᬭ
9	ᬝ	ᬮ	900	ᬝm	ᬮ

	Tamil system	Malayalam system			Tamil system	Malayalam system
TENS 10	ᬯ	ᬮ	**THOUSANDS** 1000	ᬘ	ᬘᬭ
20	உᬯ	ᬮ	2000	உᬘ	ᬓᬘᬭ
30	ᬫᬯ	ᬮᬯ	3000	ᬫᬘ	ᬫᬘᬭ
40	ᬆᬯ	ᬯ	4000	ᬆᬘ	ᬯᬘᬭ
50	ᬎᬯ	ᬚ	5000	ᬎᬘ	ᬖᬘᬭ
60	ᬝᬯ	ᬮᬯ	6000	ᬝᬘ	ᬫᬘᬭ
70	ᬏᬯ	ᬮᬯ	7000	ᬏᬘ	ᬖᬘᬭ
80	ᬛᬯ	ᬮᬯ	8000	ᬛᬘ	ᬗᬘᬭ
90	ᬝᬯ	ᬮᬯ	9000	ᬝᬘ	ᬮᬘᬭ

Fig. 25–20. The Tamil and Malayalam numerations, southern India.

Hybrid numerations can be divided into four types (fig. 25–21) whose mathematical characteristics will be described below.

FIRST TYPE. Systems of base m.

Basic signs: $1, m, m^2, m^3, m^4 \ldots$

Expression of the units of these consecutive orders:

1st order: 1 $1 + 1$ $1 + 1 + 1$ $1 + 1 + 1 + 1 \ldots$
2nd order: m $m + m$ $m + m + m$ $m + m + m + m \ldots$

3rd order: $1 \times m^2$ $(1 + 1) \times m^2$ $(1 + 1 + 1) \times m^2$ $(1 + 1 + 1 + 1) \times m^2 \ldots$
4th order: $1 \times m^3$ $(1 + 1) \times m^3$ $(1 + 1 + 1) \times m^3$ $(1 + 1 + 1 + 1) \times m^3 \ldots$
5th order: $1 \times m^4$ $(1 + 1) \times m^4$ $(1 + 1 + 1) \times m^4$ $(1 + 1 + 1 + 1) \times m^4 \ldots$

and so on to the last numerical sign available in practice.

SECOND TYPE. Systems of base m.

Basic signs: $1, 2, 3, \ldots m - 1$ (units of the 1st order)
$m, 2m, 3m, \ldots (m - 1)m$ (units of the 2nd order)
$m^2, m^3, m^4, \ldots$

Expression of units of orders equal to or greater than the square of the base:

3rd order: $1 \times m^2$ $2 \times m^2$ $3 \times m^2$ $\ldots$ $(m - 1) \times m^2$
4th order: $1 \times m^3$ $2 \times m^3$ $3 \times m^3$ $\ldots$ $(m - 1) \times m^3$
5th order: $1 \times m^4$ $2 \times m^4$ $3 \times m^4$ $\ldots$ $(m - 1) \times m^4$, etc.

THIRD TYPE. Systems of base m^2.

Basic signs: $1, 2, 3, \ldots m - 1, m, 2m, 3m, \ldots (m - 1)m$
m^2 (base)
m^4 (square of the base), etc.

Expression of units of orders equal to or higher than m^2:

2nd order (multiples of m^2 and m^3):
$1 \times m^2$ $2 \times m^2$ $\ldots$ $(m - 1) \times m^2$ $m \times m^2$ $2m \times m^2$ $\ldots$ $(m - 1)m \times m^2$

3rd order (multiples of m^4 and m^5):
$1 \times m^4$ $2 \times m^4$ $\ldots$ $(m - 1) \times m^4$ $m \times m^4$ $2m \times m^4$ $\ldots$ $(m - 1)m \times m^4$

4th order (multiples of m^6 and m^7):
$1 \times m^6$ $2 \times m^6$ $\ldots$ $(m - 1) \times m^6$ $m \times m^6$ $2m \times m^6$ $\ldots$ $(m - 1)m \times m^6$

and so on.

FOURTH TYPE. Systems of base m.

Basic signs: $1, 2, 3, \ldots m - 1$ (1st order)
m, m^2, m^3, m^4, etc.

Expression of units of orders equal to or greater than the base:

2nd order: $1 \times m$ $2 \times m$ $3 \times m$ $\ldots$ $(m - 1) \times m$
3rd order: $1 \times m^2$ $2 \times m^2$ $3 \times m^2$ $\ldots$ $(m - 1) \times m^2$
4th order: $1 \times m^3$ $2 \times m^3$ $3 \times m^3$ $\ldots$ $(m - 1) \times m^3$, etc.

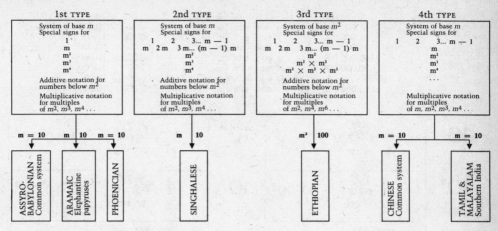

Fig. 25–21. Types of hybrid numerations.

All the hybrid written numerations that we have just considered have one thing in common: each of them is a faithful transcription of the corresponding oral numeration.

Example: 6657.

Akkadian:	*šeššu*	*limi*	*šeššu*	*me'at*	*hamšâ*	*sîbu*	
	6	× 1000 +	6	× 100 +	50	+	7

Ethiopian:	*sassá*	*ma'át*	*sadastŭ*	*ma'át*	*hamsá*	*wa*	*sab'atŭ*
(Geez)	60	× 100 +	6	× 100 +	50	&	7

Chinese:	*liù*	*qiān*	*liù*	*băi*	*wŭ*	*shí*	*qī*
	6	× 1000 +	6	× 100 +	5	× 10 +	7

A natural extension

To express large numbers, the Chinese and the Japanese seldom need any other numerical signs than the thirteen basic characters of their ordinary numeration because, using only those characters, they can indicate numbers up to at least 100,000,000,000 (10^{11}).

This method (usually limited to numbers below 10^8) is a simple extension of their ordinary numeration: it consists in taking 10,000 as a new counting unit. Figure 25–22 shows how it is used for representing powers of 10 from 10^4 to 10^{11}.

10,000	:	yī wàn (=	$1 \times 10{,}000$)
100,000	:	shí wàn (=	$10 \times 10{,}000$)
1,000,000	:	yī bǎi wàn (=	$1 \times 100 \times 10{,}000$)
10,000,000	:	yī qiān wàn (=	$1 \times 1000 \times 10{,}000$)
100,000,000	:	yī wàn wàn (=	$1 \times 10{,}000 \times 10{,}000$)
1,000,000,000	:	shí wàn wàn (=	$10 \times 10{,}000 \times 10{,}000$)
10,000,000,000	:	yī bǎi wàn wàn (=	$1 \times 100 \times 10{,}000 \times 10{,}000$)
100,000,000,000	:	yī qiàn wàn wàn (=	$1 \times 1000 \times 10{,}000 \times 10{,}000$)

10^4	一萬 yī wàn	$1 \cdot 10^4$	10^8	一萬萬 yī wàn wàn	$1 \cdot 10^4 \cdot 10^4$
10^5	十萬 shí wàn	$10 \cdot 10^4$	10^9	十萬萬 shí wàn wàn	$10 \cdot 10^4 \cdot 10^4$
10^6	一百萬 yī bǎi wàn	$1 \cdot 10^2 \cdot 10^4$	10^{10}	一百萬萬 yī bǎi wàn wàn	$1 \cdot 10^2 \cdot 10^4 \cdot 10^4$
10^7	一千萬 yī qiān wàn	$1 \cdot 10^3 \cdot 10^4$	10^{11}	一千萬萬 yī qiān wàn wàn	$1 \cdot 10^3 \cdot 10^4 \cdot 10^4$

Fig. 25–22. The ordinary Chinese notation for powers of 10.

Large numbers not multiples of the base, such as 487,390,629, for example, are expressed in this way:

四萬八千七百三十九萬六百二十九

sì wàn bā qiān qī bǎi sān shí jiǔ wàn liù bǎi èr shí jiǔ

$(4 \times 10^4 + 8 \times 10^3 + 7 \times 10^2 + 3 \times 10 + 9) \times 10^4 + 6 \times 10^2 + 2 \times 10 + 9$

That is, the number is broken down as follows:

$(4 \times 10{,}000 + 8 \times 1000 + 7 \times 100 + 3 \times 10 + 9) \times 10{,}000 + 6 \times 100 + 2 \times 10 + 9$

Or:

$$48{,}739 \times 10{,}000 + 629$$

This extension is evidently quite natural, since most other hybrid numerations have developed something like it.

In Singhalese and Tamil writings (figs. 25–19, 25–20) the number 353,549 is expressed in these forms:

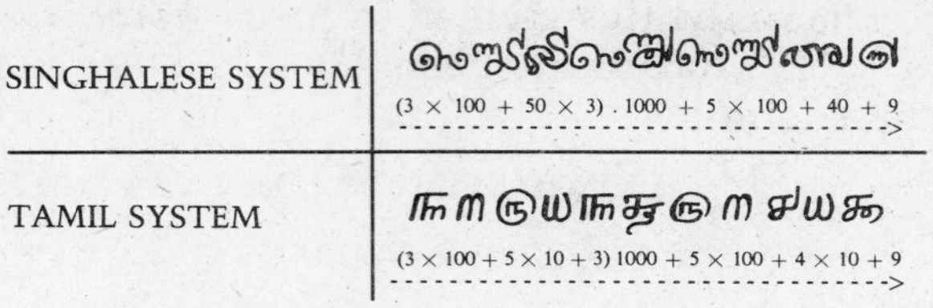

SINGHALESE SYSTEM	⟨Singhalese numerals⟩
	$(3 \times 100 + 50 \times 3) . 1000 + 5 \times 100 + 40 + 9$
TAMIL SYSTEM	⟨Tamil numerals⟩
	$(3 \times 100 + 5 \times 10 + 3) 1000 + 5 \times 100 + 4 \times 10 + 9$

In one of the Assyrian cuneiform tablets describing King Sargon II's eighth campaign in Armenia (714 B.C.) the number 305,412 is expressed in this way (fig. 23–1):

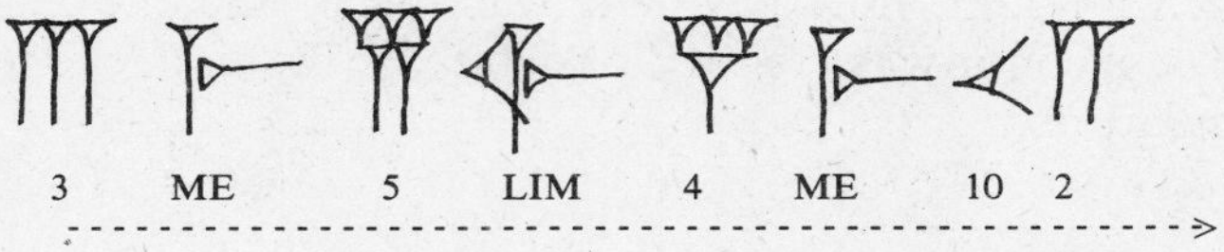

| 3 | ME | 5 | LIM | 4 | ME | 10 | 2 |

That is, in this form: $(3 \times 100 + 5) \times 1000 + 4 \times 100 + 10 + 2$.

Some written numerations based only on the additive principle have been extended to large numbers by introducing the multiplicative principle.

One example of this is an adaptation of the Roman numeration that was used occasionally in the early Christian era, and often in the Middle Ages. Instead of repeating the letters C and M to express multiples of 100 and 1000, as was done before, scribes and copyists first wrote the number of hundreds or thousands they wanted to indicate, then wrote the letter C or M in the position of a coefficient, or as a superscript (fig. 25–23).

100	C	1000	M	10,000	X^M
200	II.C or II^c	2000	II^M	15,000	XV^M
300	III.C or III^c	3000	III^M	30,000	XXX^M
400	IIII.C or IIII^c	4000	IIII^M	50,000	L^M
500	V. C or V^c	5000	V^M	100,000	C^M
600	VI. C or VI^c	6000	VI^M	300,000	CCC^M
700	VII.C or VII^c	7000	VII^M	500,000	D^M
800	VIII.C or VIII^c	8000	VIII^M	602,000	$DCLII^M$
900	IX.C or VIIII^c	9000	VIIII^M	1,000,000	M^M

Fig. 25–23. Late adaptation of the Latin numeration for indicating multiples of 100 and 1000.

Pliny the Elder used this method in his *Historia Naturalis* in the first century A.D. He wrote the numbers 83,000, 92,000, and 110,000, for example, in these forms:

LXXXIII.M for 83,000

XCII.M for 92,000

CX.M for 110,000

The register of King Philip the Fair's treasury, written in Latin and dated 1299, is among the oldest French treasury account books that have come down to us. "The king's treasurers recorded receipts and expenditures from day to day, indicating the name of the person who paid or received each amount, how payment was made, and the name of the person whose account was credited or debited" (Prou). In the register we find these notations:

col. I
1. 11

IIII^m 1(ibras). t(uronensium).

4000 livres tournois
(francs minted at Tours)

col. 1
1. 22

V^m. III^c.XVI. 1(ibras).VI.s(olidos). I. d(enarios). p(arisiensium).

5316 livres (francs), 6 sols (sous), & 1 denier parisis
(denier minted at Paris)

col. 2
1. 32

II^c.XLIIII.1(ibras).XII.s(olidos). t(uronensium).

244 livres (francs), 12 sols tournois
(sous minted at Tours)

A similar idea was exploited 3000 years earlier by Egyptian scribes who, to express large numbers, extended their hieratic notation by using the multiplicative principle (figs. 15–6, 22–14). In the Harris Papyrus (New Kingdom), the number 494,800 is represented as shown in this form:

$$800 + 4000 + \overset{10,000}{\underset{9}{\times}} + \overset{100,000}{\underset{4}{\times}}$$

And finally, the Greeks and the Arabs used the same kind of extension:

Example: 353,549

GREEK ALPHABETIC SYSTEM (See figs. 17-2 and 17-4.)	λε M 'Γ Φ M Θ 30 + 5 × + 3000 + 500 + 40 + 9 10,000 ···>
ARABIC "ABJAD" SYSTEM (See figs. 20–4 and 24–6.)	شنجغ ثمط <·· 9 + 40 + 500 + 1000. (3 + 50 + 300)

Chinese notations of large numbers in scientific works

The Chinese system already described is almost invariably used in common practice, but sometimes in works on scientific subjects, particularly astronomy, special Chinese characters are used for expressing numbers greater than 10^4. Each of these characters has a different numerical value in each of three different systems: *xià deng* ("lower degree"), *zhōng deng* ("middle degree"), and *shàng deng* ("upper degree"). The character 兆, for example, has the value of one million (10^6) in the first of these systems, one trillion (10^{12}) in the second, and ten quadrillion (10^{16}) in the third (fig. 25–24).

The lower-degree system is a direct extension of the ordinary numeration, since the values it assigns to the ten special characters are the powers of 10 from 10^5 (one hundred thousand) to 10^{14} (one hundred trillion). That is, the characters:

yì, zháo, jing, gai, . . . zheng, zài

have the values:

$10^5, 10^6, 10^7, 10^8, . . . 10^{13}, 10^{14}$

One million, 2,000,000, and 3,000,000, for example, are noted as follows in the lower-degree system:

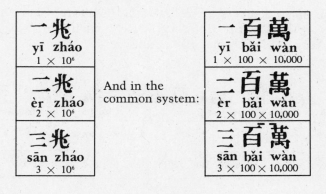

一兆 yī zháo 1×10^6	一百萬 yī bǎi wàn $1 \times 100 \times 10,000$
二兆 èr zháo 2×10^6	二百萬 èr bǎi wàn $2 \times 100 \times 10,000$
三兆 sān zháo 3×10^6	三百萬 sān bǎi wàn $3 \times 100 \times 10,000$

And in the common system:

Any number below 10^{15} can be represented without difficulty. As an illustration, here is the number 530,010,702,000,000 in this system.

五 載 三 正 一 壤 七 垓 二 兆
wǔ zái sān zheng yī ràng qī gai èr zháo

$$5 \times 10^{14} + 3 \times 10^{13} + 1 \times 10^{10} + 7 \times 10^{8} + 2 \times 10^{6}$$

In the middle-degree system, the ten special characters also correspond to powers of 10 above 10^4, but whereas in the first system the values are multiplied by 10 in going from one character to the next, in this one they are multiplied by 10,000, so that the characters:

yì, zháo, jing, . . . zheng, zài

have these values:

$10^8, 10^{12}, 10^{16}, . . . 10^{40}, 10^{44}$

Proceeding as in ordinary use, but observing the rule of never having two special numerical signs juxtaposed, any number below 10^{48} can be expressed. Example:

三 百 五 十 壤 七 千 三 百 兆 二 十 六 億
sān bǎi wǔ shí ràng qī qiān sān bǎi zháo èr shí liù yì

$$(3 \times 10^2 + 5 \times 10).10^{28} + (7 \times 10^3 + 3 \times 10^2).10^{12} + (2 \times 10 + 6).10^8$$

3,500,000,000,000,007,300,002,600,000,000

In the upper-degree system, only the first three of the ten special characters are used, with values of 10^8, 10^{16}, and 10^{32}, making it possible to express any number below 10^{64}. Example:

三京 五千三百一億二百七萬六千一百八十五兆 三億一萬
sān jing wǔ qiān sān bǎi yī yì èr bǎi qī wàn liù qiān yī bǎi bā shí wǔ zháo sān yì yī wàn

$$3 \times 10^{32} + [5 \times 10^3 + 3 \times 10^2 + 1].10^8 + [2 \times 10^2 + 7].10^4 + 6 \times 10^3 + 1 \times 10^2 + 8 \times 10 + 5] 10^{16} + 3 \times 10^8 + 1 \times 10^4$$

300,005,301,020,761,850,000,000,300,010,000

Use of such large numbers is quite infrequent, of course, but these systems are one more illustration of the high intellectual attainments of the Chinese.

		Xià deng system (lower degree)	Zhōng deng system (middle degree)	Shàng deng system (upper degree)
萬	wàn	10^4	10^4	10^4
億	yì ★	10^5	10^8	10^8
兆	zháo	10^6	10^{12}	10^{16}
京	jing	10^7	10^{16}	10^{32}
垓	gai	10^8	10^{20}	10^{64}
補	bù ★★	10^9	10^{24}	10^{128}
壤	ràng	10^{10}	10^{28}	10^{256}
冓	gou ★★★	10^{11}	10^{32}	10^{512}
澗	jián	10^{12}	10^{36}	10^{1024}
正	zheng	10^{13}	10^{40}	10^{2048}
載	zái	10^{14}	10^{44}	10^{4096}

THEORETICAL VALUES

★ Graphic variant: 亿 ★★ Equivalent word: 秭 cè
★★★ Graphic variant: 溝

Fig. 25–24. Chinese characters used for expressing large numbers.

PART VI

The Ultimate Stage of Numerical Notation

26

The First Place-Value Numeration System

Just as in alphabetic writing all the words of a language can be recorded with a limited number of graphic signs called letters, all whole numbers can be represented with the ten digits of our modern written numeration. From an intellectual standpoint, then, the numerical system that we use in our everyday lives is far superior to any of the other systems we have considered. But this is not because of the number chosen as its base (that is, the number of units required to form a unit of the next higher order), for we could use some other base, such as 2, 8, 12, 20, or 60, and obtain a perfectly rational system of representing the whole numbers with the same advantages as our decimal system. It was only an "anatomical accident" that gave most of the world's peoples the idea of counting by tens, and therefore by hundreds, thousands, etc.

The superiority of our written numeration comes from the principle that each digit has a variable value that depends on how it is placed: a given digit is associated with units, tens, hundreds, or thousands, according to whether it occupies the first, second, third, or fourth position (going from right to left) in the written expression of a number. This is the principle that makes arithmetical operations very easy to perform, comparatively speaking, in our system and those like it.

At a time that unfortunately we cannot specify exactly, though it was most likely at the beginning of the second millennium B.C., the idea of a place-value notation was conceived for the first time by the mathematicians and astronomers of Babylon.

They developed their system on the basis of the old Sumerian sexagesimal numeration, but it differs markedly from the common Assyro-Babylonian system and is superior to any of the other written numerations used in the ancient world.

It is strictly positional and has a sexagesimal base (that is, 60 units of a given order are represented by one unit of the next higher order), so that a group of signs such as 312, for example, which in our decimal place-value system corresponds to the number:

$$3 \times 10^2 + 1 \times 10 + 2 = 3 \times 100 + 10 + 2$$

corresponds in the Mesopotamian sexagesimal place-value system to:

$$3 \times 60^2 + 1 \times 60 + 2 = 3 \times 3600 + 60 + 2$$

And 1111, which in our system is:

$$1 \times 10^3 + 1 \times 10^2 + 1 \times 10 + 1 = 1000 + 100 + 10 + 1$$

is, in the Mesopotamian system:

$$1 \times 60^3 + 1 \times 60^2 + 1 \times 60 + 1 = 216{,}000 + 3600 + 60 + 1$$

Examples of this numeration were discovered early in the history of Assyriology: by Hincks in 1854, on an astronomical tablet found during excavations at Nineveh, and by Rawlinson in 1855, on a mathematical tablet found at Larsa. Since then, the existence of the numeration has been confirmed by several other scientific documents—whose deciphering and interpretation owe a great deal to the contributions of François Thureau-Dangin and Otto Neugebauer—found in various parts of Mes-

FRONT   BACK

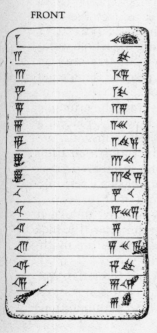
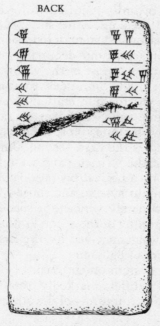

	TRANSCRIPTION		TRANSLATION (decimal place-value system)	
FRONT	1	25	1	25
	2	50	2	50
	3	1;15	3	75
	4	1;40	4	100
	5	2;05	5	125
	6	2;30	6	150
	7	2;55	7	175
	8	3;20	8	200
	9	3;45	9	225
	10	4;10	10	250
	11	4;35	11	275
	12	5;	12	300
	13	5;25	13	325
	14	5;50	14	350
	15	6;15	15	375
	16	6;40	16	400
BACK	17	7;05	17	425
	18	7;30	18	450
	19	7;45 ★	19	465 ★
	20	8;20	20	500
	30	12;30	30	750
	40	16;40	40	1 000
	50	20;50	50	1 250

*Error by the scribe

Fig. 26–1. Tablet from Susa, first half of the second millennium B.C., showing a table of multiplication by 25, with the numbers expressed in the place-value notation. It is typical of the multiplication tables used in Babylon and Susa: they usually gave the results of multiplying a given number n (below 60) by each of the first 20 whole numbers, as well as 30, 40, and 50. This made it easy to obtain the product of n and any number between 1 and 60.

Since the Babylonian system is sexagesimal, the transcriptions in Hindu-Arabic numerals can be readily understood if they are regarded as indicating hours and minutes. For example, 3;20 = 3 hours 20 minutes = 200 minutes.

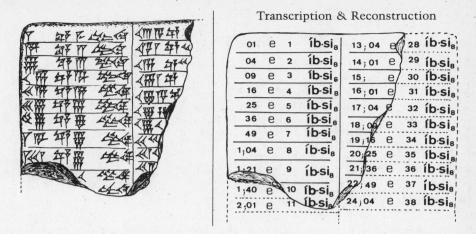

Transcription & Reconstruction

01	e	1	íb·si₈	13;04	e	28	íb·si₈
04	e	2	íb·si₈	14;01	e	29	íb·si₈
09	e	3	íb·si₆	15;	e	30	íb·si₈
16	e	4	íb·si₈	16;01	e	31	íb·si₈
25	e	5	íb·si₈	17;04	e	32	íb·si₈
36	e	6	íb·si₈	18;09	e	33	íb·si₈
49	e	7	íb·si₈	19;16	e	34	íb·si₈
1;04	e	8	íb·si₈	20;25	e	35	íb·si₈
1;21	e	9	íb·si₈	21;36	e	36	íb·si₈
1;40	e	10	íb·si₈	22;49	e	37	íb·si₈
2;01	e	11	íb·si₈	24;04	e	38	íb·si₈

Fig. 26–2. Fragment of a table of square roots, with the numbers expressed in the Babylonian place-value notation. It dates from about 1800 B.C. and was found in the ruins of Nippur (about 100 miles southeast of Babylon).

opotamia and dating from the first Babylonian dynasty to the time of the Seleucids. These documents are of several types (figs. 26–1, 26–2): tables for facilitating calculation (tables of multiplication, division, reciprocals, squares, square roots, cubes, cube roots, etc.); astronomical tables; collections of exercises in practical arithmetic or elementary geometry; lists of more or less complex mathematical problems (surveying land, solving algebraic equations, calculating areas, etc.).

There are also the mathematical tablets, found at Susa (in Elam) and dating from the end of the first Babylonian dynasty, whose publication, translation, and interpretation we owe to E. M. Bruins and M. Rutten. Because of their mathematical content, and particularly the numerical notation used in them, these tablets provide "confirmation, for a new geographical area, of an outstanding mathematical advance that had already been proven elsewhere" and constitute "still more evidence of the influence that Babylon exercised beyond its borders" (Bruins and Rutten).

The Babylonian place-value system

The first thing to be pointed out is that unlike our decimal place-value system, which uses nine distinct signs to represent numbers 1 to 9, the Babylonian sexagesimal place-value system has only two numerical signs, strictly speaking, and not fifty-nine, as one might expect. It expresses the fifty-nine units of the first order with combinations of two cuneiform signs: a vertical wedge for 1 and a crescent for 10 (fig. 26–3).

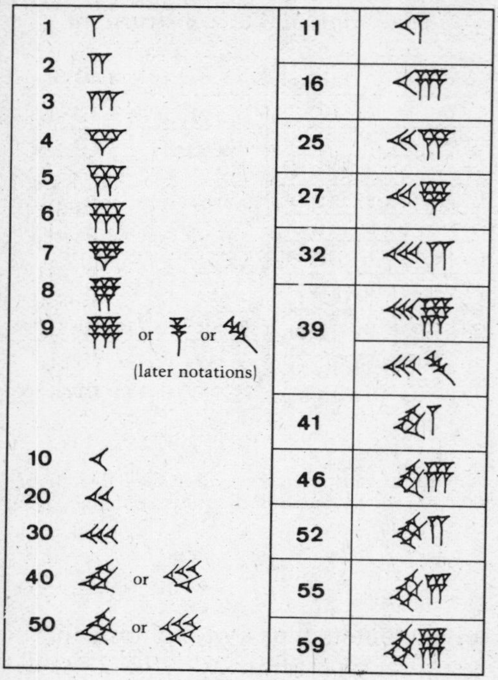

1	𒁹	11	
2		16	
3			
4		25	
5		27	
6			
7		32	
8			
9	or ... or ... (later notations)	39	
		41	
10		46	
20		52	
30		55	
40	or	59	
50	or		

Fig. 26–3. Representation of the fifty-nine units of the first sexagesimal order in the Babylonian place-value system.

Thus numbers below 60 are represented with a decimal base and the additive principle. But for numbers above 60 the place-value principle is used. The number 60, for example, is written not as:

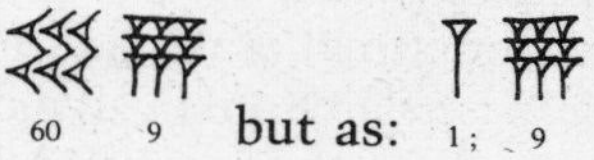

$$\underset{60 \quad 9}{\text{𒐏𒐖}} \quad \text{but as:} \quad \underset{1; \quad 9}{\text{𒁹𒐖}}$$

(From here on, when numbers expressed in the Babylonian sexagesimal place-value system are transcribed with our Hindu-Arabic numerals, a semicolon will be used to separate the different orders of units, as in the right-hand notation above.)

The second notation symbolizes $1 \times 60 + 9$ (and is analogous to our representation of sixty-nine seconds as $1'9''$), and this notation:

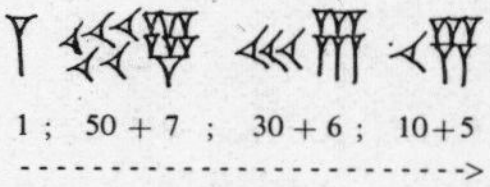

$$\underset{1; \quad 50+7 \ ; \quad 30+6 \ ; \quad 10+5}{\text{𒁹 ... ...}} \quad \text{or } 1; \ 57; \ 36; \ 15$$

symbolizes $1 \times 60^3 + 57 \times 60^2 + 36 \times 60 + 15 \ (= 423{,}375)$.

The examples at the top of the next page are from one of the oldest known Babylonian mathematical texts, a tablet containing problems that require solving second-degree equations:

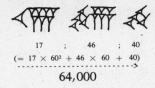

17 ; 46 ; 40

(= 17 × 60² + 46 × 60 + 40)

-------------------------------->

64,000

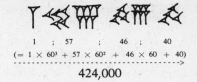

1 ; 57 ; 46 ; 40

(= 1 × 60³ + 57 × 60² + 46 × 60 + 40)

-------------------------------->

424,000

The difference between the Babylonian place-value system and the Sumerian system, based on the additive principle, can be seen in these examples, showing how each of them indicates the numbers 1859 and 4818.

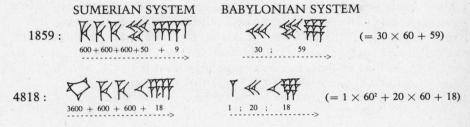

SUMERIAN SYSTEM   BABYLONIAN SYSTEM

1859 :   600 + 600 + 600 + 50 + 9   30 ; 59   (= 30 × 60 + 59)

-------------------------->   ------------------->

4818 :   3600 + 600 + 600 + 18   1 ; 20 ; 18   (= 1 × 60² + 20 × 60 + 18)

-------------------------->   ------------------->

Difficulties of this system

As we have seen, the Babylonian sexagesimal place-value system was decimal and additive within each order of units. As a result, it had ambiguities that often caused errors.

In a mathematical tablet from Susa, the number 10; 15 = 10 × 60 + 15 was noted as:

10 ; 15

-------->

which can easily be confused with:

25   10 ; 10 ; 5   (= 10 × 60² + 10 × 60 + 5)

-------->   and   -------->

Probably because they were aware of this difficulty, scribes sometimes left a blank space to indicate transition from one sexagesimal order to the next. In the same tablet from which the example above is taken, the scribe seems to have overcome the difficulty by noting the number 10; 10 = 10 × 60 + 10 in the form:

10 ; 10

-------->

that is, by distinctly separating the two crescents for 10, thus eliminating the possibility of confusion with the notation for 20.

In another mathematical text from Susa, the number 1; 1; 12 = 1 × 60² + 1 × 60 + 12 is noted in this form:

1 ; 1 ; 12

-------->

Here, the blank space clearly distinguishes this notation from:

$$\text{𝖳𝖳 ⟨𝖳𝖳} \quad (= 2 \times 60 + 12).$$

2 ; 12

Sometimes, instead of a blank space, one of the following signs was used to avoid confusion:

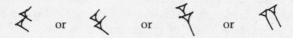

These signs were also used to indicate separation in literary as well as scientific texts. In commentaries on literary texts, a separation sign was sometimes used to distinguish words from their explanations. In bilingual or trilingual texts, it served to indicate passage from one language to another. In lists of omens, it was regularly used between two formulations, and as a sign indicating the beginning of a sentence.

Examples from a mathematical tablet found at Susa:

1.
1;10 ; | 18 ; 45
separation sign

$$(= 1 \times 60^3 + 10 \times 60^2 + 18 \times 60 + 45)$$

2.
20 ; | 3 ; 13 ; 21 ; 33
separation sign

$$(= 20 \times 60^4 + 3 \times 60^3 + 13 \times 60^2 + 21 \times 60 + 33)$$

By means of the separation sign, the first notation is clearly distinguished from that of the number:

$$1; 10 + 18; 45 = 1 \times 60^2 + 28 \times 60 + 45$$

and the second one from that of the number:

$$20 + 3; 13; 21; 33 = 23 \times 60^3 + 13 \times 60^2 + 21 \times 60 + 33.$$

The oldest zero in history

Another difficulty of the Babylonian place-value system was related to the fact that in the second millennium B.C. the use of the zero—that is, a special sign to mark the absence of sexagesimal units of a certain order—was almost certainly unknown to Mesopotamian mathematicians and astronomers.

In a mathematical tablet from Uruk (now in the Louvre Museum) dating from a period later than the first Babylonian dynasty (perhaps the second third of the second millennium B.C., according to Thureau-Dangin), there is the following:

"Reckon the square of 2 ; 27 and you will find 6 ; 9 ."

The square of the number 2; 27 = 2 × 60 + 27 = 147 is:

$$21,609 = 6 \times 60^2 + 0 \times 60 + 9 = 6; 0; 9$$

If the scribe had known the use of the zero, he would have avoided noting the square of 2; 27 in the form of 6; 9, which may be confused with:

$$6; 9 = 6 \times 60 + 9 = 369$$

Another example of the same kind is provided by a Babylonian mathematical tablet (now in the Berlin Archaeological Museum), dating from about 1700 B.C. In it, the numbers:

$$2; 0; 20 = 2 \times 60^2 + 0 \times 60 + 20$$

and:

$$1; 0; 10 = 1 \times 60^2 + 0 \times 60 + 10$$

are indicated, respectively, as:

2 ; 20 and 1 ; 10

These notations are ambiguous, since they may be confused with:

$$2; 20 = 2 \times 60 + 20 = 140$$

and:

$$1; 10 = 1 \times 60 + 10. = 70$$

In an effort to overcome this difficulty, Babylonian scribes sometimes left a blank space where a power of 60 was missing. Examples:

A : 1 ; ; 25
absence of units
of the 2nd order

$$(= 1 \times 60^2 + 0 \times 60 + 25)$$

B : 1 ; 0 ; 35

$$(= 1 \times 60^2 + 0 \times 60 + 35)$$

C : 1 ; 0 ; 40

$$(= 1 \times 60^2 + 0 \times 60 + 40)$$

D : 1 ; 27 ; 0 ; 3 ; 45

$$(= 1 \times 60^4 + 27 \times 60^3 + 0 \times 60^2 + 3 \times 60 + 45)$$

(Examples A, B, and C are from mathematical tablets found at Susa, and example D is from line 15 of the tablet shown in fig. 26–4. The interpretations of all four examples are certain because the values correspond to mathematical relations clearly indicated by the context.)

But the problem was still not solved. Careless scribes often failed to leave a blank space where it was needed. Furthermore, the absence of two or more consecutive orders of units could not be reliably indicated because blank spaces side by side formed a single wide gap and the reader could not always tell how many absent orders it was meant to represent. And finally, without a zero the sign for 4, for example, could represent 4, 4×60, 4×60^2, 4×60^3, or 4×60^4, and it was sometimes difficult or impossible to know which number was intended.

These ambiguities did not prevent Babylonian mathematicians and

TRANSCRIPTION

LINE	TA-KI-IL-TI ŠI-LI-IP-TIM ŠA IN NA-AS-SÀ-ḪU-Ú-[M]A SAG-I ...-ú	IB-SÁ SAG	IB-SÁ ŠI-LI-IP-TIM	MU-BI-IM	
3	,59, * ,15	1 ; 59	2 , 49	KI	1
4	58 , 56 , 58 , 14 , 50 , 6 , 15	56 ; 7	3 , 12 ; 1	KI	2
5	,55 , 7 , 41 , 15 , 33 , 45	1 , 16 ; 41	1 , 50 . 49	KI	3
6	53 ,10 * ,29 , 32 , 52 , 16	3 , 31 , 49	5 , 9 , 1	KI	4
7	1 , 48 , 54 , 1 , 40	1 , 5	1 , 37	KI	5
8	1 ,47 , 6 , 41 , 40	5 , 19	8 , 1	KI	6
9	1 , 43 , 11 , 56 , 28 , 26 , 40	38 , 11	59 , 1	KI	7
10	1 , 41 , 33 , 59 , 3 , 45	13 , 19	20 , 49	KI	8
11	1 , 38 , 33 , 36 , 36	9 , 1	12 , 49	KI	9
12	1 , 35 , 10 , 2 , 28 , 27 , 24 , 26 , 40	1 , 22 , 41	2 , 16 , 1	KI	10
13	1 , 33 , 45	45	1 , 15	KI	11
14	1 , 29 , 21 , 54 , 2 , 15	27 , 59	48 , 49	KI	12
15	1 , 27 , * , 3 , 45	7 , 12 ; 1	4 , 49	KI	13
16	1 , 25 , 48 , 51 , 35 , 6 , 40	29 , 31	53 , 49	KI	14
17	1 , 23 , 13 , 46 , 40	56	53	KI	15

*Blank space indicating the absence of units of a certain order

Fig. 26–4. Babylonian mathematical tablet dating from 1800–1700 B.C. Plimpton Collection, Columbia University.

astronomers from performing many sophisticated calculations with their imperfect system over a period of more than a thousand years. The reckoner himself always knew the orders of magnitude he was dealing with, and they were often clear from the context; otherwise, any confusions that the system might engender could be avoided by explanatory comments.

However, at a time that we cannot determine precisely, though it was probably earlier than the Seleucid era (fourth to first centuries B.C.), Babylonian mathematicians and astronomers developed a genuine zero to signify the absence of sexagesimal units of a certain order. Instead of a blank space, they used either of these two signs (which are forms of the separation sign):

An astronomical tablet dating from the Seleucid era, found in clandestine excavations at the site of Uruk and now in the Louvre Museum, has the number:

$$2; 0; 25; 38; 4 = 2 \times 60^4 + 0 \times 60^3 + 25 \times 60^2 + 38 \times 60 + 4$$

written in this form:

2 ; 0 ; 25 ; 38 ; 4

with the zero sign marking the absence of sexagesimal units of the fourth order.

Lines 10, 14, and 24 of the tablet shown in figure 26–5 contain these notations:

1.　　　　2 , 0;0 ; 33 ; 20　　　$(= 2 \times 60^4 + 0 \times 60^3 + 0 \times 60^2 + 33 \times 60 + 20)$

2.　　　　1 ; 0 ; 45　　　$(= 1 \times 60^2 + 0 \times 60 + 45)$

3.　　　　1 ; 0;0; 16 ; 40　　　$(= 1 \times 60^4 + 0 \times 60^3 + 0 \times 60^2 + 16 \times 60 + 40)$

4.　　　　1 ; 0 ; 7 ; 30　　　$(= 1 \times 60^3 + 0 \times 60^2 + 7 \times 60 + 30)$

5.　　　　2 ; 0 ; 15　　　$(= 2 \times 60^2 + 0 \times 60 + 15)$

No Babylonian mathematical document published so far has shown the zero sign at the beginning or end of a written number. While Babylonian mathematicians sometimes used such notations as:

1; 0; 3 and 12; 0; 5; 0; 33

they evidently never used any of this kind:

$$5; 0 = 5 \times 60 + 0$$

or:

$$17; 3; 0; 0 = 17 \times 60^3 + 3 \times 60^2 + 0 \times 60 + 0$$

In the past, some historians of science concluded from this that the Babylonians used the zero only in a medial position and that their zero was therefore not functionally identical with ours. But as we now know from the work of Otto Neugebauer, Babylonian astronomers differed from Babylonian mathematicians in this respect: they used the zero at the beginning and end of written numbers, as well as in a medial position.

In a Babylonian astronomical tablet from the Seleucic era, now in the British Museum, the number 60 is written in the following form (the value attributed to it is assured by a mathematical relation indicated in the context): 𝖳 ⧹ (= 1 × 60 + 0)
1 ; 0
---->

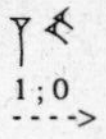

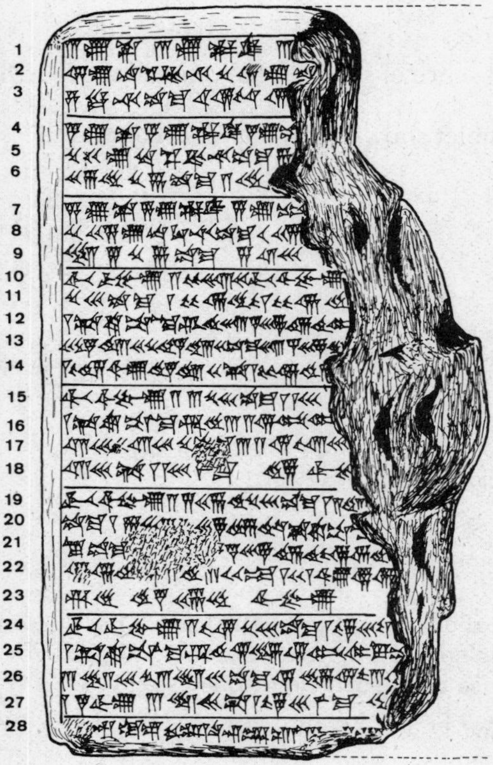

Fig. 26–5. Mathematical tablet probably dating from the late third or early second century B.C., found in clandestine excavations at Uruk. It shows some of the earliest known examples of the Babylonian zero. Louvre Museum, Paris.

in which the zero sign is used to mark the absence of units of the first order. On the back of that same tablet, the number 180 is written in this form: 𝗍𝗍𝗍 ⟨

$$\underset{3\;;\;0}{\ \ \ \ } \quad (= 3 \times 60 + 0).$$

And in another Babylonian astronomical tablet from the same period, also in the British Museum, the number:

$$2;\ 11;\ 46;\ 0 = 2 \times 60^3 + 11 \times 60^2 + 46 \times 60 + 0$$

is represented as follows:

$$\underset{2\ ;\ 11\ ;\quad 46\ ;\quad 0}{\text{𝗍𝗍 ⟨⟨ ⟨⟨𝗍𝗍𝗍 ⟨}}$$

But the final zero here is written in a rather strange way, as if it were a sign for 10 with its lower part extended. Was the upper crescent carelessly omitted, or was the sign an individual quirk of the scribe, or was it a form of rapid notation? Examination of other Babylonian astronomical tablets from the same period appears to support the latter explanation. The following example from one of those tablets seems rather eloquent:

$$\underset{3\quad ;\quad 0\ ;\ 18}{\text{𝗍𝗍𝗍 ⟨ ⟨𝗪}} \quad (= 3 \times 60^2 + 0 \times 60 + 18).$$

It must be pointed out, however, that in this same tablet the Babylonian zero is written several times in the conventional manner.

Use of the zero in the initial position enabled Babylonian astronomers to note sexagesimal fractions (that is, fractions whose denominator is equal to a power of 60) without ambiguity. Here are several examples from the tablet mentioned above:

Symbol	Value	Expansion
$\underset{0\;:\;1}{\text{⟨𝗍}}$	$= 0^\circ\ 1'$	$\left(= 0 + \dfrac{1}{60}\ \right)$
$\underset{0\;:\;4}{\text{⟨𝗪}}$	$= 0^\circ\ 4'$	$\left(= 0 + \dfrac{4}{60}\ \right)$
$\underset{0\;:\;9}{\text{⟨𝗪}}$	$= 0^\circ\ 9'$	$\left(= 0 + \dfrac{9}{60}\ \right)$
$\underset{0\;:\;53}{\text{⟨𝗍 ⟨⟨𝗍𝗍𝗍}}$	$= 0^\circ\ 53'$	$\left(= 0 + \dfrac{53}{60}\ \right)$
$\underset{0\;;\;0\;;\;30}{\text{⟨𝗍𝗍 ⟨⟨⟨}}$	$= 0^\circ\ 0'\ 30''$	$\left(= 0 + \dfrac{0}{60} + \dfrac{30}{60^2}\ \right)$
$\underset{0\;;\;6\;;\;37\;;\;40}{\text{⟨ 𝗍𝗍𝗍 ⟨⟨𝗪⟨⟨}}$	$= 0^\circ\ 6'\ 37''\ 40'''$	$\left(= 0 + \dfrac{6}{60} + \dfrac{37}{60^2} + \dfrac{40}{60^3}\ \right)$

Thus, at least as early as the first half of the second millennium B.C., Mesopotamian scholars developed a written numeration that was eminently abstract and far superior to any other system used in the ancient world; they probably devised the first strictly place-value numeration in history.

Later, they also invented use of the zero. Mathematicians seem to have used it only in a medial position; astronomers, however, used it not only in that position but also in the final position and in the initial position, which enabled them to extend their notation to sexagesimal fractions.

But to the Babylonians the zero sign did not signify "the number zero." Although it was used with the meaning of "empty" (that is, "an empty place in a written number"), it does not seem to have been given the meaning of "nothing," as in "10 minus 10," for example; those two concepts were still regarded as distinct.

In a mathematical text from Susa, the scribe, obviously not knowing how to express the result of subtracting 20 from 20, concluded in this way: "20 minus 20 . . . you see." And in another such text from Susa, at a place where we would expect to find zero as the result of a distribution of grain, the scribe simply wrote, "The grain is exhausted." These two examples show that the notion of "nothing" was not yet conceived as a number.

We have seen that the zero sign does not appear in texts from the time of the first Babylonian dynasty, and there is no direct evidence of its use before the Seleucid era, since the oldest known documents in which it appears do not go back beyond the third century B.C.

Are we to conclude that the Babylonian zero was invented in the time of the Seleucids? There is no proof of that. We must make a distinction between the time of an invention and the time of its propagation, and neither is necessarily the same as the time of its earliest known use. In some cases an invention may have been made several generations before its propagation, and the oldest known documents proving its use may date from several centuries later, either because earlier documents have perished or because they have not yet been discovered by archaeologists.

It is therefore permissible to suppose that the Babylonian zero was invented before the third century B.C., especially since we now have several reasons to believe that the tablets from the Seleucid era are copies of documents several generations older; this seems particularly clear, at least, in the case of literary texts.

But the evidence is still not conclusive and we can only hope that new archaeological discoveries will provide a final answer.

Survival of the Babylonian system

The Babylonian sexagesimal place-value system had a great influence on the scientific world in ancient times, and its influence has continued down to our own time, since in spite of our decimal numeration and systems of weights and measures we still use sexagesimal units for measuring time (hours, minutes, seconds), arcs, and angles (degrees, minutes, seconds).

At least as early as the second century B.C., Greek astronomers used this system for expressing negative powers of 60, having introduced a zero for that purpose (figs. 26–8, 26–6). But instead of the Babylonian cuneiform notation, the Greeks used an adaptation of their alphabetic numeration, noting 0°28′35″ and 0°17′49″, for example, in these forms:

$$\text{Ⲧ KH ΛΕ}$$
$$0 \; ; \; 28 \; ; \; 35 \longrightarrow \qquad (= 0 + \tfrac{28}{60} + \tfrac{35}{60^2})$$

$$\text{Ⲧ IZ MΘ}$$
$$0 \; ; \; 17 \; ; \; 49 \longrightarrow \qquad (= 0 + \tfrac{17}{60} + \tfrac{49}{60^2})$$

After the Greeks, Arab and Jewish astronomers used the same system for their astronomical tables (figs. 26–7, 26–9, 26–10), adapting it to their respective alphabetic numerations and noting the two expressions above as follows:

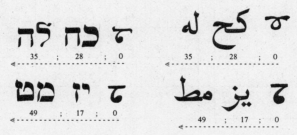

$$\overset{\text{הל כח Ⲧ}}{\underset{35 \; ; \; 28 \; ; \; 0}{\longleftarrow}} \qquad \overset{\text{ل كح σ}}{\underset{35 \; ; \; 28 \; ; \; 0}{\longleftarrow}}$$

$$\overset{\text{מט יז Ⲧ}}{\underset{49 \; ; \; 17 \; ; \; 0}{\longleftarrow}} \qquad \overset{\text{يز مط σ}}{\underset{49 \; ; \; 17 \; ; \; 0}{\longleftarrow}}$$

TRANSCRIPTION

	Greek	B (left)	B (mid)	B (right)	C (value)	C (deg)	D (value)	D (deg)
	IΘ	···			····	IB	·····	KE]
	K	···		··	····	IΓ	·····	KϚ]
	KA				····	IΔ	·····	KZ]
	KB	B	MΓ	A	[······]	IE	NB ME	[KH]
] ·	KΓ			B	KΔ ΛϚ	IϚ		KΘ
	KΔ	Δ Λ	Γ		KϚ Θ	IZ	NΔ IϚ	Λ
ϛ	KE	Θ KϚ	Δ		KZ MA	IH	**TAYPOY**	**ΔIΔYM**
] ·	KϚ	Ϛ MZ	E		KΘ IΓ	IΘ	τ KΘ NϚ	τ K[·]
	KZ	H Θ	ϛ		Λ ΛϚ	K	τ NΘ NB	τ M[·]
	KH	I NB	H		ΛB IH	KA	τ KB MH	A
	KΘ	IB Γ	Θ		ΛΓ N	KB	B NΔ ΛϚ	B
	Λ	IΓ ΛE	I		ΛE KB	KΓ	Δ KΘ	Γ
		IΔ NϚ	IA		ΛE NE	KΔ	E NΘ	Δ

TRANSLATION

	Deg	B (left)	B (mid)	B (right)	C (value)	C (deg)	D (value)	D (deg)
	19	····			···	12	····	25]
	20	····		··	····	13	····	26]
	21	··			····	14	····	27]
	22	2	43	1	···	15	52 45	[28]
] ·	23			2	24 36	16		29
	24	4	30	3	26 9	17	54 16	30
	25	9	26	4	27 41	18	**TAURUS**	**GEMINI**
6	26	6	47	5	29 13	19	0 29 56	0 20 [·]
] ·	27	8	9	6	30 36	20	0 59 52	0 40 [·]
	28	10	52	8	32 18	21	0 22 48	1
	29	12	3	9	34 50	22	2 54 36	2
	30	13	35	10	35 22	23	4 29	3
		14	56	11	35 55	24	5 59	4

Fig. 26–6.
Greek astronomical table, from a third-century papyrus.

GREEK MANUSCRIPTS

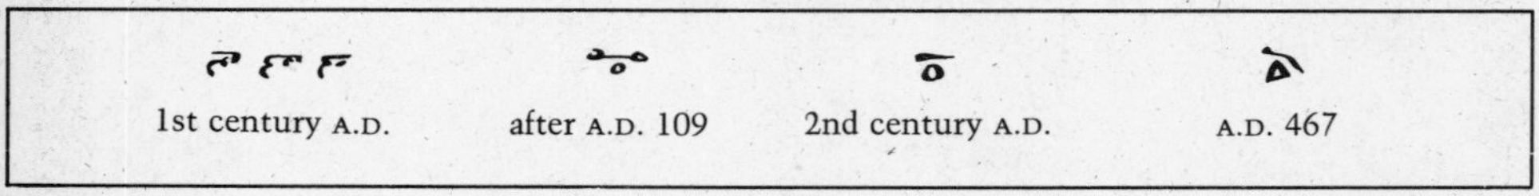

1st century A.D.	after A.D. 109	2nd century A.D.	A.D. 467

ARABIC-PERSIAN MANUSCRIPTS

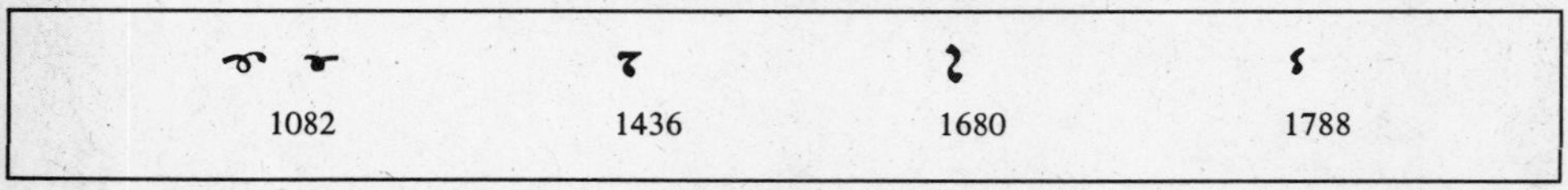

1082	1436	1680	1788

Fig. 26–7. Graphic variants of the "sexagesimal zero" used by Greek, Arab, and Jewish astronomers.

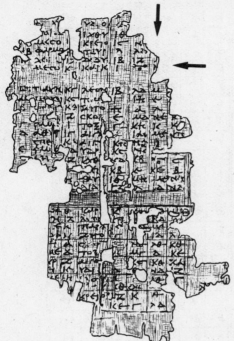

Fig. 26–8. Greek astronomical papyrus from the second century A.D. (later than 109).

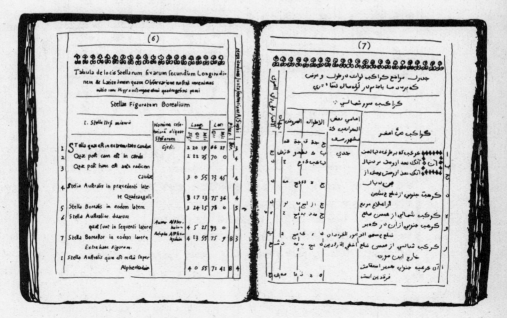

Fig. 26–9. Bilingual (Latin-Persian) astronomical table.

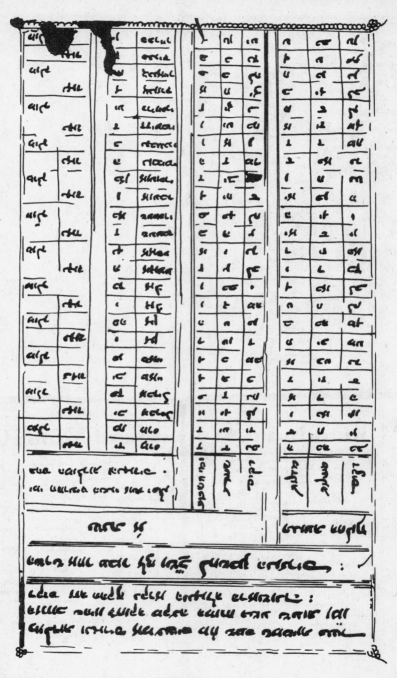

Fig. 26–10. Astronomical table by the French Jewish mathematician and philosopher Levi ben Gershon (1288–1344).

27

The Chinese Place-Value System

Chinese mathematicians and reckoners, and their Japanese counterparts, often used an ingenious written numeration whose signs, for a reason that will be explained later, are called rod numerals. It was created independently of any foreign influence and had practically no relation to the common Chinese numeration.

It is a decimal place-value system, but unlike our system, with its single series of signs for numbers 1 to 9, it has two different series of nine signs for those numbers.

In the first series, numbers 1 to 5 are represented by vertical strokes and numbers 6 to 9 by a horizontal stroke with vertical strokes below it:

FIRST SERIES

| 1 | 2 | 3 | 4 | 5 | 6 | 7 | 8 | 9 |

The signs of the second series are formed in an analogous way, but with horizontal strokes replacing vertical ones and vice versa:

SECOND SERIES

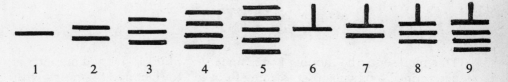

| 1 | 2 | 3 | 4 | 5 | 6 | 7 | 8 | 9 |

(Note that in the first series the horizontal stroke in the signs for numbers 6 to 9 has the symbolic value of 5, while in the second series it is the vertical stroke that symbolizes 5 in the signs for those numbers.)

To represent numbers composed of several orders of units, signs from either or both of the series can be used, with each sign, of course, having a positional value.

Here is an ancient Chinese riddle, cited by the mathematician Mei Wen-ting (1631–1721): "The character 亥 has 2 for a head and 6 for a
hai
body. Bring the 2 down to the body and you will obtain the number of days in the age of the old man of Kiang-hien."

The numeration involved in this riddle (the place-value system of rod numerals) was used at least as early as the Han dynasty (206 B.C.–A.D. 220); details of its use are not known to us any farther back than the second century B.C., but it may have originated at a much earlier time (fig. 27–1). According to Chinese sources, the riddle itself probably dates from before the Christian era, and some authors place it as early as the middle of the Zhu dynasty (eighth to seventh centuries B.C.).

To solve the riddle, the character it refers to must not be considered in its modern (*kǎishū*) form: 亥
hai

but in the form it had at the time when the riddle was composed. It was then written in the old *dà zhuàn* style (used before the Han dynasty, during the Zhu dynasty and the Warring States period), and according to Mei Wen-ting its form was: 𥘅
hai

In this form, the character *hai* has "2 for a head" (that is, its upper part is composed of the sign for 2 in the second series of rod numerals) and its lower part, or "body," is composed of three signs resembling the sign for 6 in the first series of rod numerals. By "bringing the 2 down to the body" and placing it vertically, we obtain:

‖ 𝝘𝝘𝝘
HEAD   BODY

which is closely similar to the following rod numerals of the first series:

‖ T T T
2 ; 6 ; 6 ; 6

This represents the number 2666, since the value of each rod numeral is determined by its position, reading from left to right and beginning with the highest decimal order.

But 2666 days (a little less than seven and a half years) is obviously not a satisfactory solution of the riddle: someone with that age could scarcely be called an "old man."

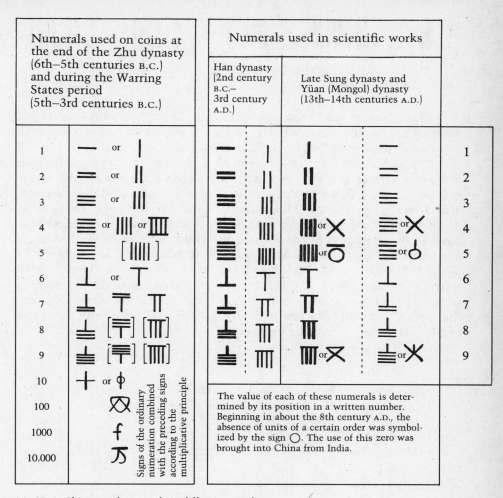

Fig. 27–1. Chinese rod numerals in different periods.

Since a zero was not used in the Chinese place-value system till the eighth century A.D., we may assume that the number indicates not days but tens of days; in other words, that it must be interpreted as "2 tens of thousands, 6 thousands, 6 hundreds, and 6 tens" of days, or 26,600 days. "There could be no question of adding more than one zero because that would have turned the old man of Kiang-hien into a Chinese Methuselah, whereas 26,600 days, about 73 years, is a perfectly reasonable age for him" (Guitel 1).

And here are some numerical expressions written by Ts'ai Keioufong, a philosopher who died in 1230:

This notation could cause confusion if the vertical strokes were not properly separated: the number 12, for example, could be confused with 3 or 21, or 25 with 7, 34, 43, 52, 214, 223, etc.

It was no doubt for this reason that, at least as early as the Han dynasty, users of the system generally agreed to alternate signs of the first series with those of the second series to make sure that decimal orders were clearly distinguished within written numbers. They either noted units of the first order with vertical strokes, units of the second order with horizontal ones, units of the third order with vertical ones, and so on (fig. 27–3), or else they began with horizontal strokes for units of the first order, used vertical ones for units of the second order, horizontal ones for units of the third order, and so on. In other words, they represented units of odd rank $(1, 10^2, 10^4, 10^6,$ etc.) with numerals of the first series and units of even rank $(10, 10^3, 10^5, 10^7,$ etc.) with numerals of the second series, or vice versa.

REPRESENTATION OF THE NUMBER 3764

on a counting board, with rods | with rod numerals combined according to the place-value principle

$3 \times 10^3 + 7 \times 10^2 + 6 \times 10 + 4$

Fig. 27–2. Origin of the Chinese rod numerals, showing how a calculating device engendered a written place-value numeration.

The system is particularly interesting because its signs reproduce arrangements of rods on a counting board (fig. 27–3), which explains the term "rod numerals." These rods were used in China, Korea, and Japan for performing arithmetical operations and solving algebraic equations. When they were laid out on a counting board, a square was left empty for each missing order of units.

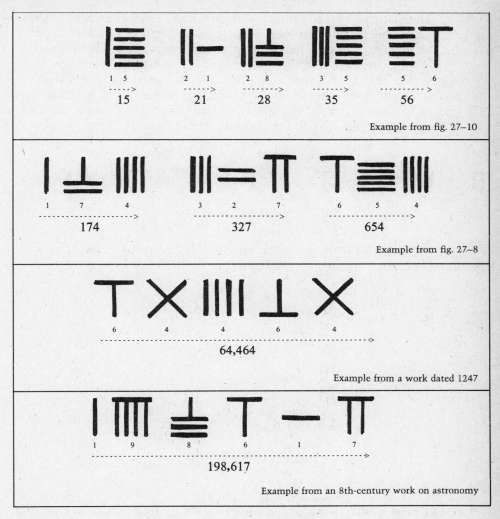

Fig. 27–3. Rod-numeral representations of numbers not multiples of the base (see fig. 27–1).

The Chinese place-value numeration originated in a system that used physical objects (rods), became stabilized when the rods were placed on a counting board, then eliminated the counting board and became a written numeration. "The fact that we have direct evidence of all three phases makes this a unique case. Intellectually speaking, the system was positional from the start: the counting board was only a material support for it which facilitated the development of techniques for performing operations" (Guitel 1).

The history of the rod-numeral system proves that the Chinese knew the place-value principle at least as early as the beginning of the Christian era. But for centuries they had no zero, and this caused serious confusions.

During the stage when rods were used on a counting board, an empty square clearly indicated that a decimal order was missing. When arrangements of rods were transformed into written numerical signs, however, simply leaving an empty space was unsatisfactory, since numbers such as 764, 70,064, and 76,400, for example, could easily be confused.

At first the Chinese overcame this difficulty by expressing such numbers either in the common system:

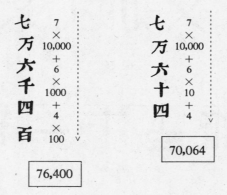

or with rod numerals placed in squares, exactly like rods on a counting board:

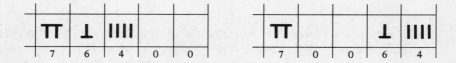

It was not till the eighth century A.D. that Chinese scholars, no doubt influenced by Indian mathematicians, began using a special sign (a circle) to mark the absence of units of a certain order (fig 27–4).

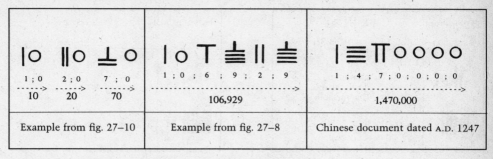

Fig. 27–4. Use of the zero in the Chinese rod-numeral system.

From then on, progress was rapid: all the arithmetical and algebraic rules concerning whole, mixed, and irrational numbers reached a degree of development equivalent to the forms in which they are taught today (figs. 27–5, 27–6, 27–7).

Fig. 27–5. How Chinese mathematicians extended their place-value notation to decimal fractions. Examples reconstructed on the basis of a work from the Yüan dynasty.

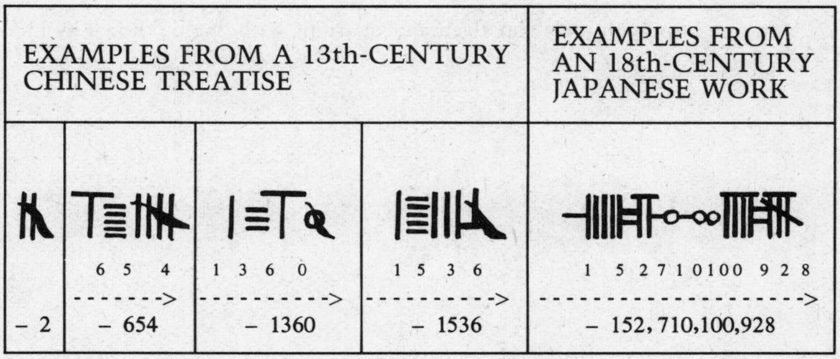

EXAMPLES FROM A 13th-CENTURY CHINESE TREATISE				EXAMPLES FROM AN 18th-CENTURY JAPANESE WORK
6 5 4	1 3 6 0	1 5 3 6		1 5 2 7 1 0 1 0 0 9 2 8
– 2	– 654	– 1360	– 1536	– 152, 710,100,928

Fig. 27–6. Extension of the bar-numeral system to negative numbers: to indicate that a given number was negative, Chinese and Japanese mathematicians often drew a slanting stroke through the last sign of its representation.

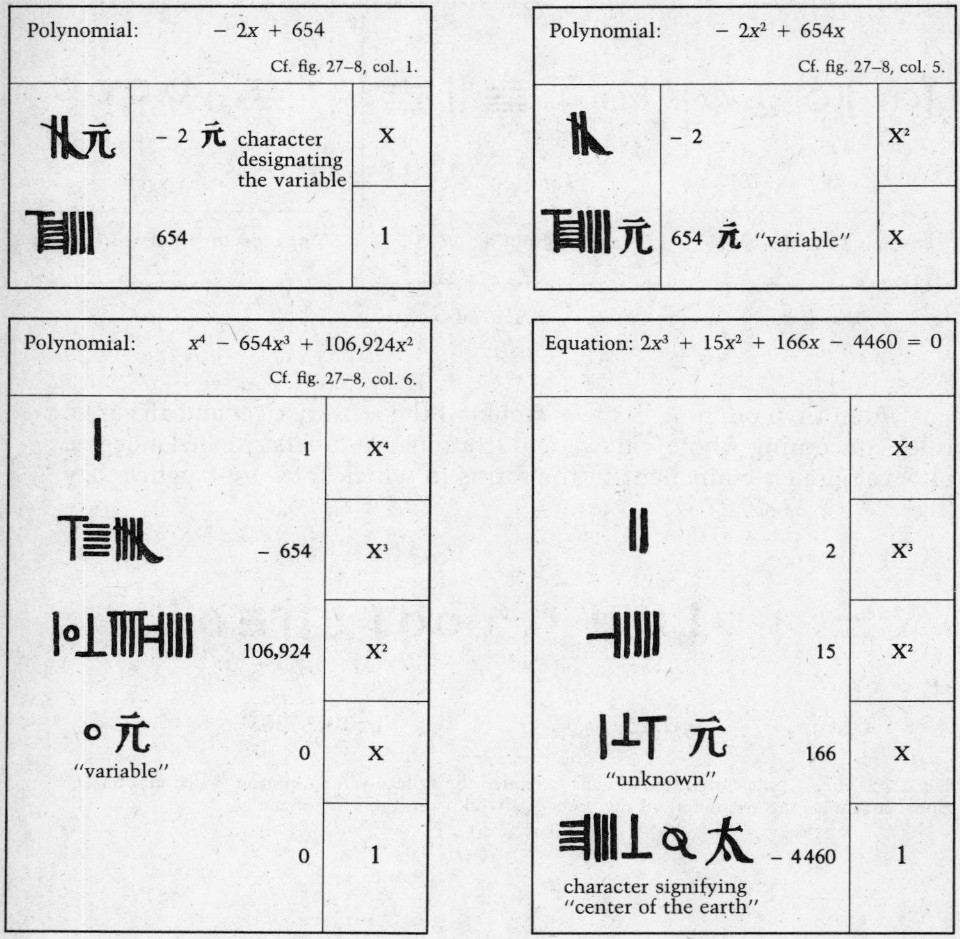

Polynomial: $-2x + 654$

Cf. fig. 27–8, col. 1.

九元	-2 元 character designating the variable	X
654	654	1

Polynomial: $-2x^2 + 654x$

Cf. fig. 27–8, col. 5.

	-2	X^2
	654 元 "variable"	X

Polynomial: $x^4 - 654x^3 + 106{,}924x^2$

Cf. fig. 27–8, col. 6.

	1	X^4
	-654	X^3
	106,924	X^2
元 "variable"	0	X
	0	1

Equation: $2x^3 + 15x^2 + 166x - 4460 = 0$

		X^4
	2	X^3
	15	X^2
元 "unknown"	166	X
太 character signifying "center of the earth"	-4460	1

Fig. 27–7. Notation of polynomials and algebraic equations with one unknown by Li Yeh (1178–1265).

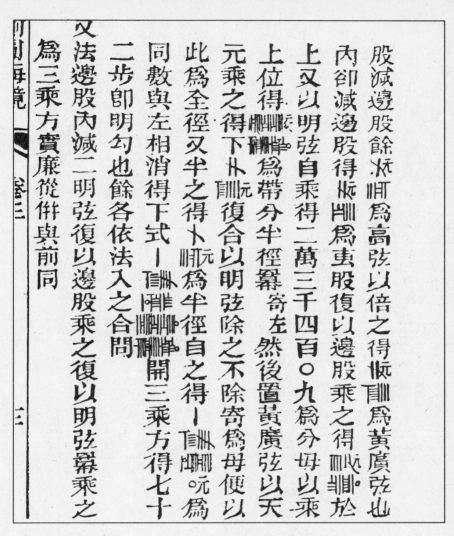

Fig. 27–8. From *Tshe Yuan Hai Ching*, written in 1248 by the mathematician Li Yeh.

‖‖‖	‖‖丁	丅‖‖	丌‖‖‖‖
Ⅰ ⟂ Ⅲ	Ⅲ = π	丅 ≡ ‖‖‖	— 丌 ≡ ‖‖‖ — Ⅰ ⟂ 丅 ⟂ 〇
1 7 4	3 2 7	6 5 4	1 9 5 5 1 1 9 6 8 0
174	327	654	1,955,119,680

Fig. 27–9. Chinese rod numerals are usually written and printed in "compressed" forms like those shown here. These examples are from figure 27–8.

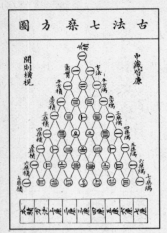

Fig. 27–10. Page from *Ssu Yuan Yü Chien*, written in 1303 by the Chinese mathematician Chu Shih-Chieh.

COMMENTS ON FIG. 27-10

In the West, Blaise Pascal (1623–1662) is often regarded as having been the first to devise the triangular arrangement (known as "Pascal's arithmetical triangle" or "the Pascal triangle") giving the numerical coefficients of the expansion of $(a + b)^n$, where n is 0 or any positive whole number:

EXPANSIONS OF BINOMIALS	"THE PASCAL TRIANGLE"
$(a + b)^0 = 1$ $(a + b)^1 = a + b$ $(a + b)^2 = a^2 + 2ab + b^2$ $(a + b)^3 = a^3 + 3a^2b + 3ab^2 + b^3$ $(a + b)^4 = a^4 + 4a^3b + 6a^2b^2 + 4ab^3 + b^4$ $(a + b)^5 = a^5 + 5a^4b + 10a^3b^2 + 10a^2b^3 + 5ab^4 + b^5$ $(a + b)^6 = a^6 + 6a^5b + 15a^4b^2 + 20a^3b^3 + 15a^2b^4 + 6ab^5 + b^6$ - >	1 1 1 1 2 1 1 3 3 1 1 4 6 4 1 1 5 10 10 5 1 1 6 15 20 15 6 1 - >

Actually, however, as can be seen from the following transcription of the table reproduced in fig. 27–10, the Chinese knew this triangle long before the famous French mathematician:

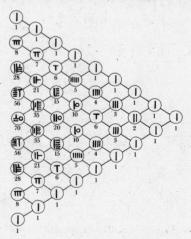

28

Amazing Achievements of a Vanished Civilization

Several dozen dead cities and towns, lost for centuries in the forests of Central America, bear witness to one of the most mysterious episodes in history.

"In their [the Mayas'] imposing temples perched atop pyramids that sometimes rose to a height of more than 150 feet, ritual and initiatory ceremonies took place, ceremonies now only dimly reflected by a few enigmatic bas-reliefs. The architectural structures of those forgotten cities, the magnificently carved stone altars and stelae, the multicolored pottery, the mysterious hieroglyphic signs engraved on the monuments— these things show that the people who made them had a high degree of civilization" (Ivanoff).

Greatness and decline of Mayan civilization

This civilization was well developed by the third century A.D. and reached great intellectual and artistic heights long before Columbus arrived in the New World.

Archaeologists call its initial phase the Formative Period and place its beginning in the fifth century B.C. For a long time it was believed that this period lasted till about A.D. 320, because that was the date of the oldest known Mayan document: a carved jade plaque found in 1864 (fig. 28–20). But the end of the Formative Period was pushed back some thirty years in 1959, when a stela bearing a date that corresponds to A.D. 292 was discovered in the ruins of the city of Tikal (fig. 28–1).

The Formative Period was followed by what is known as the Classic Period, during which Mayan civilization reached its peak. According to the view generally accepted today, this period ended in 925. While it lasted, the Mayas had a brilliant and original civilization in such fields as art, architecture, education, commerce, religion, mathematics, and astronomy.

But then, in the ninth and tenth centuries, something happened that

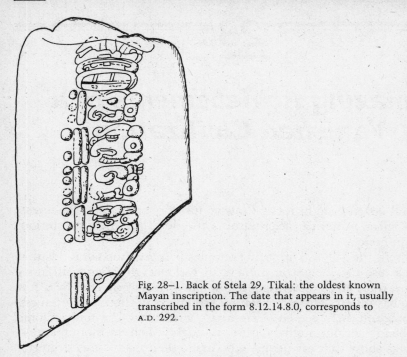

Fig. 28–1. Back of Stela 29, Tikal: the oldest known Mayan inscription. The date that appears in it, usually transcribed in the form 8.12.14.8.0, corresponds to A.D. 292.

is still mysterious to us because archaeologists have not yet succeeded in explaining it convincingly: the Mayas abandoned their ceremonial centers and the cities in the central region of their territory, sometimes so abruptly that they left unfinished buildings.

All sorts of theories have been suggested to account for this exodus. Although there is still not enough evidence for a firm conclusion, the fact remains that "the Maya did abandon work on their ceremonial and religious centers, and a number, but not the whole of the population did move away. The most probable explanation is that the peasants revolted against the hierarchy, a revolt brought about by reasons often repeated in other pages of the world's history—the inequality of rights between two classes growing greater until the masses called a halt. This cessation in ceremonial city life and the things of aesthetic value that attended it, was a contributory cause of the decline and finally close of the Classic period of Maya history" (Cottrell).

Sources of our knowledge

By the time of the Spanish conquest in the sixteenth century, Mayan civilization had already been dead for several generations. Most of its magnificent cities had fallen into ruin and been swallowed up by the jungle, forgotten by the Indians of the region, who no longer showed any traces of their ancestors' culture. This explains why the first Spanish chroniclers, probably dazzled by the brilliance of Aztec civilization, made so little mention of the Mayan centers that still existed.

Furthermore, because of their desire to convert the Indians to Christianity—and also because of their greed—the Spanish conquerors relentlessly tried to destroy all vestiges of Mayan civilization. Yet it is to a Spaniard that we owe a considerable part of our knowledge of the history, customs, and institutions of the Mayas.

"In 1869 Charles Etienne Brasseur de Bourbourg, an indefatigable and eccentric French priest, discovered *Relación de las cosas de Yucatán* in the library of the Spanish Royal Academy in Madrid. It was a work written shortly after the Spanish conquest by Friar Diego de Landa, first Bishop of Mérida, in Yucatán, containing invaluable cultural information, descriptions, and drawings of the hieroglyphic writing used by the Indians of Yucatán in the sixteenth century. It is one of the ironies of history that Landa boasted of having burned all native books in that writing, to make it easier to bring the Indian population of Yucatán into the bosom of the Catholic Church. His fanaticism had destroyed the precious painted codices that depicted a whole civilization, but a desire to explain his criminal act led him to write his chronicle and, in doing so, he involuntarily saved from oblivion the basic elements of one of the most important Indian cultures in America" (Ivanoff).

Although Landa's *Relación* is an essential source of information on Mayan civilization, it is not the only one, for we also have information from Indian chronicles.

"After the Conquest, the natives were taught to transpose their own language into Spanish. Once the Maya were able to read and write, the propagation of Christian teachings would move rapidly; it was to that end alone that the friars painstakingly instructed their subjects. It was inevitable, however, that this knowledge was put to use in recording matters of concern to native chroniclers who were justly alarmed by the rapid passing of their heritage.

"A few of these post-Conquest documents have survived to convey with eloquent simplicity the reflections of their authors concerning their history and traditions" (Gallenkamp).

And finally, we have the three ancient Mayan manuscripts, or codices, which miraculously escaped the ravages that followed the coming of the Spanish conquistadores. They were brought to Europe, probably by soldiers or missionaries soon after the conquest, and are now sometimes known by the names of the cities where they are preserved: the Dresden Codex, or Codex Dresden, or Codex Dresdensis; the Codex Tro-Cortesianus, or Codex Madrid, or Codex Madridensis; and the Codex Peresianus, or Codex Parisiensis, or Codex Paris (fig. 28–2).

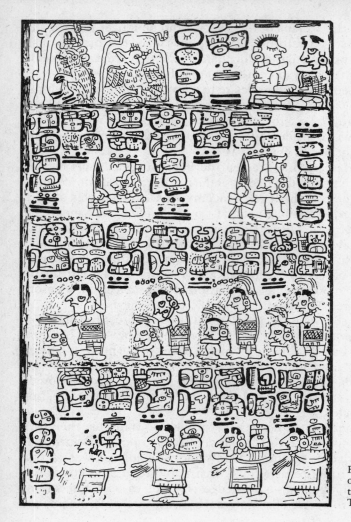

Fig. 28–2. Lower part of page 93 of the Codex Tro-Cortesianus.

Six centuries of intellectual and artistic achievement

Archaeologists have gradually succeeded in bringing to light the contours of Mayan civilization and the main aspects of its greatness. But many gaps remain in our knowledge of it. The research that is still being done may eventually make us alter certain conclusions that now seem well established.

"One of the most remarkable features of Mayan art is the skill with which sculptors collaborated with architects: the harmony of their decoration, the proportions of their figures, and the way they used the interplay of light and shadow place those sculptors among the best" (Lehmann). Yet the simple tools they used—made of stone, bone, or wood—would not have been out of place in the Neolithic.

As for Mayan painting, "the frescoes at Bonampak show the high degree of perfection which that art also attained. These frescoes are so beautiful that they have been compared with those of the Italian Renaissance." And Mayan pottery "is equally remarkable for its elegance and the variety of its multicolored decoration" (Lehmann). The Mayas discovered cement, used the technique of corbeled vaulting, and built vast, magnificent cities and a system of roads, yet they never invented the wheel and did not use draft or pack animals (fig. 28–3).

Fig. 28–3. Temple I at Tikal, built in about A.D. 702.

In mathematics, they developed a true place-value written numeration that enabled them to express numbers going beyond 100,000,000. And they used a special sign to indicate the absence of units of a certain order, which amounts to saying that they discovered the zero concept.

They also made remarkable advances in astronomy and the calendar, which were particularly important to them. They amassed impressive records of their observations of celestial phenomena and created calendars whose accuracy surpassed that of the Gregorian calendar we use today (figs. 28–4, 28–5).

They knew the corrections that had to be made in the 365-day year if it was to correspond exactly to the time taken by one of the earth's revolutions around the sun. The duration of the solar year as determined by modern astronomy is 365.242198 days. Its duration in our Gregorian calendar, 365.242500 days, therefore has an error of 3.02 ten-thousandths of a day. But its duration in the Mayan calendar, 365.242000 days (the

Mayas did not express it in that form, since they lacked a notation for fractions), has an error of only 1.98 ten-thousandths of a day.

The average period between two successive full moons is 29.53059 days, according to modern computations made with precision instruments; the astronomers of the city of Copán evaluated it at 29.53020 days (4400 days for 149 lunar months), and those of Palenque at 29.53086 days (2392 days for 81 lunar months).

Fig. 28–4. Alone in the darkness of night, a Mayan astronomer observes the stars. Detail of the Codex Tro-Cortesianus.

Fig. 28–5. Astronomic observation, as shown by the Nuttall Codex and the Selden Codex. On the left, an astronomer is observing the sky with the aid of two crossed sticks; the drawing on the right shows an eye in the angle formed by the sticks.

To the Mayas, however, "time was never a purely abstract denominator by which events were arranged into orderly sequences: rather it was an infinite beyond-world inhabited by omnipotent forces of creation and destruction. Its alternating cycles—the days, months, and years—were believed to bring with them the benevolence or evil of gods who bore them along their endlessly recurring cycles. Each bearer was the patron of a sacred number and assumed a form by which it could be portrayed in hieroglyphic inscriptions, and divisions of the calendar were regarded as 'burdens' carried on the backs of those divine guardians of time.* If a deity of malevolent intent happened to assume the burden of

*"The burdens were carried on the back, the weight supported by tumplines across the forehead. In terms of our own calendar it is as though for December 31, 1952, there were six bearers: The god of number thirty-one with December on his back; the god of number one carries the millennium; the god of number nine bears the centuries; the god of number five, the decades; and the god of number two, the years. At the end of the day there is a momentary pause before the procession restarts, but in that moment the god of number one with the burden of January replaces the god of thirty-one with his December load, and the god of number three relieves the god of number two as bearer of the year" (Thompson).

a particular cycle, grievous consequences could be expected until it was relinquished to a more provident bearer at the end of its natural course. Whether a certain month or year held promise of good or bad fortune was a matter predetermined by the temperament of the god on which it was borne.

"It is a curious belief, and one which explains in part the far-reaching power of the priesthoods over the populace who must surely have deemed survival impossible without learned mediators to interpret the irascible tendencies of their gods. Only the astronomer-priests stood between normal continuation of life and catastrophes brought about by misjudging the inclinations of the divinities. Having recognized the varying aspects of the gods and plotted their restless paths along the highways of time and space, they alone could determine when it was that beneficial gods were in possession of specific periods; or, as was more often the case, when the greatest number of benevolent deities marched in conjunction with less sympathetic ones. Thus the Mayan obsession with time was very largely tantamount to a grand-scale quest for lucky and unlucky periods in the hope that once forewarned of future prospects their destiny could be guided along the most favorable possible course" (Gallenkamp).

And so we understand that "Mayan astronomy was essentially different from what that word means to us, since its fundamental purpose was mythical interpretation of the magic powers that ruled the universe [fig. 28–6]" (Girard).

Fig. 28–6. The cyclical concept of events in Mayan mystical thought: Chac, the god of rain, plants a tree; behind him, Ah Puch, the god of death, breaks it; and finally Yum Kaax, the god of corn and farming, repairs it. Detail of the Codex Tro-Cortesianus.

Writing, arithmetic, and astronomy

Another outstanding intellectual achievement of the Mayas—and, more generally, the pre-Columbian peoples of Middle America (Central America and Mexico)—is the development of a form of hieroglyphic writing. There are still whole walls covered with this writing and it is also found on stelae and in the codices. Its characters (usually called glyphs rather than hieroglyphs, though the two terms are interchangeable) are arranged vertically in columns and were probably both ideographic and symbolic.

Some scholars believe that the Mayas' writing was beginning to be phonetic when their civilization reached its peak.

Unfortunately, Mayan writing is still largely undeciphered. Specialists have succeeded in identifying only signs and glyphs related to numeration, the calendar, directions, and some of the Mayan gods, as well as the "emblem glyphs" of some of the main cities in the central region (Palenque, Quiriguá, Tikal, and Yaxchilán, for example). It seems likely that other glyphs, the elements of Mayan writing in the strict sense of the term, will long continue to resist the efforts of the best decipherers, who would no doubt give a great deal for the discovery of a Mayan equivalent of the Rosetta Stone (fig. 28-7).

The Mayas did not count according to a decimal base, as we do, but by twenties and powers of 20, probably because at an early stage they developed the habit of counting on their toes as well as their fingers. As we have seen in Chapter 2, for numbers 39, 40, 80, 120, 400, 800, 8000, 16,000, and 24,000, for example, they said something like the following:

nineteen after twenty (39 = 19 + 20)
two twenties (40 = 2 × 20)
four twenties (80 = 4 × 20)
six twenties (120 = 6 × 20)

MAYAN GODS

Fig. 28-7. Some of the Mayan glyphs that have been deciphered.

HUNAB-KU
The god who created the world; supreme deity of the Mayan pantheon

AH PUCH
God of death

YUM KAAX
God of corn

CHAC
God of rain

"EMBLEM GLYPHS"
of three Mayan cities

PIEDRAS NEGRAS   **TIKAL**   **COPÁN**

DIRECTIONS

LIKIN
east

ČIKIN
west

OTHER GLYPHS

KIN, "DAY"
Stylized images of the solar disk, evoking the idea of "sun" and, by extension, "day"

UINAL, "20-DAY MONTH"
This glyph, an abstract representation of the moon, is the Mayan symbol of the number 20.

one "four hundred" $(400 = 1 \times 20^2)$
two "four hundreds" $(800 = 2 \times 20^2)$
one "eight thousand" $(8000 = 1 \times 20^3)$
two "eight thousands" $(16,000 = 2 \times 20^3)$
three "eight thousands" $(24,000 = 3 \times 20^3)$

To record the results of all sorts of counts, they must have used a purely vigesimal written numeration based on the additive principle. The number 1 must have been represented by a bar or a dot (a sign common to many Middle American peoples; it probably came from the cacao bean, which was used as a medium of exchange). There must also have been special signs for 20, 400 $(= 20^2)$, 8000 $(= 20^3)$, and so on, and these signs must have been repeated as often as necessary.

This is only a conjecture, however, and cannot be verified in the present state of our knowledge. We have no trace of the system that the Mayas used in ordinary practice, since the manuscripts that probably would have given us valuable information on the subject were all destroyed by the idiotic fanaticism of Spanish inquisitors.

The only known surviving examples of Mayan numerical notation are related to either astronomy or the computation of time.

The Mayan calendar

The Mayas had two distinct calendars, which they used concurrently: one based on the sacred year (*tzolkin*) and the other on the civil year (*haab*). The latter was a solar calendar.

The sacred year was composed of twenty cycles of thirteen days each, making a total of 260 days. It had a series of twenty days whose names and order of succession were as follows:

Imix	Cimi	Chuen	Cib
Ik	Manik	Eb	Caban
Akbal	Lamat	Ben	Eznab
Kan	Muluc	Ix	Cauac
Chicchan	Oc	Men	Ahau

Each of these names was associated with a glyph, whose style might vary from one inscription to another (fig. 28–8).

"The twenty days which formed the Maya 'month' were regarded as gods and were the recipients of prayers. The days were in a way embodiments of gods, such as the sun and the moon, the maize deity, the death god, and the jaguar god, which were drawn from their various categories to be reassembled in this series" (Thompson). The day named Kan, for example, was associated with the corn god; Cimi, with the death god;

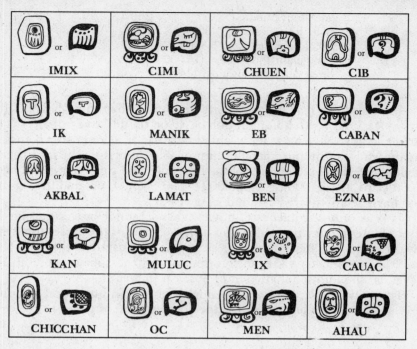

Fig. 28–8. Glyphs of the twenty days of the Mayan calendar, with their names in the Yucatec language.

Oc, with the dog; Ahau (the name means "flower" or "lord"), with the sun, etc.

Each of the twenty days was also associated with a number that varied cyclically from 1 to 13. That is, considering the series for the first time, the first day was associated with the number 1, the second with 2, the thirteenth with 13. Then the sequence of thirteen numbers began again, so that the fourteenth day was associated with 1, the fifteenth with 2, the twentieth with 7. Then the sequence of twenty days began again: the first day was associated with 8, the second with 9, and so on. This was continued through successive cycles of thirteen numbers assigned to twenty days (fig. 28–9).

It took 260 days for the process of pairing numbers with days to return to its initial point. (It can be shown mathematicaly that the number of possible pairs is $13 \times 20 = 260$.)

Each of the 260 days in the sacred year was characterized by its name and its corresponding number, and those two elements situated it perfectly. The following days, for example:

	I	II	III	IV	V	VI	VII	VIII	IX	X	XI	XII	XIII
IMIX	1	8	2	9	3	10	4	11	5	12	6	13	7
IK	2	9	3	10	4	11	5	12	6	13	7	1	8
AKBAL	3	10	4	11	5	12	6	13	7	1	8	2	9
KAN	4	11	5	12	6	13	7	1	8	2	9	3	10
CHICCHAN	5	12	6	13	7	1	8	2	9	3	10	4	11
CIMI	6	13	7	1	8	2	9	3	10	4	11	5	12
MANIK	7	1	8	2	9	3	10	4	11	5	12	6	13
LAMAT	8	2	9	3	10	4	11	5	12	6	13	7	1
MULUC	9	3	10	4	11	5	12	6	13	7	1	8	2
OC	10	4	11	5	12	6	13	7	1	8	2	9	3
CHUEN	11	5	12	6	13	7	1	8	2	9	3	10	4
EB	12	6	13	7	1	8	2	9	3	10	4	11	5
BEN	13	7	1	8	2	9	3	10	4	11	5	12	6
IX	1	8	2	9	3	10	4	11	5	12	6	13	7
MEN	2	9	3	10	4	11	5	12	6	13	7	1	8
CIB	3	10	4	11	5	12	6	13	7	1	8	2	9
CABAN	4	11	5	12	6	13	7	1	8	2	9	3	10
EZNAB	5	12	6	13	7	1	8	2	9	3	10	4	11
CAUAC	6	13	7	1	8	2	9	3	10	4	11	5	12
AHAU	7	1	8	2	9	3	10	4	11	5	12	6	13

Fig. 28–9. The 260 consecutive days of the Mayan sacred year.

 13 CHUEN

 4 IMIX

occupied the 91st and 121st positions, respectively, in the sacred year that began with 1 Imix.

In the religion of the Mayas—and of other Middle American peoples who had calendars similar to theirs—time flowed inexorably and indefinitely in cycles of 260 days. Each of those days was either favorable or unfavorable. A wedding could take place only on a propitious day; a person's future depended on the nature of the day when he was born; a military expedition could not begin on a malignant day, and so on. The sacred calendar was so important that it still survives among the descendants of the Mayas. "Today, in the Guatemalan highlands, a child takes the name of his birthday, as well as its character. Imix symbolizes the hidden forces of the universe that are manifested in madness. A child born on 1 Imix will bear that name and will always be regarded as an abnormal person with unpredictable reactions" (Ivanoff).

Some authors—who rightly maintain that choice of the number 260 could have resulted only from an association between 20 and 13—believe that the origin of the 260-day cycle must have been essentially related to religious considerations and that the sacred calendar represented a kind of symbolic alliance between human beings (associated with 20) and the thirteen gods of the Upper World. To explain the part played by the number 20 in the sacred calendar, these authors point out that 20 was used as a numeration base in Middle America and that the word *uinal*, which designates a period of twenty days, has the same root as *uinic*, "man," which the Mayas used in *hun uinic*, "twenty." They explain the presence of the number 13 by the important place it occupied in Middle American religions: the Mayas believed that the world had thirteen heavens, each ruled by one of the thirteen gods of the Upper World who composed the *Oxlahuntiku* (as opposed to the *Bolontiku*, the nine gods of the Lower World who were ruled by the death god and regulated the succession of nights).

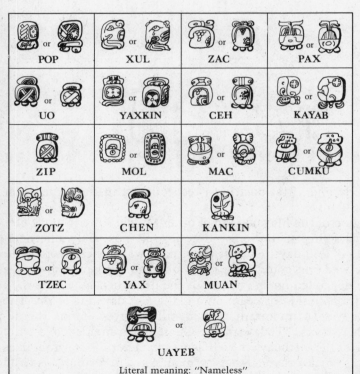

Fig. 28–10.
Glyphs and names of the eighteen twenty-day "months" of the Mayan civil calendar.

The Mayas also had a secular solar calendar. The corresponding year had 365 days and was composed of eighteen *uinals* (twenty-day "months") with an extra five-day period following the last *uinal*.

Here are the names and order of the eighteen *uinals*, or months:

Pop	Yaxkin	Mac
Uo	Mol	Kankin
Zip	Chen	Muan
Zotz	Yax	Pax
Tzec	Zac	Kayab
Xul	Ceh	Cumku

"The months were named according to agricultural or devotional activities of natural phenomena, and each was dedicated to a certain deity. The Maya used name hieroglyphs of patron gods or animals for the month names [fig. 28–10]" (Peterson).

Uayeb, the extra five-day period, was represented by a glyph associated with the ideas of chaos, disaster, and corruption. Those five days, called "phantom" or "useless" days, were regarded as empty, sad, and ill-omened (like the epagomenal days of the ancient Egyptians and Greeks). Those born in that period of the year, it was believed, would be worthless, unlucky, miserable, and poor all their lives. "During those days they did not comb or wash themselves, nor did the men nor women free themselves from lice, nor did they undertake any mechanical or fatiguing work, for fear some misfortune should happen to them" (Diego de Landa, quoted by Peterson).

In the civil calendar, the first day of each month, and of Uayeb, was represented by a combination of the corresponding glyph and the following one:

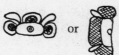

This glyph, usually transcribed by our sign O, indicated that the period in question was beginning, and in the Mayan religion it indicated the day when the god who had carried the preceding month was relinquishing his "temporal burden" and the god of the beginning month was taking possession of it. Since Zip and Zotz, for example, were consecutive months (fig. 28–10), the notation:

0   ZOTZ

evoked the idea of a burden passing from the patron god of Zip to the patron god of Zotz.

The other days of each month were numbered from 1 to 19, the second day bearing the number 1, the third the number 2, and so on to the last one, which bore the number 19. And the last four days of Uayeb were numbered from 1 to 4 (fig. 28–11).

POP	UO	ZIP	ZOTZ	TZEC	XUL	YAXKIN	MOL	CHEN	YAX	ZAC	CEH	MAC	KANKIN	MUAN	PAX	KAYAB	CUMKU	UAYEB
0	0	0	0	0	0	0	0	0	0	0	0	0	0	0	0	0	0	0
1	1	1	1	1	1	1	1	1	1	1	1	1	1	1	1	1	1	1
2	2	2	2	2	2	2	2	2	2	2	2	2	2	2	2	2	2	2
3	3	3	3	3	3	3	3	3	3	3	3	3	3	3	3	3	3	3
4	4	4	4	4	4	4	4	4	4	4	4	4	4	4	4	4	4	4
5	5	5	5	5	5	5	5	5	5	5	5	5	5	5	5	5	5	
6	6	6	6	6	6	6	6	6	6	6	6	6	6	6	6	6	6	
7	7	7	7	7	7	7	7	7	7	7	7	7	7	7	7	7	7	
8	8	8	8	8	8	8	8	8	8	8	8	8	8	8	8	8	8	
9	9	9	9	9	9	9	9	9	9	9	9	9	9	9	9	9	9	
10	10	10	10	10	10	10	10	10	10	10	10	10	10	10	10	10	10	
11	11	11	11	11	11	11	11	11	11	11	11	11	11	11	11	11	11	
12	12	12	12	12	12	12	12	12	12	12	12	12	12	12	12	12	12	
13	13	13	13	13	13	13	13	13	13	13	13	13	13	13	13	13	13	
14	14	14	14	14	14	14	14	14	14	14	14	14	14	14	14	14	14	
15	15	15	15	15	15	15	15	15	15	15	15	15	15	15	15	15	15	
16	16	16	16	16	16	16	16	16	16	16	16	16	16	16	16	16	16	
17	17	17	17	17	17	17	17	17	17	17	17	17	17	17	17	17	17	
18	18	18	18	18	18	18	18	18	18	18	18	18	18	18	18	18	18	
19	19	19	19	19	19	19	19	19	19	19	19	19	19	19	19	19	19	

Fig. 28–11. The 365 days of the Mayan civil year.

The following date, for example:

4   XUL

referred not to the fourth day of Xul but to the fifth.

Each of the twenty days always had the same position in each of the eighteen *uinals* of a given year. If the year began with the day named Eb, for example, all the other months of that year began with Eb, and if Eznab had the seventh position in the first month, it kept that position all through the year.

Because of the five additional days, however, each day had a different number from the one it had had the year before. If Ahau had the number 8 in one year, for example, in the next four years it had the numbers 3, 18, 13, and, again, 8 (fig. 28–12).

Each day thus moved back five places each year, and returned to its initial position every fifth year. Only four days could correspond to the date:

0   POP

The 20 basic days	1st year 𝒰	1st year UAYEB	2nd year 𝒰	2nd year UAYEB	3rd year 𝒰	3rd year UAYEB	4th year 𝒰	4th year UAYEB	5th year 𝒰	5th year UAYEB
Eb	0	0	15		10		5		0	0
Ben	1	1	16		11		6		1	1
Ix	2	2	17		12		7		2	2
Men	3	3	18		13		8		3	3
Cib	4	4	19		14		9		4	4
Caban	5		0	0	15		10		5	
Eznab	6		1	1	16		11		6	
Cauac	7		2	2	17		12		7	
Ahau	8		3	3	18		13		8	
Imix	9		4	4	19		14		9	
Ik	10		5		0	0	15		10	
Akbal	11		6		1	1	16		11	
Kan	12		7		2	2	17		12	
Chicchan	13		8		3	3	18		13	
Cimi	14		9		4	4	19		14	
Manik	15		10		5		0	0	15	
Lamat	16		11		6		1	1	16	
Muluc	17		12		7		2	2	17	
Oc	18		13		8		3	3	18	
Chuen	19		14		9		4	4	19	

U: any one of the 18 20-day months

Fig. 28–12. Successive positions of the twenty basic days in the Mayan civil calendar.

In other words, only four could occupy the first position in the civil year. These four, Eb, Caban, Ik, and Manik, were therefore called the Year Bearers.

Since the Mayas used their 260-day sacred calendar concurrently with their 365-day civil calendar, the complete expression of a date required taking both calendars into account, as shown in figure 28–13.

Because the days recurred cyclically in a fixed order, a complete date (one expressed in terms of both calendars) necessarily recurred at the end

Position of the day in the sacred year

Position of the day in the civil year

Fig. 28–13. Example of a complete date, expressed with reference to both calendars, sacred and civil.

13 AHAU ;   18 CUMKU

of a certain time, which calculation shows to be 18,980 days, or fifty-two civil years. That is, when a certain day of the sacred calendar coincided with a certain day of the civil calendar, those same two days would coincide again in fifty-two civil years (or seventy-three sacred years).

To form a concrete idea of this cycle, imagine two intermeshing cogwheels, A and B, with cogs numbered from 1 to 365 and from 1 to 260, respectively. For the two wheels to return to a given position in relation to each other (the position in which the two cogs numbered 1 coincide, for example), wheel A must make fifty-two revolutions while wheel B makes seventy-three.

To put it in mathematical terms, the number of days in the cycle is equal to the least common multiple of 260 and 365, and since 5 is their greatest common divisor, the number of days in the cycle is equal to:

$$\frac{260 \times 365}{5} = 18,980 \text{ days} = 52 \text{ civil years} = 73 \text{ sacred years}$$

Such is the basis of the fifty-two-year cycle called the Calendar Round, which played an important part in the religious lives of the Mayas, Aztecs, and other pre-Columbian peoples of Middle America. The Aztecs, for example, believed that the end of each sacred cycle would be accompanied by all sorts of catastrophes, and as it approached they offered great numbers of human sacrifices in the hope of inducing their gods to let them live through another sacred cycle.

Chronology and numeration

The Mayas also used a time-counting system, called the Long Count system, in which the basic unit was the day and, for practical reasons, there was a "year" of 360 days. This is the system used in chronological inscriptions, particularly in the lavishly carved stelae that the Mayas periodically erected during the Classic Period, evidently to commemorate important dates and honor the gods (fig. 28–14).

These dated stelae—the oldest one that has been discovered was erected at Tikal in A.D. 292 (fig. 28–1)—provide some of the most interesting chronological inscriptions that the Mayas have left to us. By comparing their extreme dates, we can form an idea of the duration of the great Mayan cities. For example, the earliest known stela erected at Tikal dates from A.D. 292 and the latest from 869; for Uaxactún, the dates are 328 and 889; for Copán, 469 and 800; for Yaxchilán, 509 and 771; for Piedras Negras, 509 and 830; for Palenque, 538 and 785.

In the Long Count system, time was counted in *kins* (days), *uinals* (twenty-day "months"), *tuns* (360-day "years"), *katuns* (cycles of twenty "years"), *baktuns* (cycles of 400 "years"), etc. (fig. 28–15).

Fig. 28–14. Stela A, Quiriguá, erected in A.D. 775. Gods are carved on its front and back, and glyphs (calendrical, astronomic, and others) on its sides.

Orders of units	Names and definitions	Equivalents	Numbers of days
1st	kin DAY		1
2nd	uinal 20-DAY "MONTH"	20 kins	20
3rd	tun 360-DAY "YEAR"	18 uinals	360
4th	katun CYCLE OF 20 "YEARS"	20 tuns	7200
5th	baktun CYCLE OF 400 "YEARS"	20 katuns	144,000
6th	pictun CYCLE OF 8000 "YEARS"	20 baktuns	2,880,000
7th	calabtun CYCLE OF 160,000 "YEARS"	20 pictuns	57,600,000
8th	kinchiltun CYCLE OF 3,200,000 "YEARS"	20 calabtuns	1,152.000,000
9th	alautun CYCLE OF 64,000,000 "YEARS"	20 kinchiltuns	23,040,000,000

Fig. 28–15. Units of the Long Count system.

Since the *tun* had only 360 days, the *katun* was not equal to twenty of our years but to twenty years minus 104.842 days; the *baktun* was equal to 400 years minus 2096.84 days, and so on.

The Mayas used a strictly vigesimal system when they counted persons, animals, or objects, but not when they counted time. The Long Count system has an irregularity at the third order of units. If they had kept the system vigesimal at this point (that is, if they had made the *tun* equal to twenty *uinals* instead of eighteen), they would have had a 400-day "year" even farther removed from the true solar year than the 360-day *tun*.

Each unit of time was represented by a glyph that, like most other Mayan glyphs, occurred in two and sometimes even three different forms, depending on whether it was used in a painted codex, for example, or in a carefully executed carving on a stela or a building. Besides the relatively simple normal form, there was what is called the head variant, showing the head of a god, a person, or an animal. And occasionally (as at Quiriguá and Palenque, where such exceptions were more frequent) there was a form showing the full figure of a god, a person, or an animal (fig. 28–16).

NORMAL	HEAD VARIANT			FULL FIGURE

Fig. 28–16. Different forms of the glyph for *kin* ("day").

Numbers associated with time units in the Long Count system were also indicated by glyphs that could have two or three different forms.

One way of writing numbers 1 to 19 was with head-variant signs: for example, 5 was indicated by the head of the corn god, 10 by the head of the death god, and so on (fig. 28–17).

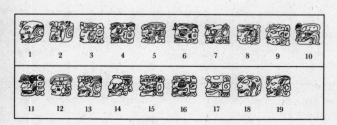

Fig. 28–17. Mayan head-variant numerals for the first nineteen whole numbers.

The glyphs for numbers 1 to 13 were representations of the *Oxla-huntiku*, the thirteen gods of the Upper World. Those for numbers 14 to 19 were derived from those for numbers 4 to 9, as shown in this example:

VARIANTS OF THE GLYPH FOR 9

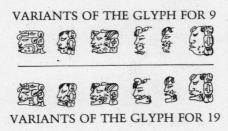

VARIANTS OF THE GLYPH FOR 19

The rule followed here is that the lower jaw is shown fleshless for numbers 14 to 19 (fig. 28–18), the fleshless jaw indicating the death's head which symbolizes 10:

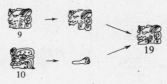

Fig. 28–18. Detail of Lintel 48, Yaxchilán, showing a singular representation of "sixteen days." The sitting monkey is a full-figure glyph sometimes used to indicate the *kin* (day). On his hands he holds the head of the god of the number 6 and, on his legs, the death's head associated with the number 10.

This system, however, appears in only a limited number of inscriptions, and was used even less often in manuscripts.

For numbers 1 to 19, the Mayas much more frequently used a very simple system with only two basic symbols: a dot or a circle for 1 and a vertical or horizontal bar for 5. Numbers 1 to 4 were represented by as many dots or circles as necessary; the number 5 by a bar; numbers 6 to 9 by one, two, three, or four dots placed above or beside the bar; numbers 10 and 15 by two and three bars, and so on (fig. 28–19).

The Mayas could use either of their two graphic systems for Initial Series (Long Count) dates. A date was expressed not in solar or lunar years but in multiples of recurrent periods whose totals indicated the number of days that had passed since a date that we transcribe as 13.0.0.0 4 Ahau 8 Cumku. The Mayas must have chosen that starting date for mythological or religious reasons, but we do not know what they were. It has

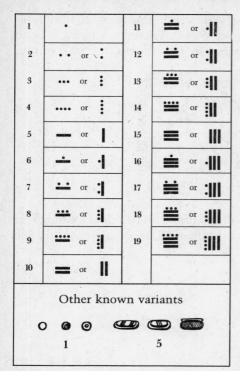

Fig. 28–19. The common Mayan numerals for the first nineteen whole numbers.

been calculated (in what is known as the Thompson Correlation) at August 12, 3113 B.C.

An example of a date represented in this way is provided by the Leyden Plate, shown in figure 28–20. The inscription begins with the following glyph, which introduces the Initial Series:

YAXKIN

This is the name glyph of the patron god of the month (in the civil calendar) containing the last day of the date expressed.

Farther down, the position of this day is specified in relation to both the civil year and the sacred year:

1   Eb   0 Yaxkin

BACK   FRONT

Fig. 28–20. The Leyden Plate, a jade plaque about eight and a half inches long, found near Puerto Barrios, Guatemala, and thought to have been carved at Tikal. On one side a richly dressed figure, probably a god, stands on a captive; on the other side is a date corresponding to A.D. 320. Rijksmuseum voor Volkenkunde, Leyden, Netherlands.

The number of days that have passed since the starting date of the Mayan era is expressed in this form:

8 baktuns

14 katums

3 tuns

1 uinal

12 kins

This date, transcribed as 8.14.3.1.12, is interpreted as follows:

8 *baktuns*	= 8 × 144,000 days	=	1,152,000 days
14 *katuns*	= 14 × 7200 days	=	100,800 days
3 *tuns*	= 3 × 360 days	=	1,080 days
1 *uinal*	= 1 × 20 days	=	20 days
12 *kins*	= 12 × 1 day	=	12 days
		Total:	1,253,912 days

Converting 1,253,912 days into 3433 of our calendar years (using the close approximation of 365.24 days per year) and taking 3113 B.C. as the Mayas' starting date, we can calculate that the date on the Leyden Plate corresponds to A.D. 320.

Fig. 28–21. Initial Series date from the Hieroglyphic Stairway of Palenque, expressed with head-variant numerals.

Let us now consider a date taken from the Hieroglyphic Stairway of Palenque and shown in figure 28–21. Like the preceding date, this one begins with the name glyph of the patron god of the month containing the last day of the date:

POP

This day is expressed by its position in the civil calendar and the sacred calendar:

8 Ahau   13 Pop

And the date is indicated in this way:

9 baktuns

14 katuns   3 tuns

1 uinal   0 kins

INTERPRETATION

GLYPH INTRODUCING THE INITIAL SERIES

The grotesque head in the center is the name glyph of the patron god of the month (here Cumku) in which the last day of the Initial Series falls.

9 BAKTUNS
(9 × 144,000 days = 1,296,000 days)

17 KATUNS
(17 × 7200 days = 122,400 days)

0 TUNS
(0 × 360 days = 0 days)

0 UINALS
(0 × 20 days = 0 days)

0 KINS
(0 × 1 day = 0 days)

13 AHAU

Name glyph of the patron god of the ninth day in the nine-day series (the nine gods of the Lower World)

Undeciphered glyph

Phases of the moon on the last day of the Initial Series (here, new moon)

Position of the current lunar month in the lunar half-year period (here, second position)

Undeciphered glyph

Undeciphered glyph

Current lunar month (here, 29 days in length)

18 CUMKU

Fig. 28–22. Initial Series date on Stela E, Quiriguá, with a Supplementary Series giving various details of it. Transcribed as 9.17.0.0.0 13 *Ahau* 18 *Cumku*, this date corresponds to January 24, A.D. 771. The interpretation is taken from Morley (see bibliography).

Transcribed as 9.8.9.13.0, it is interpreted as follows:

9 *baktuns*	=	9 × 144,000 days	=	1,296,000 days
8 *katuns*	=	8 × 7200 days	=	57,600 days
9 *tuns*	=	9 × 360 days	=	3,240 days
13 *uinals*	=	13 × 20 days	=	260 days
0 *kins*	=	0 × 1 day	=	0 days
		Total:		1,357,100 days

Calculation shows that the date corresponds to A.D 603.

Finally, the date on Stela E, Quiriguá (fig. 28–22), transcribed 9.17.0.0.0, is interpreted:

9 *baktuns*	=	9 × 144,000 days	=	1,296,000 days
17 *katuns*	=	17 × 7200 days	=	122,400 days
0 *tuns*	=	0 × 360 days	=	0 days
0 *uinals*	=	0 × 20 days	=	0 days
0 *kins*	=	0 × 1 day	=	0 days
		Total:		1,418,400 days

This stela, commemorating the date 13 Ahau 18 Cumku, was therefore erected 1,418,400 days after the starting point of Mayan chronology, which corresponds to January 24, A.D. 771.

Discovery of the place-value principle and the zero in the New World

In their inscriptions, the Mayas symbolized the absence of *kins*, *uinals*, *tuns*, etc. with a glyph that could take many different forms (figs. 28–23, 28–24).

Fig. 28–23. Detail of a plaque from Palenque, showing a full-figure representation of "0 kins." (See figs. 28–16, 28–24.)

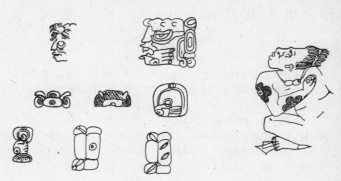

Fig. 28–24. Forms of the zero sign in Mayan carved inscriptions.

It may seem that they introduced a zero sign into a system that had no need for it. The date on Stela E, Quiriguá (fig. 28–22), for example, was recorded in the form of 9 *baktuns*, 17 *katuns*, 0 *tuns*, 0 *uinals*, 0 *kins*, but could it not have been recorded more simply as 9 *baktuns*, 17 *katuns*?

Guitel suggests that the use of a special sign to indicate the absence of units of a certain order was dictated by practical, religious, and aesthetic reasons. Because of their stately, ritual character, which required special care in the "layout" of the glyphs and divine images on them, Mayan stelae were like "large stone checkerboards [figs. 28–14, 28–22] on which glyphs were ordered as rigorously as if they had been tokens on a counting board" (Guitel 1). The time units, from the *kin* to the *baktun* (and sometimes even from the *kin* to the *alautun*), always appeared on the stelae in order of magnitude (fig. 28–15). When a certain time unit was missing, it was necessary—for aesthetic reasons and to avoid error—to fill in the corresponding empty space with a sign that, combined with the sign for the time unit, indicated its absence.

And because of that, the Mayas were led to devise a remarkable place-value written numeration with a genuine zero.

It is striking to see yet another example of how peoples far removed from each other in both space and time, but proceeding by trial and error under similar conditions, took similar paths and arrived at similar results.

Let us return for a moment to the common Chinese hybrid numeration (fig. 25–21) in which numbers composed of several orders of units are traditionally expressed by placing the sign for 10 between the sign for units of the first order and the sign for units of the second order, the sign for 100 between the sign for units of the second order and the sign for units of the third order, and so on. But in some Chinese works (such as a table of logarithms published in 1713 by Emperor Kăngshi, and a work entitled *Ting Chü Suan-fa*, dating from 1355), the signs for powers of 10 are often omitted, so that the number 67,859, for example, is expressed in this form:

六 七 八 五 九   instead of   六 萬 七 千 八 百 五 十 九

6   7   8   5   9   6 × 10,000 + 7 × 1000 + 8 × 100 + 5 × 10 + 9

This shift toward a strictly place-value notation is in the normal line of development of systems based, at least partially, on the hybrid principle.

That is probably why the sign for 100 is occasionally omitted in some Aramaic numerical representations from the early Christian era. In a Hatran inscription dated 436 of the Seleucid era (A.D. 124–125), for example, the date is noted in this shortened form:

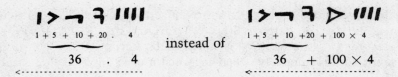

1 + 5 + 10 + 20 . 4   instead of   1 + 5 + 10 + 20 + 100 × 4

36 . 4   36 + 100 × 4

In the Middle Ages, some Jewish scholars who ordinarily wrote numbers such as 5845 in this form:

5 + 40   HUNDRED   8   THOUSAND   5

Powers of 10 written out

sometimes abridged it as follows:

5 + 40 , 8 . 5

The Greek mathematician Diophantus of Alexandria (c. A.D. 250), after representing numbers such as 98,610,732 in this form:

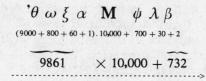

(9000 + 800 + 60 + 1) . 10,000 + 700 + 30 + 2

9861   × 10,000 + 732

later adopted this simplified notation:

$$'\theta\ \omega\ \xi\ \alpha\ .\ \ \psi\ \lambda\ \beta$$

(9000 + 800+60+1) 700+30+2

9861 . 732

-------------------------------->

And, like the Chinese, users of the Tamil and Malayalam systems (in southern India) often omitted signs for tens, hundreds, thousands, etc. (fig. 25–20):

Example: 5843

	Complete notation	Abridged notation
Tamil system	௫ ௲ ௮ ௱ ௪ ௰ ௩ 5 1000 8 100 4 10 3 -------------------->	௫ ௮ ௪ ௩ 5 8 4 3 -------------------->
Malayalam system	൫ ൲ ൮ ൱ ൪ ൰ ൩ 5 1000 8 100 4 10 3 -------------------->	൫ ൮ ൪ ൩ 5 8 4 3 -------------------->

The above examples show that partially hybrid numerations (fig. 25–21, types 1, 2, and 3) have been able to evolve only toward partially place-value notations, while completely hybrid numerations (fig. 25–21, type 4) have evolved toward strictly place-value notations.

Generally speaking, users of a hybrid numeration may become inclined to shorten it, when there is no danger of confusion, by omitting signs for orders of units, leaving only the corresponding coefficients.

The Mayan Long Count system of recording dates and durations can be regarded as a hybrid numeration, for two reasons:

1. The time units are always written in the same order (in order of decreasing value, going from top to bottom).
2. Multiples of the time units are expressed by using the multiplicative principle, and more complex dates are expressed by using the additive as well as the multiplicative principle.

The expression 5 *pictuns,* 17 *baktuns,* 11 *katuns,* 8 *tuns,* 0 *uinals,* 6 *kins,* for example, corresponds to this number of days:

$$5 \times 2{,}880{,}000 + 17 \times 144{,}000 + 11 \times 7200 + 8 \times 360 + 0 \times 20 + 6,$$

which in turn corresponds to:

$$5 \times (18 \times 20^4) + 17 \times (18 \times 20^3) + 11 \times (18 \times 20^2) + 8 \times (18 \times 20) + 0 \times 20 + 6.$$

Like all hybrid systems, this one had within it the germ of the discovery of a true place-value numeration. In the codices, Long Count expressions of dates were often simplified by omitting the glyphs for time units, leaving only the corresponding numerical coefficients. In the Dresden Codex, for example, this period of time:

5 *baktuns,* 17 *katuns,* 6 *tuns,* 11 *uinals,* 19 *kins*
$$(= 5 \times 144{,}000 + 17 \times 7200 + 6 \times 360 + 11 \times 20 + 19 = 844{,}799 \text{ days})$$

is expressed in the form:

5   17   6   11   19,

with those five numerical coefficients written in the places associated respectively with *baktuns, katuns, tuns, uinals,* and *kins.*

The glyph used in carved inscriptions to indicate the absence of units of a certain order was replaced in the codices with an equivalent glyph that was usually easier to reproduce (figs. 28–24, 28–25).

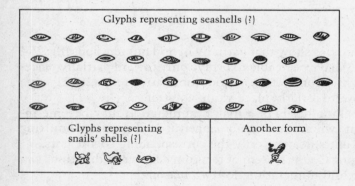

Glyphs representing seashells (?)

Glyphs representing snails' shells (?)

Another form

Fig. 28–25.
Forms of
the zero sign
in the codices.

In their manuscripts, the Mayas thus used a written numeration of base 20, in which the value of a sign depended on its position in a written number. Besides the zero, the system had only two basic signs (fig. 28–19), a dot for 1 and a bar for 5, and the units of each order were indicated by combinations of them:

1   2   4   6   9   13   18

Each number greater than 20 was written in a column with as many horizontal lines as there were orders of units, and was read from top to bottom in decreasing order of values, the bottom line being associated with units of the first order, the second line with multiples of 20, the third line with multiples of 360 = 18 × 20 (and not with multiples of 400 = 20 × 20), the fourth line with multiples of 7200 = 18 × 20² (and not with multiples of 8000 = 20 × 20 × 20), and so on.

Example: 13,495

1 .... 1 × 7200

17 .... 17 × 360

8 .... 8 × 20

15 15

When units of a certain order were missing in the expression of a number, a form of the zero sign (fig. 28–25) was placed on the line associated with that order. The numbers 20 (= 1 × 20 + 0) and 1,087,200 (= 7 × 144,000 + 11 × 7200 + 0 × 360 + 0 × 20 + 0), for example, were represented in these forms, respectively:

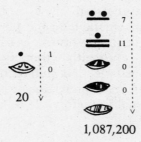

1,087,200

This numeration was therefore based on the place-value principle and had a genuine zero. It was not, however, strictly vigesimal; if it had been, the successive units would have corresponded to 1, 20, 20², 20³, 20⁴, and so on. But beginning with the third order it had an irregularity, since its units were 1, 20, 18 × 20, 18 × 20², 18 × 20³, and so on.

Because of this irregularity, the Mayan zero—which was used in a medial as well as a final position (fig. 28–26)—did not function as an operator, since placing a zero at the end of a numerical expression did not multiply the corresponding number by 20:

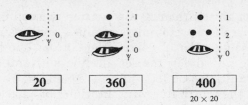

In a vigesimal place-value system with no irregularities, the zero functions as an operator because a notation in the form 10 corresponds to the number $1 \times 20 + 0 = 20$, and a notation in the form 100 (derived from the previous one by adding a zero) corresponds to the number $1 \times 20^2 + 0 \times 20 + 0 = 400 = 20 \times 20$. But in the Mayan system the forms 10 and 100 correspond respectively to $1 \times 20 + 0 = 20$ and $1 \times 360 + 0 \times 20 + 0 = 360$, and the number $20 \times 20 = 400$ was written in the form 120 $(= 1 \times 360 + 2 \times 20 + 0)$, not in the form 100.

This irregularity is not surprising, however, because the system was designed not for the needs of ordinary calculation but for those of astronomy and time counting.

Conclusion

It is true that the main concerns of Mayan scholars were mystical and divinatory, but it is also true that, in the history of civilizations, astrology and religion have often opened the way for philosophy and science. In any case, we must admire that brilliant line of astronomer-priests who were able to develop such remarkable concepts and achieve such amazingly precise results in astronomy on the basis of observations that, as far as we know, were made only with extremely rudimentary equipment. The Mayas' intellectual, artistic, and architectural achievements make their civilization one of the most prestigious in pre-Columbian America, and one of the most unusual in history.

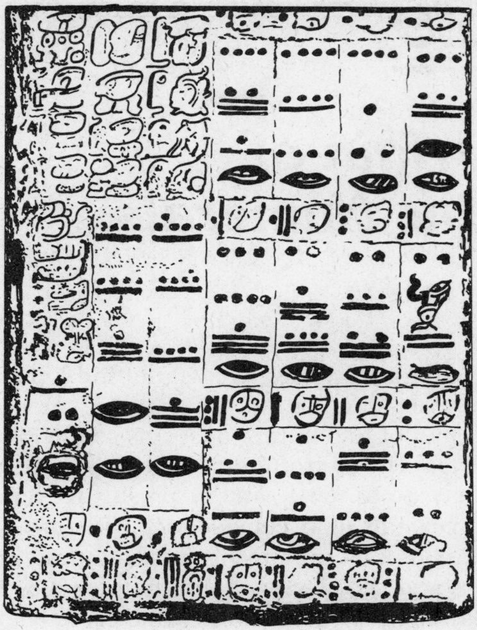

The tables (right side) contain Mayan numeral columns labeled L, K, J, I / H, G, F, E / D, C, B, A:

L	K	J	I
4	4	4	3
17	9	1	13
6	4	2	0
0	0	0	0
35,040	**32,120**	**29,200**	**26,280**

H	G	F	E
3	2	2	2
4	16	8	0
16	14	12	10
0	0	0	0
23,360	**20,440**	**17,520**	**14,600**

D	C	B	A
1	1	16	8
12	4	4	2
8 (*)	6	0	0
0	0		
11,680	**8760**	**5840**	**2920**

*Omission of the three dots by the scribe

The numbers recorded are part of a table concerning the synodical revolution of Venus (which was reckoned at 584 days) and can be expressed as follows:

$$
\begin{aligned}
A &= 2920 &&= 1 \times 2920 \; [= 5 \times 584 \text{ days}] \\
B &= 5840 &&= 2 \times 2920 \; [= 10 \times 584 \text{ days}] \\
C &= 8760 &&= 3 \times 2920 \; [= 15 \times 584 \text{ days}] \\
D &= 11{,}680 &&= 4 \times 2920 \; [= 20 \times 584 \text{ days}] \\
E &= 14{,}600 &&= 5 \times 2920 \; [= 25 \times 584 \text{ days}] \\
F &= 17{,}520 &&= 6 \times 2920 \; [= 30 \times 584 \text{ days}] \\
G &= 20{,}440 &&= 7 \times 2920 \; [= 35 \times 584 \text{ days}] \\
H &= 23{,}360 &&= 8 \times 2920 \; [= 40 \times 584 \text{ days}] \\
I &= 26{,}280 &&= 9 \times 2920 \; [= 45 \times 584 \text{ days}] \\
J &= 29{,}200 &&= 10 \times 2920 \; [= 50 \times 584 \text{ days}] \\
K &= 32{,}120 &&= 11 \times 2920 \; [= 55 \times 584 \text{ days}] \\
L &= 35{,}040 &&= 12 \times 2920 \; [= 60 \times 584 \text{ days}]
\end{aligned}
$$

Fig. 28–26. Page 24 of the Dresden Codex (detail).

29

The Origin of Hindu-Arabic Numerals

"If we knew history better," said Emile Mâle, "we would find a great intelligence at the origin of every innovation."

Our modern written numeration, of Indian origin, seems so obvious to us that it is difficult for us to realize its profundity and importance. Our civilization uses it unthinkingly, so to speak, and as a result we tend to be unaware of its merits. But no one who considers the history of numerical notations can fail to be struck by the ingenuity of our system, because its use of the zero concept and the place-value principle gives it an enormous advantage over most of the other systems that have been devised through the centuries.

The keystone of our modern numeration system

As the first step in tracing the line of development that led to our system, I will summarize and compare the main numerations considered so far. As Guitel has shown, they can all be divided into three basic types (fig. 29–1).

The first type is based on the additive principle, and its numerical signs are entirely independent of each other.

The Egyptian hieroglyphic numeration, which is undoubtedly one of the most primitive, is a written version of the ancient concrete systems that consist in representing a given number by bringing together as many material objects as necessary. It has a special sign only for 1 and each power of the base 10, and repeats these signs as often as necessary; to write the number 7897, for example, 31 signs are needed. Because of the tedious repetitions of signs that it requires, such a system is clumsy to use. To alleviate this difficulty, some peoples introduced additional signs.

The Greeks and the Sabaeans, for example, attributed a special sign not only to each of the numbers 1, 10, 100, 1000, 10,000, etc., but also to each of the supplementary units 5, 50, 500, 5000, 50,000, etc. The Etruscan and Roman numerations had this feature, but they had it from their inception, since they were a natural extension of the use of notches

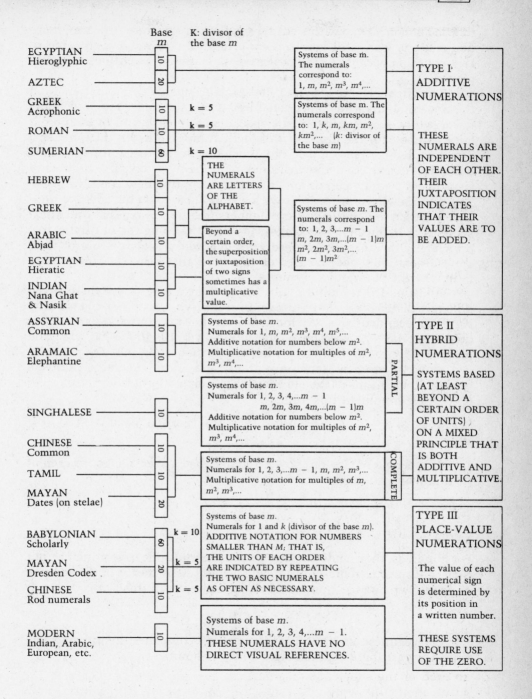

Fig. 29–1. Types of written numerations. Classification by G. Guitel, revised by the author.

in wood or bone. Even with signs for supplementary units, however, these numerations were still cumbersome. And some of them even came to lack cohesion because of the multitude of graphic conventions they employed.

The Greek, Hebrew, Syriac, and Arabic alphabetic numerations, and the ancient Indian and Egyptian hieratic numerations, required many fewer symbols: they attributed a special sign to each unit of each order (nine symbols for the nine units of the first order, nine for the nine tens, nine for the hundreds, and so on). But while such procedures shortened numerical expressions, they required memorizing a very large number of signs and, beyond a certain order of units, adopting new conventions.

That is why some peoples with a system based on the additive principle were led to represent units of higher orders not by repeating identical symbols or by means of special symbols but by using the multiplicative principle. Other peoples, probably influenced by the structure of their oral numerations, had the idea of extending that principle to representations of the other orders of units.

And this led to numerations of the hybrid type, based on a mixed principle that is both additive and multiplicative, with signs that are only partly independent. The Aramaeans and the Assyro-Babylonians attributed a special sign to each of the numbers 1 10, 100, 1000, etc., and proceeded by multiplication for hundreds, thousands, etc. They continued, however, to represent units, tens, and, more generally, numbers below 100 by means of the old additive principle; they wrote the number 6345, for example, in this form:

$$(1 + 1 + 1 + 1 + 1 + 1) \times 1000 + (1 + 1 + 1) \times 100 + (10 + 10 + 10 + 10) + (1 + 1 + 1 + 1 + 1)$$

The people of Ceylon (now Sri Lanka) had a special sign for each of the following numbers:

1	2	3	4	5	6	7	8	9
10	20	30	40	50	60	70	80	90
100	1000,	etc.						

For hundreds, thousands, etc., they used the multiplicative principle, but they continued to represent numbers below 100 by means of the additive principle, writing the number 6345, for example, in this form

$$6 \times 1000 + 3 \times 100 + 40 + 5$$

The Chinese and the Tamils created "complete" hybrid systems, since they attributed special signs to these numbers:

1	2	3	4	5	6	7	8	9
10	100	1000	10,000,	etc.				

and expressed tens, hundreds, thousands, etc. by using only the multi-plicative principle, writing the number 6345 in this form:

$$6 \times 1000 + 3 \times 100 + 4 \times 10 + 5$$

To see why place-value systems are superior to all others, we can begin by considering the Greek alphabetic numeration. It has very short notations for commonly used numbers: no more than four signs are needed for any number below 10,000. But that is not the main criterion for judging a written numeration. What matters most is the ease with which it lends itself to arithmetical operations. Although Greek and Byzantine mathe-maticians succeeded in performing those operations with their alphabetic numerals, their procedures were much more complicated than ours and they had to use all sorts of artifices.

Now let us take the case of the common Chinese numeration. It is simple up to a certain point but, practically speaking, it is not a means of representing all whole numbers because it would require an unending process of creating new symbols.

Using a small number of signs, a properly conceived place-value numeration permits not only simple representation of any number, how-ever large, but also easy performance of arithmetical operations. And that is why our present system is one of the basic intellectual tools of the modern world. To realize its efficiency, we have only to try to perform an addition by means of Roman numerals, for example:

CCLXVI	266
MDCCCVII	1807
DCL	650
MLXXX	1080
MMMDCCCIII	3803

It does not take long to see that without converting the Roman numerals into our modern system the problem is difficult if not impossible to solve. And this is only an addition—multiplication or division would be even worse. Such systems do not lend themselves to calculation because of the static nature of their basic numerals, which are essentially only ab-breviations for recording the results of calculations already done by means of a counting board or abacus.

"That is why, from the beginning of history until the advent of our modern *positional* numeration, so little progress was made in the art of reckoning.

"Not that there were no attempts to devise rules for operating on these numerals. How difficult these rules were can be gleaned from the great awe in which all reckoning was held in these days. A man skilled in the art was regarded as endowed with almost supernatural powers. . . .

"And to a certain extent this awe persists to this day. The average man identifies mathematical ability with quickness with figures. 'So you are a mathematician? Why, then you must have no trouble with your income-tax return!' What mathematician has not at least once in his career been so addressed?" (Dantzig).

The following story about a German merchant illustrates the situation as it still continued in the fifteenth century. The merchant "had a son whom he desired to give an advanced commercial education. He appealed to a prominent professor of a university for advice as to where he should send his son. The reply was that if the mathematical curriculum of the young man was to be confined to adding and subtracting, he could perhaps obtain the instruction in a German university; but the art of multiplying and dividing, he continued, had been greatly developed in Italy, which in his opinion was the only country where such advanced instruction could be obtained.

"As a matter of fact, multiplication and division as practiced in those days had little in common with the modern operations bearing the same names. Multiplication, for instance, was a succession of *duplations*, which was the name given to the doubling of a number. In the same way division was reduced to *mediation*, i.e. 'halving' a number. . . .

"We begin to understand why humanity so obstinately clung to such devices as the abacus or even the tally. Computations which a child can now perform required then the services of a specialist, and what is now only a matter of a few minutes meant in the twelfth century days of elaborate work.

"The greatly increased facility with which the average man today manipulates number has been often taken as proof of the growth of the human intellect. The truth of the matter is that the difficulties then experienced were inherent in the numeration in use, a numeration not susceptible to simple, clear-cut rules. The discovery of the modern positional numeration did away with those obstacles and made arithmetic accessible even to the dullest mind" (Dantzig).

The place-value (positional) principle was certainly not obvious, since its discovery eluded most peoples, including those of Europe, who had to wait till the Arabs transmitted it to them after receiving it from the Indians.

It was independently conceived only four times in history. Three of those conceptions were by the Babylonians (probably in the early second millennium B.C.), the Mayas (probably in the Classic Period, third to ninth centuries A.D.), and the Chinese (shortly before the beginning of the Christian era). But the Babylonian, Mayan, and Chinese place-value systems were defective in comparison with the Indian numeration, which is the ancestor of ours (fig. 29–1).

The Babylonian system was of base 60, had special signs only for 1 and 10, and proceeded by additive repetitions of those two signs for all other numbers below 60, which means that the auxiliary base 10 and the additive principle were used for representing the units of each sexagesimal order.

The Mayan system of base 20 followed the same pattern: it had special signs only for 1 and 5, repeated those two signs for other numbers below 20, and therefore used the auxiliary base 5 and the additive principle for representing the units of each order. Moreover, since it was adapted to the Mayan method of time counting, it had an irregularity beginning at the third order: instead of attributing the values 1, 20, $20^2 = 400$, $20^3 = 8000$, etc. to the consecutive orders of units, as would normally be done in a vigesimal system, it attributed to them the values 1, 20, $18 \times 20 = 360$, $18 \times 20^2 = 7200$, etc. And it was this irregularity that deprived the Mayan system of any operative possibilities.

The Chinese place-value system was decimal, like ours, but it represented units of the first order by means of the additive principle, with a special sign for 1 and a symbolic indication of 5.

We can now understand the ingenuity and superiority of place-value systems of Indian origin, in which the basic numerals are signs with no direct visual reference and the place-value principle is used consistently with powers of the base 10.

In conjunction with the place-value principle, discovery of the zero marks the decisive stage in a process of development without which we cannot imagine the progress of modern mathematics, science, and technology. The zero freed human intelligence from the counting board that had held it prisoner for thousands of years, eliminated all ambiguity in the written expression of numbers, revolutionized the art of reckoning, and made it accessible to everyone. "And the influence of this great discovery was by no means confined to arithmetic. By paving the way to a generalized number concept it played just as fundamental a rôle in practically every branch of mathematics" (Dantzig).

We appreciate the importance of that discovery when we realize that it was made only three times in history: by Babylonian scholars, by Mayan astronomer-priests, and by Indian mathematicians and astronomers. (Although Chinese mathematicians conceived the place-value principle, the zero concept was evidently still unknown to them for many centuries: it was not introduced in China till the eighth century A.D., no doubt from India.)

In their Initial Series dates the Mayas used a special sign to mark the absence of time units of a certain order, and thus created a genuine zero sign that they used in medial and final positions in numerical representations (fig. 29–2). But because of the irregularity in their numeration, this zero had no operative possibilities.

For nearly fifteen centuries Mesopotamian scholars did not use a zero in their numerical representations, and this sometimes caused confusion. In an effort to overcome the difficulty, they first left an empty space to indicate the absence of units of a certain order (it was as if we were to write the number 106 as 1 6). But the problem was not solved, because the empty space was often omitted. Furthermore, it was difficult to indicate the absence of two or more consecutive orders in that way, since the difference between one empty space and two or more of them was not easily discernible. It was not until a relatively late period, probably about the fourth century B.C., that a special sign was introduced for that purpose. Mathematicians evidently used this zero sign only in a medial position, but astronomers used it in both medial and final positions in their numerical representations and were therefore probably aware of its operative possibilities. In no case, however, was the Babylonian zero conceived as a number, that is, as a synonym of "nothing" or "null quantity."

Our modern zero, a direct descendant of the Indian zero, is obviously superior to the two others (fig. 29–2).

The zero and the place-value principle were not discovered through the unaided efforts of some solitary genius: they appeared after a long process of trial and error, groping, reflection, and invention. "The passage of time enabled some scholars to succeed, when it was possible, in perfecting the primitive instrument they had inherited from their ancestors. They did it because they had a kind of passion for large numbers and wanted to record them. Other scholars, realistic and persistent, succeeded in making the reckoners of their time accept that revolutionary innovation. We are the heirs of both groups" (Guitel 1).

A significant anecdote

The Arab poet al-Sabhadi, who lived in Baghdad in the Middle Ages, wrote in one of his works that there were three things on which the Indian nation prided itself: its methods of reckoning, the game of chess, and the book titled *Kalila wa Dimna* (a collection of legends and fables rather similar to those of Aesop).

He tells us that in his time and region numbers were recorded and reckoning was done by means of the following ten signs, which he calls "Indian figures":

He explains that these signs have variable values, depending on their position, and that the dot is used to mark absent units. After praising this method for its superiority over the system of Arabic numeral letters and for the great ease and elegance of its use in reckoning, he gives several

	MAYAN	BABYLONIAN	INDIAN	MODERN
SYSTEMS	Base 20 with an irregularity beginning at the third order	Base 60	Base 10	Base 10
	Place-value principle			
	Basic numerals, excluding zero, formed by additive combinations of these signs: • — 1 5	Y ⟨ 1 10	Basic numerals, excluding zero, with no direct visual reference: ୧ ୨ ୩ ୪ ୫ ୬ ୭ ୮ ୯ 1 2 3 4 5 6 7 8 9	1 2 3 4 5 6 7 8 9
ZEROS	◉	⪦ or 𐏓 or ⪨	೦	0

This sign (originally meaning "empty") marks the absence of units of a certain order in a numerical representation.

Attested:	Attested:	Attested:	Used:
– In medial position 9 0 0 7 $9 \times 7200 + 0 \times 360 + 0 \times 20 + 7$	– In medial position 9 0 0 7 ---------> $9 \times 60^3 + 0 \times 60^2 + 0 \times 60 + 7$	– In medial position 9 0 0 7 ------------> $9 \times 10^3 + 0 \times 10^2 + 0 \times 10 + 7$	– In medial position 9 0 0 7
– In final position 6 4 9 0 $6 \times 7200 + 4 \times 360 + 9 \times 20 + 0$	– In final position (evidently used only by astronomers) 6 4 9 0 --------> $6 \times 60^3 + 4 \times 60^2 + 9 \times 60 + 0$	– In final position 6 4 9 0 ----------> $6 \times 10^3 + 4 \times 10^2 + 9 \times 10 + 0$	– In final position 6 4 9 0

When this zero is placed at the end of a numerical representation, it indicates that the number represented before it is to be multiplied by the base (e.g., 120 = 12 × base).

Example:
640 = 64 × 10

	This sign eventually came to be regarded as representing a number, the "number zero." It then became a synonym of "nothing."	Sign representing the "zero value" or "null number."

Fig. 29–2. Comparative table of the different zeros in history.

This zero is fundamental to all of modern mathematics.

examples of it. Then, in the course of presenting the Indian view of the origin of chess, he relates an anecdote that I will repeat here in a freely adapted version.

One day an Indian scholar named Sessa invented chess. Later, this game was presented to the King of India. He was so amazed by its ingenuity and the great variety of its possible combinations that he sent for Sessa so that he could reward him in person.

"For your remarkable invention," the king said enthusiastically, "I'm willing to give you any reward you want. To you, my generosity will be boundless."

"I'm deeply grateful to you, Your Majesty," replied Sessa, "but I won't ask for a great reward. I'd simply like you to give me as many grains of wheat as it would take to fill the sixty-four squares of my chessboard if the first square held one grain, the second one two, the third one four, the fourth one eight, and so on, doubling the number of grains from one square to the next."

"Your request is much too modest!" exclaimed the king, surprised. "You offend me by asking for something so unworthy of my benevolence and so insignificant compared with the magnificence of what I can give you."

But Sessa insisted till the king resigned himself and ordered his vizier to bring the "bag of wheat" that Sessa wanted.

A few days later, the king asked his vizier if that fool Sessa had taken possession of his meager reward.

"The reckoners attached to your court haven't yet determined the number of grains," said the vizier. "They hope to finish tonight, before dawn."

But the next morning the royal reckoners had not yet calculated the amount of wheat that had to be given to Sessa.

The king was exasperated.

"How can they take so long to solve such a simple problem? I order them to finish before tomorrow morning."

The problem was still unsolved the next day. Judging them to be incompetent, the king angrily dismissed his reckoners and replaced them with others. But the new ones made no faster progress than the old ones.

After several days of ceaseless work, the leader of the reckoners came to the king to announce the final result.

"Well, have you given Sessa his reward?" the king asked impatiently.

"Forgive me for saying this, Your Majesty, but in spite of all your wealth and power it's impossible for you to provide that reward. Even if you emptied all the granaries in your kingdom, you'd have only a minute fraction of the amount of wheat you promised to Sessa. There's not that much wheat in the granaries of all the kingdoms on earth."

"Be more specific! Exactly how much is it?"

"Your Majesty, the number of grains of wheat that Sessa wants is eighteen quintillion, four hundred and forty-six quadrillion, seven hundred and forty-four trillion, seventy-three billion, seven hundred and nine million, five hundred and fifty-one thousand, six hundred and fifteen."

The number can be written in our modern notation as:

18,446,744,073,709,551,615.

Al-Sabhadi claims to have obtained this result in a very short time, using an Indian method. He expresses it in the following notation, which he rightly praises for its concision and precision, and which differs from ours only in the forms of its signs:

۱	*۲*	*۳*	*۴*	*8*	*۶*	*۷*	*۸*	*۹*	*•*
1	2	3	4	5	6	7	8	9	0

"Sessa showed as much ingenuity in making his request as he did in inventing the game of chess!" said the king. "Since it's impossible for me to pay my debt to him, what do you advise me to do?"

"Tell him that he must count the grains of wheat himself," replied the reckoner. "Even if he worked constantly, night and day, counting them at the rate of one per second, and lived another fifty or sixty years, it wouldn't take a very large room to hold all the grains he could count before he died."

The cradle of our modern system

That system is as important as the invention of agriculture, the wheel, writing, or the steam engine. To which people are we to attribute the honor of first devising it? A long-lived tradition attributes it to the Arabs, but in all likelihood the first written numeration of base 10 with the same structure as ours, and graphic signs that prefigured our modern digits, was born in southern India some fifteen centuries ago. This is supported by abundant evidence and it has often been stated by Arab as well as European authors.

The earliest known examples are provided by documents on copperplate dating from the sixth to tenth centuries A.D. Written in Sanskrit, these engraved documents are deeds concerning property given to Brahmanist religious authorities by kings or wealthy individuals. They give the name of the donor, describe the property, and record the date in accordance with one of the Indian eras then in use (Chedi, Shaka, Vikramaditya, etc.). These dates are expressed either in words or in signs whose values depend on their position (fig. 29–4). The deed shown in figure 29–3 is regarded as the oldest known example of the Indian place-value numeration. It is dated 346 Chedi (A.D. 595) and indicates the number 346 as follows:

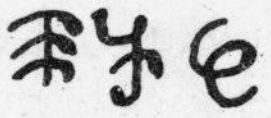

$$\text{ॠ} \quad \text{ヰ} \quad \text{ℓ}$$

3   4   6

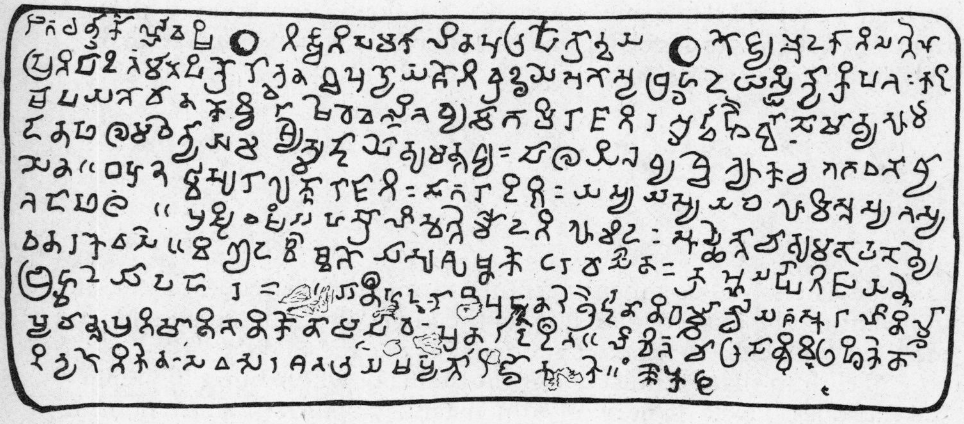

Fig. 29–3. Copperplate deed dated 346 Chedi (A.D. 595), from Sankheda, near the modern city of Borach in western India.

Other examples are in two stone inscriptions from the reign of Bhojaveda (second half of the ninth century A.D.), discovered in the nineteenth century near Gwalior, about 180 miles south of Delhi, in the temple of Vaillabhatta Svamin consecrated to Vishnu.

The first of these inscriptions is dated (with the number words written out) 932 Vikramaditya, which corresponds to A.D. 875. It contains a Sanskrit poem whose twenty-six verses are numbered as follows:

| 1 | 2 | 3 | 4 | 5 | 6 | 7 | 8 | 9 | 10 | 11 | 12 | 19 | 20 | 26 |

The second inscription is dated, in numerical signs, 933 Vikramaditya (A.D. 876). Written in incorrect Sanskrit, it is a deed of property given to the temple of Vaillabhatta Svamin by the inhabitants of Gwalior. It describes a piece of land, 270 *hastas* long and 187 *hastas* wide, to be used as a flower garden from which the temple will be provided with fifty garlands of flowers each day. The numbers 933, 270, 187, and 50 are indicated in these notations:

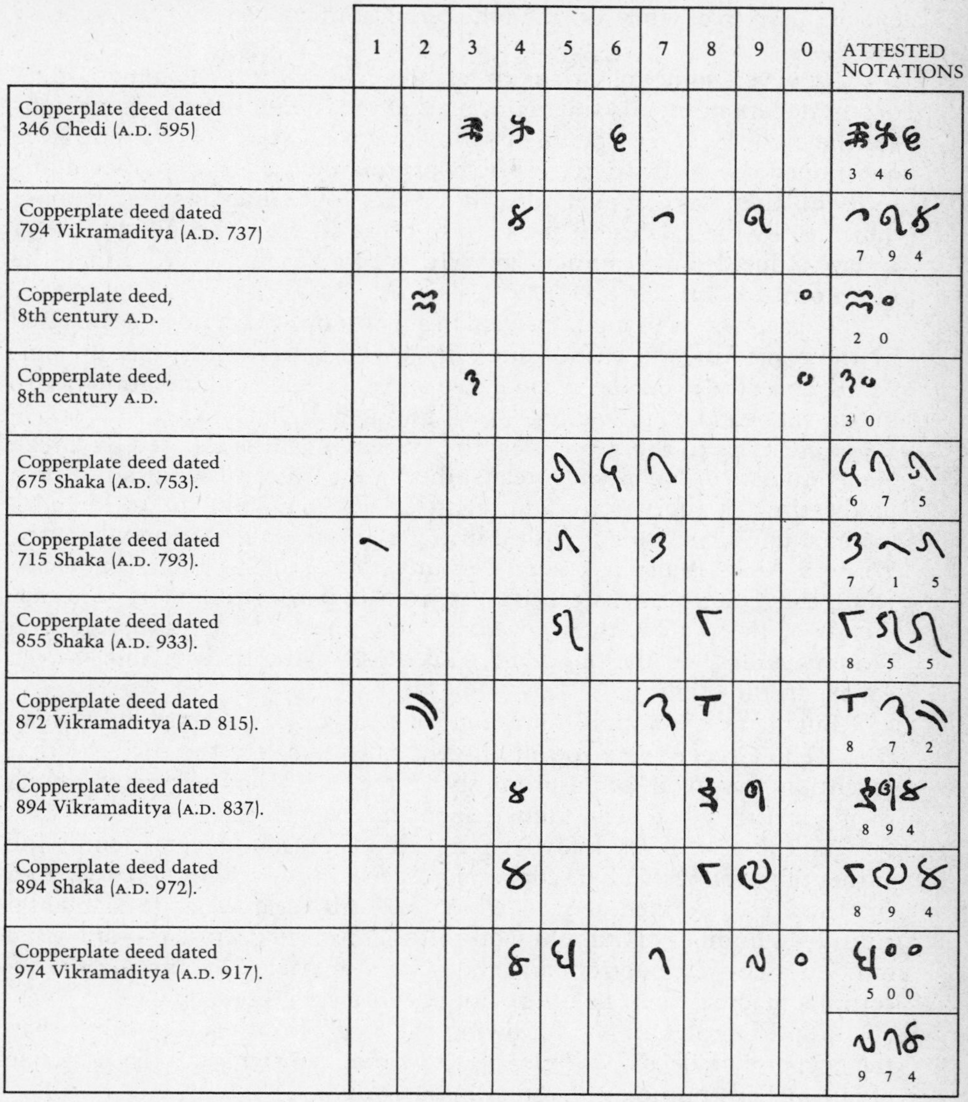

	1	2	3	4	5	6	7	8	9	0	ATTESTED NOTATIONS
Copperplate deed dated 346 Chedi (A.D. 595)			𑀳	𑀴		𑁆					𑀳𑀴𑁆 3 4 6
Copperplate deed dated 794 Vikramaditya (A.D. 737)				𑀵			𑀶	𑀷			𑀶𑀷𑀵 7 9 4
Copperplate deed, 8th century A.D.		≋								०	≋० 2 0
Copperplate deed, 8th century A.D.			𑀳							०	𑀳० 3 0
Copperplate deed dated 675 Shaka (A.D. 753).				𑀵	𑁆	𑀶					𑁆𑀶𑀵 6 7 5
Copperplate deed dated 715 Shaka (A.D. 793).	𑀶				𑀵		𑀳				𑀳𑀶𑀵 7 1 5
Copperplate deed dated 855 Shaka (A.D. 933).					𑀵			𑀶			𑀶𑀵𑀵 8 5 5
Copperplate deed dated 872 Vikramaditya (A.D 815).		𑀶					𑀳	𑀶			𑀶𑀳𑀶 8 7 2
Copperplate deed dated 894 Vikramaditya (A.D. 837).				𑀵				𑀳	𑀶		𑀳𑀶𑀵 8 9 4
Copperplate deed dated 894 Shaka (A.D. 972).				𑀵				𑀶	𑀷		𑀶𑀷𑀵 8 9 4
Copperplate deed dated 974 Vikramaditya (A.D. 917).				𑀵	𑀶		𑀷		𑀲	०	𑀵०० 5 0 0
											𑀲𑀷𑀵 9 7 4

Fig. 29–4. Some of the oldest known examples of the Indian place-value numeration.

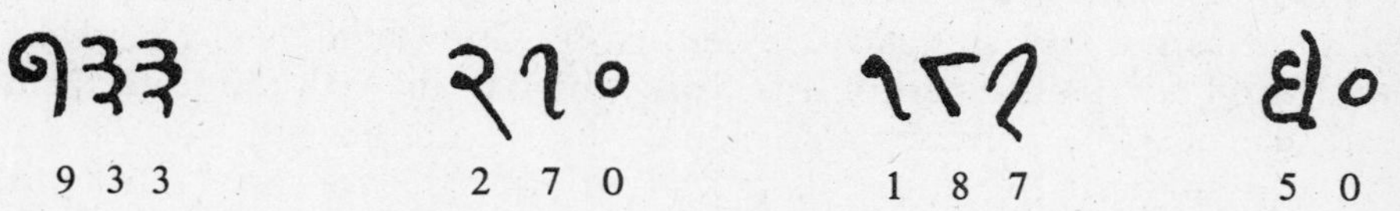

933　　270　　187　　50

　Judging solely from the documents we have just considered, a place-value system with signs for numbers 1 to 9 was used in India before the end of the sixth century A.D., and the zero was in use by the eighth century (fig. 29–3).

However, some historians of science, who contest the Indian origin of our modern numerals and maintain that the Gwalior stone inscriptions are the earliest Indian documents attesting the use of this system, have questioned the authenticity of the copperplate deeds because, according to them, they were copied, altered, or forged at times later than those indicated by their dates. In the view of these authors, the Indians' use of a place-value decimal numeration with a zero began no earlier than the ninth century A.D.

This view was first put forward in support of certain theories inspired by the conjectures of earlier humanists who believed that the Indians, Arabs, and Europeans owed their zero and place-value numeration not to Indian scholars but to ancient Greek mathematicians.

According to these theories, the system was first devised by Greek Neoplatonists or Neopythagoreans, then went from Alexandria to Rome in the time of the empire and, a little later, was brought to India by merchants. From Rome it passed into the provinces in Africa, where the Western Arabs found it several centuries later, during their conquests, while the Eastern Arabs received it from Indian astronomers. The numerals of the system then developed into those of the Europeans and Western Arabs, on the one hand, and those of the Indians and Eastern Arabs, on the other.

But there is a simple, common-sense argument against these theories: "The Greeks were too intelligent not to realize the merit of that invention; they would have promptly adopted it if it had originated among them, or if they had even known about it" (Montluca).

To the best of my knowlege, no serious reason for questioning the authenticity of the Indian copperplate deeds has ever been stated. And the theories described above—which are still regarded as fundamental truth by a number of authors who call themselves occultists—are based only on a desire to corroborate the "Greek miracle" by any means; assertions made in their favor are not supported by any evidence.

The examples in the following pages, all from sources other than the copperplate deeds, will clearly show that the signs and the principle of our modern numeration are of Indian origin.

First, an example from the eighth century A.D. in China. "A zero symbol is mentioned in the *Khai-Yuan Chan Ching*, that great compendium of astronomy and astrology edited by Chhüthan Hsi-Ta between +718 and +719. The part of this work which deals with the Chiu Chih calendar of +718 contains a section on Indian methods of calculation. After saying that the numerals are all written cursively with only one stroke each, the writer goes on to say, 'When one or other of the 9 numbers [is to be used to express a multiple of] 10 [lit. reaches 10], then it is entered in a column in front of [the unit digit] [*chhien wei*]. Whenever there is an empty space in a column [i.e, a zero], a dot is always placed [to signify it]' " (Needham). The author was not Chinese: "Chhüthan Hsi-Ta himself was an Indian Buddhist, the most eminent member of one of three clans of Buddhist astronomers and calendar experts originating from India and resident at the Chinese capital during Thang times. . . . [His] name certainly transliterates Gautama Siddhartha" (Needham). This example confirms the influence that the spread of Buddhism had on the propagation of Indian science in the Far East, and it suggests that by the beginning of the eighth century the modern place-value numeration was already being used by Indian mathematicians and had even been brought to China.

The next example comes from Mesopotamia and goes back to the seventh century. We owe it to Severus Sebokht, a Syrian bishop who lived in the Kenneshre monastery on the Euphrates and studied philosophy, mathematics, and astronomy. "In one of the fragments of his works that have come down to us, of date 662, he directly refers to the Hindu numerals. He seems to have been hurt by the arrogance of certain Greek scholars who looked down on the Syrians, and in defending the latter he claims for them the invention of astronomy. He asserts the fact that the Greeks were merely the pupils of the Chaldeans of Babylon, and he claims that these same Chaldeans were the very Syrians whom his opponents condemn. He closes his argument by saying that science is universal and is accessible to any nation or to any individual who takes the pains to search for it. It is not, therefore, a monopoly of the Greeks, but is international. . . .

"It is in this connection that he mentions the Hindus by way of illustration, using the following words: 'I will omit all discussion of the science of the Hindus, a people not the same as the Syrians; their subtle discoveries in this science of astronomy, discoveries that are more ingenious than those of the Greeks and the Babylonians; their valuable methods of calculation; and their computing that surpasses description. I wish only to say that this computation is done by means of nine signs. If those who believe, because they speak Greek, that they have reached the limits of science should know these things they would be convinced that there are also others who know something' " (Smith).

This example shows that by the middle of the seventh century the numerals of Indian origin were already known and appreciated beyond the borders of India.

Southeast Asia provides us with several examples that are especially valuable because they are in stone inscriptions whose authenticity is guaranteed by a whole set of facts. The ancient civilizations of Indochina and Indonesia, particularly those of the Khmers in Cambodia and the Chams on the southeast coast of what is now Vietnam, underwent a strong Indian influence in the early Christian era, partly because they were intermediaries in the trade in spices, silk, and ivory between India and China, and partly because of the expansion of Shivaism and Buddhism.

Among the stone inscriptions that these civilizations have left to us, those written in Khmer, Old Malay, Cham, and Old Javanese (that is, in vernacular languages) show the concurrent use of two distinct systems of recording numbers.

The first of these systems was for numbers used in such common activities as measuring lengths, areas, and volumes, or counting slaves, objects, animals, offerings presented to gods and temples, and so on. It usually consisted simply in writing the words for the corresponding numbers in the vernacular language. The Khmers, however, often preferred to use the numerals of their "vernacular numeration," which bore traces of the base 20 and was essentially additive (fig. 29–5).

But neither local number words nor the Khmer vernacular written numeration was used for expressing the dates of inscriptions: another way of recording numbers was exclusively reserved for that purpose.

1	**I**	10	𝟖 or 𝟗 or 𝟗	
		20	ō or ᵕᵕ	
2	**II**	30	ōꝗ	= 20 + 10
3	**III**	40	ꝉ ou ᵕᵗ	= 20 + 1 × 20 Stroke added to sign for 20
4	**IIII** or ꝉ	50	ōꝗ	= 40 + 10
5	ꝗ or ꝉ or ꝗ	60	ꝉ or ꝉ	= 20 + 2 × 20 Two strokes added to sign for 20
		70	ōꝗ	= 60 + 10
6	ꝉ or ꝉ or ꝉ	80	ꝉ or ꝉ	= 20 + 3 × 20 Three strokes added to sign for 20
7	ꝉ or ꝉ or ꝉ	90	ꝉꝗ	= 80 + 10
		100	ꝉ or ꝉ	
8	ꝗ or ꝗ or ꝗ	200	ꝉ	= 100 + 1 × 100 Stroke added to sign for 100
9	ꝗ or ꝗ	300	ꝉ	= 100 + 2 × 100 Two strokes added to sign for 100

ATTESTED EXAMPLES						
ꝗII	ꝗIII	ōꝗꝗ	ꝉōꝗ	ꝉōII	ꝉ	ꝉ
10 + 2	10 + 3	20+10+5	80 + 7	100+80+2	200+10+6	300+80+10+6
---->	---->	------>	------>	------->	--------->	----------->
12	13	35	87	182	216	396

Fig. 29–5. Signs of the Khmer vernacular numeration which was used till the thirteenth century A.D. for expressing ordinary numbers in inscriptions. Table revised by Claude Jacques.

At first, the number words of the Sanskrit language were used. Then dates were expressed by means of a decimal place-value numeration with a true zero. This is attested from the second half of the seventh century in the vernacular inscriptions of Cambodia and Sumatra, and from the early ninth century in those of Champa (figs. 29–6, 29–7, 29–8). It is also attested in Java, where the earliest inscription date expressed in numerals is 682 Shaka (A.D. 760).

The numerals used in these inscriptions are variants of the Indian numerals adapted to the local scripts (Khmer, Malay, Cham, Javanese) and aesthetic standards.

Moreover, the inscriptions show that for many generations dates were indicated with references to an era whose origin is unquestionably Indian: the Shaka era of Indian astronomers. (It began in A.D. 78; 605 Shaka, for example, is A.D. 683 or 684.)

All this is important because it shows the great influence that Indian astronomers had on the ancient civilizaions of Southeast Asia, and in

		Christian era	Shaka era	
EARLIEST KHMER INSCRIPTION DATED IN NUMERALS	Province of Sambór, Cambodia	683	605	 6 0 5
EARLIEST OLD MALAY INSCRIPTIONS DATED IN NUMERALS	Palembang, Sumatra	683	605	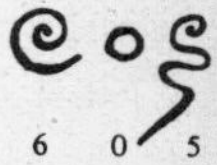 6 0 5
	Palembang, Sumatra	684	606	 6 0 6
	Island of Bangka, Sumatra	686	608	 6 0 8
EARLIEST CHAM INSCRIPTIONS DATED IN NUMERALS	Champa	813	735	 7 3 5
	Champa	829	751	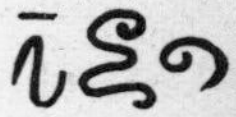 7 5 1

Fig. 29–6. Shaka dates in Southeast Asian vernacular inscriptions, expressed by means of numerals, including a zero, of Indian origin.

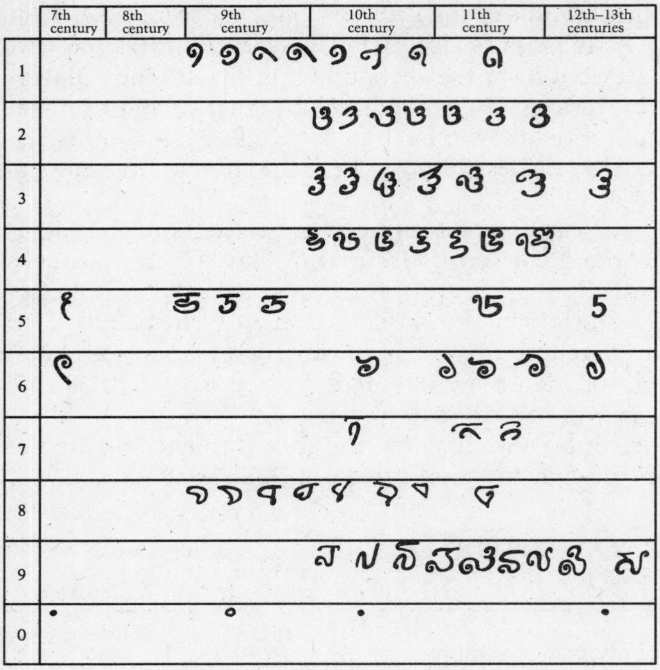

	7th century	8th century	9th century	10th century	11th century	12th–13th centuries
1						
2						
3						
4						
5						
6						
7						
8						
9						
0						

Fig. 29–7. Dated variants of numerals in the place-value system used exclusively for expressing Shaka dates in Khmer inscriptions. Table revised by Claude Jacques.

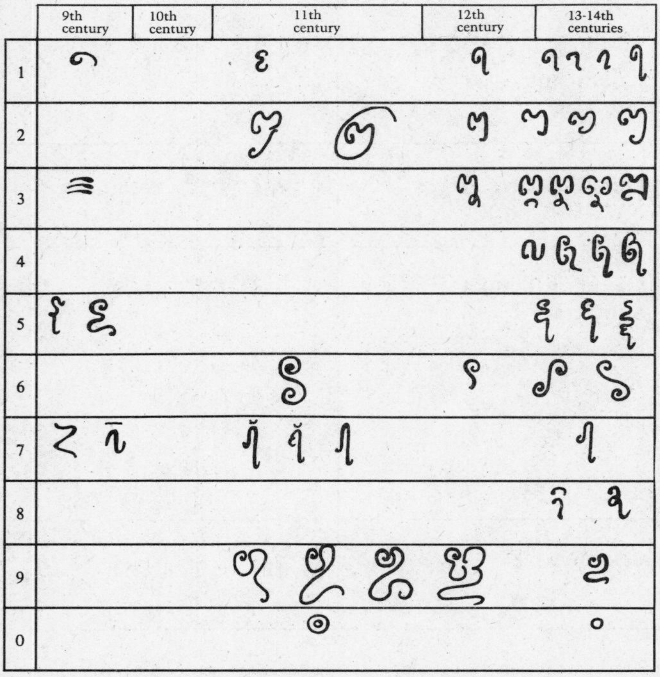

	9th century	10th century	11th century	12th century	13-14th centuries
1					
2					
3					
4					
5					
6					
7					
8					
9					
0					

Fig. 29–8. Dated variants of numerals in the place-value system used for expressing Shaka dates in Cham inscriptions. (In the nineteenth century, misinterpretation of the values of some of these numerals caused serious errors in the dating of Cham inscriptions, which in turn led to misinterpretation of the corresponding events.)

particular it shows that Khmer, Cham, and Javanese scholars (who were deeply impregnated with Indian culture) borrowed elements of Indian astronomy and strictly conformed to the corresponding arithmetical rules for several centuries.

"It is interesting to find it [the Indian numeration] attested in Indochina and the Malay Archipelago in the seventh century A.D., which, if G. R. Kaye's pessimistic views on the evidence of Indian epigraphy are correct, is two centuries before its appearance in India itself." But "unless one claims that Hindu-Arabic numerals came from the Far East, their use in early Indian colonies is clearly in favor of their use in India at a still earlier time" (Coedès).

The symbolic number words of Indian astronomers

An unusual method of expressing whole numbers that is unquestionably of Indian origin was commonly used in Indian astronomical texts. As we will see, it holds an important place in the history of the zero and our modern place-value numeration.

Most of the information that follows was graciously given to me by Roger Billard, who has recently done some remarkable studies of the numerical data and Sanskrit texts of Indian scientific astronomy.

First, let us recall that in a consistent oral numeration of base 10—in which numbers 1 to 9 and the powers of 10 have individual names—numbers are expressed according to the hybrid principle, following either a descending or an ascending order of powers of the base. For example, the number 4568:

$$4 \times 10^3 + 5 \times 10^2 + 6 \times 10 + 8$$

is spoken as:

> four thousand, five hundred, six-ten, (and) eight

in descending order of powers of 10, and as:

> eight, six-ten, five hundred, (and) four thousand

in ascending order.

In the Indian method being considered here, only words for the corresponding units were used, taken in ascending order of powers of 10—that is, all words for the base and its powers were omitted, so that the number 4568 was spoken in an abridged form that can be translated as "eight, six, five, four." And "two, eight, nine, three, one," for example, corresponded to the number:

$$2 + 8 \times 10 + 9 \times 10^2 + 3 \times 10^3 + 1 \times 10^4 = 13,982$$

The method thus constituted a genuine place-value oral numeration of base 10. It is all the more noteworthy because such a numeration has evidently never been devised anywhere but in India.

To designate numbers 1 to 9, it used not only the ordinary Sanskrit words for them:

eka	dvi	tri	catur	pañca	ṣaṭ	sapta	aṣṭa	nava
1	2	3	4	5	6	7	8	9

but also a large collection of other Sanskrit words, each with a meaning that evoked a certain numerical value—that is, the things, ideas, persons, or animals denoted by the words were taken as number symbols. Such words will be referred to as symbolic number words. The incomplete list in figure 29–9 will give some idea of the system:

ONE

pitāmaha : "first father" (Brahma) ādi : "beginning" rūpa : "form" tanū : "body"	warā , mahī go , pṛthivī dhara , kṣiti "earth"	abja , mṛgāṅka indu , candra soma, śaśāṅka "moon"

TWO

Aśvin : "the twin gods" Yama : "the primordial couple" yamala , dasra , yugma yugala , nāsatya , dvaya Words designating twins or couples	netra , nayana : "eyes" bahū : "arms" gulpha : "ankles" pakṣa : "wings" Words designating paired parts of the body

THREE

guṇa triguṇa "The 3 primordial properties"	loka bhuvana "The 3 worlds"	kāla trikāla "The 3 divisions of time"	agni, vaiśvānara vahni, dahana "Fire" (the 3 Vedic fires)	trinetra Haranetra "The 3 eyes of Shiva"

FOUR

sāgara abdhi sindhu ambudhi jaladhi "ocean"	Veda (sacred book divided into 4 parts) diś : "cardinal point" yuga : "cosmic cycle" (There are 4 of them.) kṛta (name of the first of the 4 cosmic cycles) irya : "the (4) positions of the human body" Haribahū : "the (4) arms of Vishnu" Brahmāsya : "the (4) faces of Brahma"

Fig. 29–9.

FIVE

bāna	*Pāṇḍava*	:"the (5) brother kings"
śara	*indriya*	:"the (5) senses"
iṣu	*Rudrāsya*	:"the (5) faces of Shiva"
"the (5) arrows" (of Kama)	*bhūta*	:"elements"
	mahāyajña	:"sacrifices"
	prāṇa	:"breaths"

SIX

aṅga	: "the (6) members" (head, trunk, 2 arms, 2 legs)
rasa	: "the (6) flavors"
ṛtu	: "season"
ṣaṇmukha *kumāravadana*	} "the (6) faces of Kumara"

SEVEN

aśva	: "the (7) horses" (of Surya)
naga	: "mountain"
ṛṣi	: "sage"
bhaya	: "fear"
svara	: "vowel"

EIGHT

Vasu	: (There are 8 Vasu gods.)
gaja	: "elephant".
nāga	: "snake" (8 types)
maṅgala	: "good omen"
mūrti	: "the (8) forms of Shiva"

NINE

aṅka	: "the (9) numerals"
chidra	: "the (9) orifices" (of the human body)
graha	: "planets"
Aja	: "the god Brahma"

Furthermore, this system had words to indicate the absence of units of a certain order; some of them are in figure 29–10.

ZERO

kha , *ambara* , *ākāśa* , *antarikṣa* *gagana*, *abhra* , *viyat* , *nabhas* "sky, atmosphere, space"	*śūnya* "void"	*bindu* "dot"

Fig. 29–10.

Writing in A.D. 629, the Indian scholar Bhaskara (not to be confused with Bhaskara Acharya, "Bhaskara the Learned," who lived in the twelfth century) expressed the number 4,320,000 as follows:

viyadambarākāśaśaśūnyayamarāmaveda

sky/atmosphere/space/void/primordial couple/Rama/Veda
0   0   0   0   2   3   4

Bearing in mind that, in this system, numbers are represented in ascending order of powers of 10, we see that the number in question is:

$$0 + 0 \times 10 + 0 \times 10^2 + 0 \times 10^3 + 2 \times 10^4 + 3 \times 10^5 + 4 \times 10^6 = 4,320,000$$

The system of symbolic number words thus shows a firm grasp of the zero concept and the place-value principle.

Here is another example from Bhaskara:

khagnyadrirāmārkarasavasurandhrendavaḥ

space/fire/mountains/Rama/sun/flavors/Vasu/orifices/moon

0	3	7	3	12	6	8	9	1

This expression of the number 1,986,123,730 contains a symbolic word for a number greater than 10: *ārka* ("sun") is used to indicate 12.

Several other Indian astronomical texts show the use of a symbolic word to indicate a number of two or more digits: for example, 10 is sometimes indicated by *anguli* ("fingers") or *Rāvaṇaśiras* ("the 10 heads of Rāvaṇa"); 11 by one of the eleven names or attributes of Shiva (Rudra, Shiva, Ishvara, etc.); 12 by *raśi* ("sign of the zodiac"), *cakra* ("zodiac") or *ārka* ("sun"), an allusion to the twelve suns of tradition; 13 by one of the names of Kama, who presides over the thirteenth lunar month; 15 by *pakṣa* ("wing"), an allusion to the two wings, black and white, of the luna month; 20 by *nakha* ("nails of the fingers and toes"); 27 by *nakṣatra* ("lunar mansions"); 32 by *danta* ("teeth"); 49 by *tāna* ("tone"), an allusion to the seven octaves of seven notes each; etc.

Such usages (which were motivated only by the requirements of versification in Indian astronomical texts) were not an exception to the rule of place-value. In the example above, the expression:

space/fire/mountains/Rama/sun/flavors/Vasu/orifices/moon

0	3	7	3	12	6	8	9	1

is a symbolic designation of the number $0 + 3 \times 10 + 7 \times 10^2 + 3 \times 10^3 + 12 \times 10^4 + 6 \times 10^6 + 8 \times 10^7 + 9 \times 10^8 + 1 \times 10^9 = 1,986,123,730$ (this is assured by the context), but the word for "sun," which in isolation designates the number 12, acquires the value it has here ($12 \times 10^4 = 120,000$) solely by the position it occupies in the expression.

We can therefore say that the essential features of our modern numeration are at least as old, in India, as the system of symbolic number words.

The system was widely used in Southeast Asia for expressing Shaka dates in Sanskrit inscriptions, beginning in the late sixth century in Cambodia, the late seventh in Champa, and the early eighth in Java (fig. 29–11).

In India, where the system originated, its use by mathematicians and astronomers began at least as early as the second half of the sixth century and continued for hundreds of years. It is found, for example, in the words of Putumanasomayajin in the eighteenth century, Paramesh-

		Christian era	Shaka era
EARLIEST DATED SANSKRIT INSCRIPTIONS IN CAMBODIA	Province of Kompon Thom	598	520 KHADVIŚARA sky/two/arrows 0 2 5
	Phnom Bayang Stela	604	526 RASADASRAŚARAIŚ flavors/twins/arrows 6 2 5
EARLIEST SANSKRIT INSCRIPTIONS IN CHAMPA DATED BY MEANS OF SYMBOLIC NUMBER WORDS	Mī-so'n Stela	687	609 ĀNANDĀMVARAṢAṬŚATA* (the 9) Nandas/space/six hundred* 9 0 6 × 100
	Mī-so'n Stela	731	653 RĀMĀRTTHAṢATKAIŚ Rama/sense objects/six 3 5 6
EARLIEST DATED SANSKRIT INSCRIPTION IN JAVA	Čangal Stela	732	654 ŚRUTĪNDRIYARASAIR Veda/sense organs/flavors 4 5 6

*This seems to show a certain lack of experience in using symbolic number words.

Fig. 29–11. Shaka dates from Sanskrit inscriptions of highly Indianized civilizations in Southeast Asia.

It is important to distinguish between Southeast Asian inscriptions in vernacular languages and those in Sanskrit. At first, dates in both kinds of inscriptions were expressed with ordinary Sanskrit number words. Then, beginning in the late seventh century, dates in vernacular inscriptions were expressed with numerals of the Indian place-value written numeration. But during this second phase Sanskrit symbolic number words, governed by the place-value principle, were used for expressing dates in Sanskrit inscriptions (which, being nearly always in verse, provided no occasion for using numerals).

vara (fifteenth century) Bhaskara Acharya (twelfth), Bhoja and Sripati (eleventh), Lalla (ninth), Haridatta, Brahmagupta, and Bhaskara (seventh), and Varahamihira (c. 575).

Bhaskara often felt the need to express a number not only in symbolic number words but also, at the same time, in numerals, as in the following example from his commentary on the *Aryabhatiya*, by the mathematician and astronomer Aryabhata:

śūnyāmbarodadhiviyadagniyamākāśaśaraśarādriśūnyendurasāmbarāṅgāṅkādrisvarendu /

aṅkair api 1779606107550230400 /

void/sky/ocean/sky/fire/couple/space/arrows/arrows/mountains/void/moon /
 0   0   4   0   3   2   0   5   5   7   0   1

flavors/atmosphere/members/numerals/mountains/horses/moon
 6   0   6   9   7   7   1

in numerals: 1,779,606,107,550,230,400

In this example, Bhaskara uses the Sanskrit word *anka* not only in its ordinary meaning of "numeral" but also as a symbolic designation of the number 9, alluding to the nine numerals (the zero sign was not regarded as a numeral) of the Indian place-value system.

The example proves that use of the nine numerals and the zero sign according to the place-value principle was already established in India by the time of Bhaskara, that is, the early seventh century.

In the early sixth century, Aryabhata showed that he was acquainted with the Indian numerals and zero sign as well as the system of symbolic number words, though he used a third system in his writings (one in which numerical values were assigned to syllables of the Indian alphabet). In his *Aryabhatiya* (c. 510) there are two examples of symbolic number words: "form" for 1 and "zodiac" for 12.

In the second chapter of this work (a chapter on arithmetic) he mentions a remarkable procedure that can be practiced only in writing: it uses a decimal base, the place-value principle, nine numerals, and a tenth sign that functions as a zero. This part of the *Aryabhatiya* also gives (in versified form) a rule for extracting square and cube roots that closely resembles the one we know today.

The *Lokavibhaga*, a Jain work on cosmology bearing a date that corresponds to August 25, 458, in the Julian calendar, reveals that the system of Sanskrit symbolic number words was known before Aryabhata's time. The expression "minus 1" is indicated by *rūponaka* ("diminished by form"). The concept of zero is designated sometimes by *gagana* or *ambara* ("sky") and sometimes by *śūnya* ("void"). And the number 13,107,200,000 is indicated in this interesting way:

> *pañcabhyaḥ   khalu   śūnyebhyaḥ   param   dve   sapta   cāmbaram*
> *ekaṃ   trīni   ca   rūpam   ca*
> *five voids then two and seven sky one and three and form*
>   00000   2   7   0 1   3   1

As can be seen in this example, the author of the *Lokavibhaga* made sparing use of symbolic number words and usually limited himself to the ordinary Sanskrit words for numbers 1 to 9. This may have been because, in his time, symbolic number words were not very well known outside of scholarly circles and, wanting to praise the science of the Jain religious movement, he felt it would be better to make his work accessible to as many people as possible (personal communication from Roger Billard).

He expressed some numbers wihout any symbolic words at all; for example, 14,236,713:

> *trīny   ekaṃ   sapta   ṣaṭ   trīni   dve   catvāry   ekakam*
> three   one   seven six three two   four   one

Moreover, he sometimes felt obliged to add *kramāt* ("in order"), *ankak-ramena* ("in order of the numerals"), or *sthānakramād* ("in order of position") to numerical representations of this kind.

The *Lokavibhaga* shows that the zero concept and the place-value principle were known in India by the middle of the fifth century, which means that our modern numeration goes back at least that far.

To understand why Indian astronomers generally preferred to use symbolic number words rather than numerals, we must first realize that the difference between the two systems was merely a matter of graphic conventions.

This can be illustrated by the change that the Javanese made in their method of recording numbers. Beginning in the eighth century they wrote their language in Kavi characters, which were derived from an old Pali alphabet and have now fallen into disuse. In that writing, they used a decimal place-value numeration with the following numerals (of Indian origin):

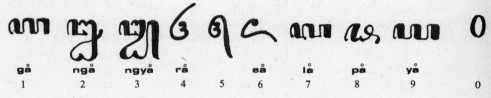

Then, at a relatively recent time, they replaced the first nine of these numerals with letters or combinations of letters from the modern Javanese alphabet:

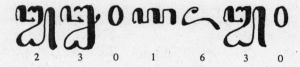

But they kept the Indian zero sign and continued to use a place-value notation, expressing the number 2,301,630, for example, in this form:

Thus the essence of a decimal numeration with the same structure as ours is independent of its symbolization. The nature of the symbols chosen (graphic signs with or without direct visual reference, letters of the alphabet, words with or without evocative meanings, etc.) does not matter, provided they are never ambiguous and the corresponding system consistently uses the zero concept and place-value principle.

Indian scientific works were often written in verse. Of the different kinds of numerical notation that satisfied the requirements of Sanskrit verse, the one most commonly used was the system of symbolic number words. "This procedure proved to be extremely effective in preserving numbers and it was no doubt designed for that purpose." The numerical expressions in the works of Indian astronomers "have been accurately preserved through time and many handwritten copies." This preservation "is all the more striking because Indian manuscripts are never very old, physically speaking: our copies of them were usually made no more than two or three centuries ago. Numerical data recorded in numerals would surely have come down to us in an unusable state" (Billard).

The values of Indian numerals varied not only in the course of time but also from one region of India to another. For this reason, Indian astronomers avoided using numerals in their writings and generally preferred to use symbolic number words instead.

Origin of the Indian place-value system

When and how did Indian scholars discover the zero concept and the place-value principle? Was the discovery entirely their own, or did it result from a foreign influence? We can only speculate on those questions because we do not have enough evidence for conclusive answers. One thing seems certain, however: the forms of the first nine numerals used in the Indian place-value written numeration are of strictly Indian origin.

All Indian scripts, and some of those used in Central and Southeast Asia, have a common origin in the ancient Indian writing known as Brahmi, which was in all likelihood derived from the Phoenician alphabet. Brahmi was used in India at least as early as the third century B.C., since its first known appearance is in the edicts of King Asoka, who reigned during that century. It also appears, slightly altered, in the inscriptions of Nana Ghat (second century B.C.) and Nasik (first or second century A.D.), and in the contemporary Sunga, Saka, Andhra, Mathura, Ksatrapa, and Kusana inscriptions. It later gave rise to several distinct types of writing from which the main groups now in use are derived. The differences among these groups resulted either from the characteristics of the languages to which they were adapted or from the nature of the writing materials used (fig. 29–12).

As for the numerical signs used in India and Central and Southeast Asia (for the moment, we will consider only the signs for numbers 1 to 9), the great diversity of their forms does not place their common origin in doubt because they are all derived from the first nine signs of the Brahmi numeration (fig. 29–13A).

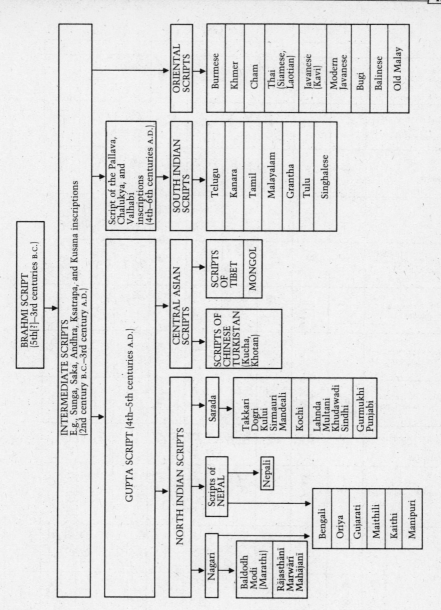

Fig. 29–12. Some of the scripts derived from Brahmi.

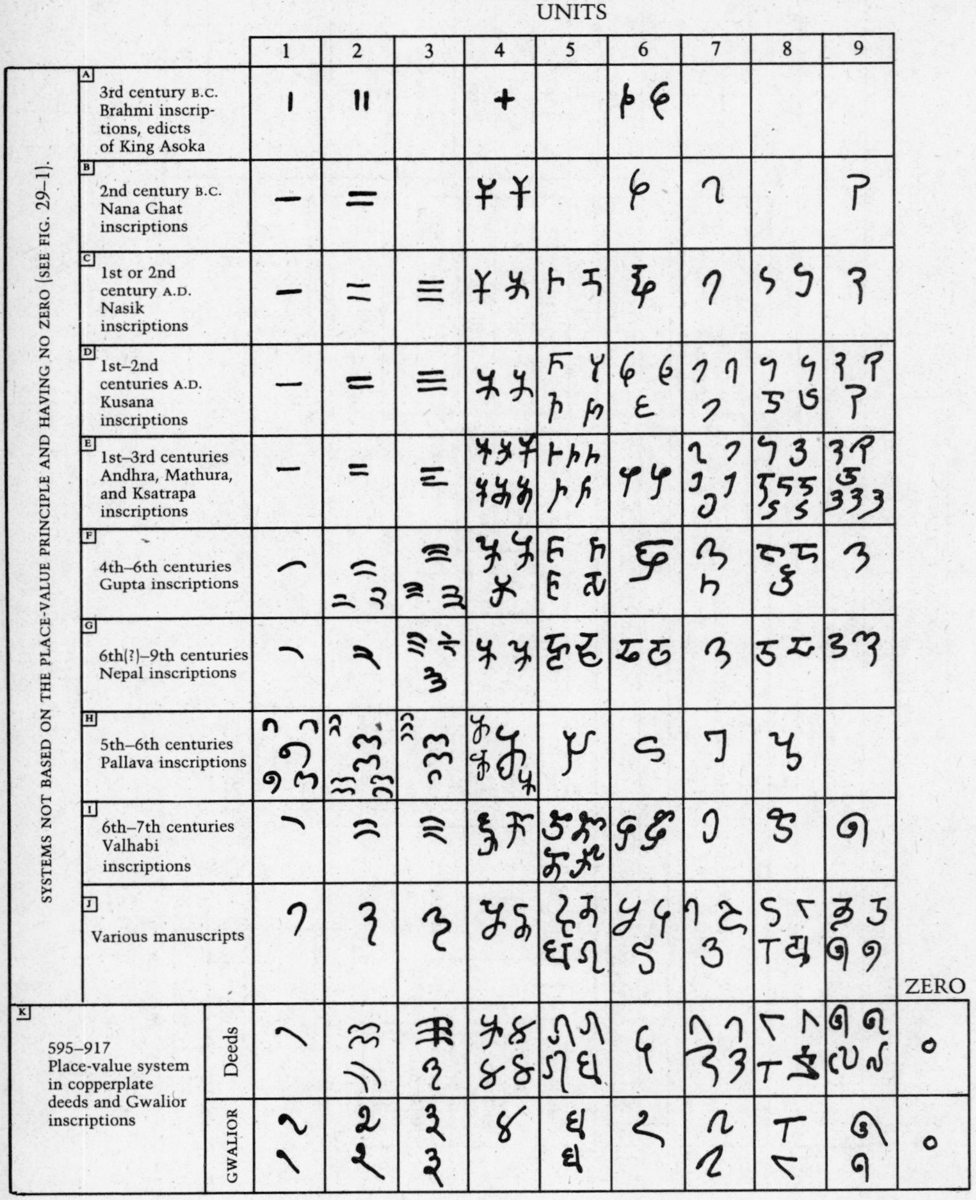

UNITS

		1	2	3	4	5	6	7	8	9
A	3rd century B.C. Brahmi inscriptions, edicts of King Asoka									
B	2nd century B.C. Nana Ghat inscriptions									
C	1st or 2nd century A.D. Nasik inscriptions									
D	1st–2nd centuries A.D. Kusana inscriptions									
E	1st–3rd centuries Andhra, Mathura, and Ksatrapa inscriptions									
F	4th–6th centuries Gupta inscriptions									
G	6th(?)–9th centuries Nepal inscriptions									
H	5th–6th centuries Pallava inscriptions									
I	6th–7th centuries Valhabi inscriptions									
J	Various manuscripts									

(left vertical label: SYSTEMS NOT BASED ON THE PLACE-VALUE PRINCIPLE AND HAVING NO ZERO (SEE FIG. 29–1).)

ZERO

| **K** | 595–917 Place-value system in copperplate deeds and Gwalior inscriptions | Deeds | | | | | | | | | | 0 |
| | | GWALIOR | | | | | | | | | | 0 |

Fig. 29–13A. Comparison of the ancient Indian numerations directly related to the Brahmi system (A to J) with the first strictly place-value Indian notations (K).

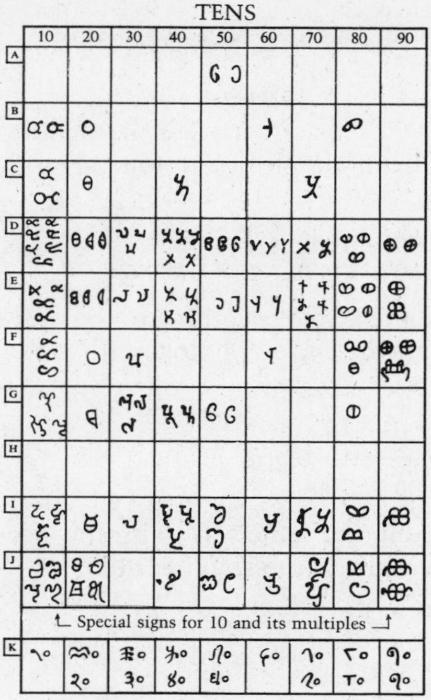

TENS

	10	20	30	40	50	60	70	80	90

(Fig. 29–13B — table of Brahmi tens numeral forms, rows A through K)

⌐ Special signs for 10 and its multiples ⌐

Fig. 29–13B.

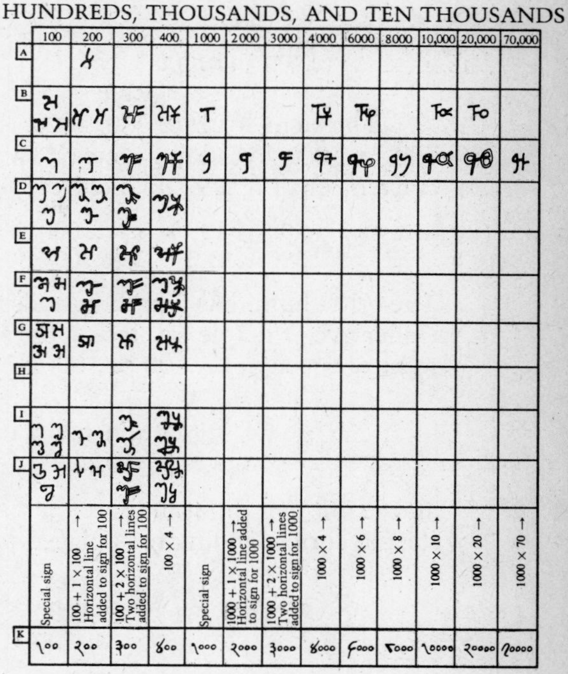

HUNDREDS, THOUSANDS, AND TEN THOUSANDS

	100	200	300	400	1000	2000	3000	4000	6000	8000	10,000	20,000	70,000

(Fig. 29–13C — table of Brahmi hundreds, thousands, and ten thousands numeral forms, rows A through K)

Bottom annotations (left to right):
Special sign | 100 + 1 × 100 Horizontal line added to sign for 100 | 100 + 2 × 100 Two horizontal lines added to sign for 100 | 100 × 4 | Special sign | 1000 + 1 × 1000 Horizontal line added to sign for 1000 | 1000 + 2 × 1000 Two horizontal lines added to sign for 1000 | 1000 × 4 | 1000 × 6 | 1000 × 8 | 1000 × 10 | 1000 × 20 | 1000 × 70

Fig. 29–13C. Signs drawn with solid lines are actually attested; those drawn with double lines are reconstructions made on the basis of comparative study.

The first nine Indian numerals (in their various forms) were not always used according to the place-value principle, as can be seen from an examination of the written numeration that was employed in India, with graphic variations, from the third century B.C. to the eighth century A.D. It is partially revealed to us by the Brahmi inscriptions of King Asoka and, more completely, by the Sunga, Saka, Andhara, and Ksatrapa inscriptions, and others (fig. 29–13). At first sight, it seems to be a decimal and essentially additive numeration with special signs for these numbers:

1	2	3	4	5	6	7	8	9
10	20	30	40	50	60	70	80	90
100	200	300	400	500	600	700	800	900
1000	2000	3000	4000	5000	6000	7000	8000	9000

and so on. But if we look at the signs more closely we see that they are not all independent of each other, and that the only numerals with special signs are the following:

1	2	3	4	5	6	7	8	9
10	20	30	40	50	60	70	80	90

100 and 1000

The signs for 200, 300, 2000, and 3000 are derived from the sign for 100 or 1000 by adding one or two horizontal lines. The four graphic conventions (fig. 29–13C) correspond to:

$$200 = 100 + 1 \times 100 \qquad 2000 = 1000 + 1 \times 1000$$
$$300 = 100 + 2 \times 100 \qquad 3000 = 1000 + 2 \times 1000$$

The other hundreds and thousands are represented according to the multiplicative principle by placing the sign for the corresponding units to the right of the sign for 100 or 1000:

$$400 = 100 \times 4 \qquad 4000 = 1000 \times 4$$
$$500 = 100 \times 5 \qquad 5000 = 1000 \times 5$$
$$600 = 100 \times 6 \qquad 6000 = 1000 \times 6$$

and so on. Ten thousands are represented in the same way, with the sign for the corresponding tens placed to the right of the sign for 1000:

$$10,000 = 1000 \times 10$$
$$20,000 = 1000 \times 20$$
$$30,000 = 1000 \times 30$$

and so on.

It is clear that the structure of the old Indian written numeration was modeled on that of the Sanskrit oral numeration:

1 eka	10 daśa	100 śata		1000 sahasra	
2 dvi	20 viṃśati	200 dviśata	(2×100)	2000 dvisahasra	(2×1000)
3 tri	30 triṃśati	300 triśata	(3×100)	3000 trisahasra	(3×1000)
4 catur	40 catvāriṃśati	400 caturśata	(4×100)	4000 catursahasra	(4×1000)
5 pañca	50 pañcāśat	500 pañcaśata	(5×100)	....	
6 ṣaṭ	60 ṣaṣṭi	600 ṣaṭśata	(6×100)	10,000 daśasahasra	(10×1000)
7 sapta	70 sapti	700 saptaśata	(7×100)	20,000 viṃśatsahasra	(20×1000)
8 aṣṭa	80 aśīti	800 aṣṭaśata	(8×100)	30,000 triṃśatsahasra	(30×1000)
9 nava	90 navati	900 navaśata	(9×100)	....	

For example, the notation for 4769:

$$1000 \times 4 \;+\; 100 \times 7 \;+\; 60 \;+\; 9$$

corresponded to the Sanskrit oral expression:

nava	ṣaṣṭi	saptaśata	ca	catur	sahasra
nine	sixty	seven-hundred	and	four	thousand

Expressions of numbers in signs (which were generally written from left to right, in descending order of powers of 10) were therefore faithful transcriptions of the corresponding oral expressions (in which Sanskrit number words were generally spoken in the opposite order, that is, in ascending order of powers of 10).

All this shows that among certain scholarly groups in India, beginning at a time that was surely earlier than A.D. 458, the first nine numerals of the ancient Brahmi numeration were, like the first nine ordinary Sanskrit number words, entirely adapted to the base 10 and the place-value principle, so that the number 4769 was written either in this form:

$$\not{\!\!f}\ \ 7\ \ \varsigma\ \ ?$$

$$4 \quad 7 \quad 6 \quad 9$$

or in this one:

nava	ṣaṭ	sapta	catur
nine	six	seven	four

Was this a purely Indian development? It is not impossible that it originated in a decimal adaptation of the Babylonian sexagesimal place-value numeration, since we now know that some of the elements and procedures of Babylonian astronomy were brought into India between the third century B.C. and the first century A.D.

There is a second possibility. As we saw in Chapter 27, beginning in the Han period (206 B.C.–A.D. 220) Chinese rod numerals were used in a decimal place-value system, the number 4769 being written in this form:

$$\equiv \pi \perp \text{Ⅲ}$$

$$4 \quad 7 \quad 6 \quad 9$$

As a result of contacts between India and China beginning in the second century A.D., Indian scholars may have developed their system by adapting the Chinese notation to the first nine Indian numerals or the first nine Sanskrit number words.

But another hypothesis seems to me more plausible, though there is no conclusive evidence in its favor.

At an early stage, the Indians undoubtedly performed arithmetical operations by means of a concrete numeration, using fingers, pebbles, or other objects or some sort of counting board. There are good reasons—notably, later accounts of Indian reckoning by Arab, Persian, and European authors—for believing that at some point Indian reckoners must have begun inscribing the first nine signs of their written numeration in columns marked off in advance (on the ground, a wax tablet, or a slate) to separate the decimal units, going from right to left in ascending order, so that the columns formed a counting board on which numerals were used as counters. To note the number 6483, for example, which was expressed as *tri aśīti catur śata ṣaṭ sahasra* ("three, eighty, four hundred, six thousand") in Sanskrit number words and as follows in the old written numeration:

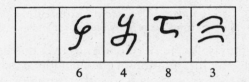

$$1000 \times 6 \ + \ 100 \times 4 \ + \ 80 \ + \ 3$$

Indian reckoners must have written the sign for 3 in the first column on the right, the sign for 8 (indicating tens) in the second column, the sign for 4 (hundreds) in the third, and the sign for 6 (thousands) in the fourth, something like this (adopting the style of the old Indian numerals):

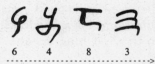

| | 6 | 4 | 8 | 3 |

This kind of counting board must have provided a remarkable means of writing whole numbers (no matter how large, theoretically at least) with the first nine signs of the ancient Indian numeration, each of them having a value that depended on its position, and without using a zero sign, the absence of units of a certain order being indicated simply by an empty column. Later, when the habit of considering orders of units in that invariable order had become ingrained, reckoners must have eliminated the columns and represented 6483 in this strictly place-value form:

$$6 \qquad 4 \qquad 8 \qquad 3$$

They must then have felt the need of a new sign to indicate the absence of units of a certain order, that is, a sign to replace the empty column that had formerly indicated such an absence. And that must have been the origin of the Indian zero sign (either a small circle or a dot, designated by various Sanskrit words: *śūnya* ["void"], *kha* ["sky"], *gagana* ["space"], *ambara* ["atmosphere"], etc.). The number 1,024,607,130,000, which must formerly have been represented on the counting board in this form:

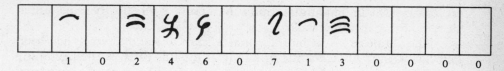

1	0	2	4	6	0	7	1	3	0	0	0	0

was represented with the first nine numerals and the new sign in the following way, which is the same as ours, except for graphic differences in numerals:

1	0	2	4	6	0	7	1	3	0	0	0	0

and it was often expressed in this literary form:

kha	*ambara*	*śūnya*	*gagana*	*agni*	*rūpa*	*sapta*	*śūnya*	*rasa*	*catur*	*netra*	*kha*	*eka*
sky	atmosphere	void	space	fire	form	seven	void	flavors	four	eyes	sky	one
0	0	0	0	3	1	7	0	6	4	2	0	1

It is therefore quite possible that the zero and the place-value principle were discovered in India independently of any foreign influence.

The zero sign later came to be viewed as the sign of a numerical value, the "null number." We know, for example, that in his *Brahmasiddhānta* (A.D. 628) the mathematician and astronomer Brahmagupta told how to perform the six basic operations (addition, subtraction, multiplication, division, raising to powers, and extraction of roots) on positive numbers, negative numbers, and the null number. The zero originally meant "void," an empty column on the counting board. When and how did it become enriched by acquiring the meaning of "nothing," as in "10 minus 10"? That question is one of the most interesting in the history of science, but unfortunately we cannot answer it in the present state of our knowledge.

Worldwide propagation of the Indian system

The Indian place-value numeration with a zero sign ranks among humanity's fundamental discoveries. Through the centuries it has been propagated even more widely than the alphabet of Phoenician origin, and it has now become the only real universal language.

When its advantages became apparent to the scholars and reckoners of civilizations in contact with India, they gradually abandoned the imperfect systems transmitted to them by their ancestors. But this transition occurred differently in different places. Some cultures adopted only the structure of the Indian numeration and improved or radically transformed their traditional numerations.

The Chinese, with their decimal system of rod numerals, had been familiar with the place-value principle at least as early as the beginning of the Christian era, but for centuries they had no zero. At first they overcame the difficulty by expressing a number such as 1,470,000, for example, either by writing it with characters of their ordinary numeration (which is equivalent to a notation in number words) or by placing rod numerals in squares, like those on a counting board, leaving an empty square for each missing order of units:

	I	☰	TT				
	1	4	7	0	0	0	0

It was not until the eighth century, probably under the influence of Indian Buddhists, that the Chinese introduced a zero sign (a circle) into their place-value numeration. They now represented 1,470,000 in this form:

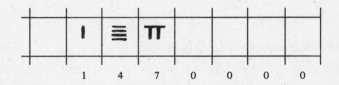

The traditional Chinese numeration uses these thirteen signs according to the hybrid principle (Chapter 25):

一	二	三	四	五	六	七	八	九	十	百	千	萬
1	2	3	4	5	6	7	8	9	10	100	1000	10,000

Normally the number 7829, for example, is represented in this system as follows:

$$7 \times 1000 + 8 \times 100 + 2 \times 10 + 9$$

In some Chinese works, however, the signs indicating powers of 10 are often omitted, so that 7829 is represented in this abridged form (Chapter 28):

七 八 二 九

7   8   2   9

But this use of the first nine numerals of the ordinary Chinese numeration according to the place-value principle does not necessarily reflect a foreign influence. Users of the system may very well have simplified it on their own, to make notation more rapid; as we have seen, such simplification is in the normal line of development of hybrid systems. Without a zero, however, the abridged notation could be used consistently only for numbers that had no missing orders of units. To avoid confusion between 3605 and 365, for example, the sign for 100 had to be used in noting 3605:

三 大 百 五   三 大 五
3   6 × 100 5   3   6   5
-------------->   --------------->
    3605   365

It was not until they had come under the influence of the Indian place-value numeration that users of the simplified Chinese notation were able to eliminate the classical characters for 10, 100, 1000, and 10,000 by introducing a circular zero sign to indicate the absence of units of a certain order:

9,420,279,060   九 四 二 〇 二 七 九 〇 大 〇
                  9   4   2   0   2   7   9   0   6   0
                 -->

(The example above is from a table of logarithms published in 1713.)

The traditional Chinese numeration now became perfectly adapted to arithmetical operations because its first nine signs acquired a dynamic character that they had lacked before. This is illustrated by the example below, from a Chinese work of 1355 entitled *Ting Chü Suan-fa* ("Ting Chü's Computation"). It shows how to multiply 3069 by 45, using traditional Chinese numerical signs and the Indian zero in a method identical with ours:

		TRANSLATION
三 〇 大 九		3 0 6 9
四 五		4 5
一 五 三 四 五		1 5,3 4 5
一 二 二 七 大		1 2,2 7 6
一 三 八 一 〇 五		1 3 8,1 0 5

Abraham ben Meir ibn Ezra, also known as Rabbi Ben Ezra, a Spanish Jewish scholar born in Toledo in 1092, traveled extensively in the Orient, beginning in 1139. He later lived in Italy and southern France and finally emigrated to England, where he died in 1167. He learned the Indian com-

putation methods and presented their basic rules in a Hebrew work entitled *Sefer ha Mispar* ("Book of Number"). But instead of using the Indian
numerals to represent numbers 1 to 9, he preferred to use the first nine
letters of the Hebrew alphabet, with which he had been familiar since
childhood:

ALEPH	BETH	GIMEL	DALETH	HE	VAV	ZAYIN	HETH	TETH
1	2	3	4	5	6	7	8	9

He kept the circular Indian zero sign, calling it either *galgal* ("wheel" in
Hebrew) or *sifra* (probably from an Arabic word meaning "empty"). In
adapting the old Hebrew alphabetic numeration (based on the additive
principle) to the place-value principle, he transformed it into a decimal
system with the same structure as ours. Instead of expressing the number
200,733 in the traditional form:

ג ל ש ת ֹר

3 + 30 + 300 + 400 + 200,000

he used this notation:

3   3   7   0   0   2

Some cultures adopted not only the structure of the Indian numeration but also its numerals. This was done by various peoples in India,
of course, and also in Central and Southeast Asia. But in the course of
their propagation the first nine numerals were gradually changed until
finally they took on a wide variety of forms, each appropriate to one of
the regional writing styles of India and its neighboring countries (figs.
29–3, 29–7, 29–8, 29–13, 29–14).

The Arabs, and later the peoples of Europe, also borrowed the numerals of the Indian system as well as its structure, then adapted them
to their respective forms of writing.

Adoption of the Indian system by the Eastern Arabs

In the vast empire that they built within less than a century after Mohammed's death, the Moslems forced conquered peoples to adopt the
Arabic language and its writing, which soon became a means of com-

		1	2	3	4	5	6	7	8	9	0
NORTHERN INDIA	Modern Nagari										o
	Marathi										o
	Nepali										o
	Bengali										o
	Oriya										o
	Gujarati										o
	Sarada										•
	Sindhi										o
	Punjabi										•
CENTRAL ASIA	Tibetan										o
	Mongol										o
SOUTHERN INDIA	Telugu (11th century)										o
	Kanara (16th century)										o
	Modern Telugu										o
	Modern Kanara										o

Fig. 29–14. Various forms of the numerals in the decimal place-value numeration now used in India and central Asia. Their variations do not cast doubt on their common origin, since they are all derived from the first nine signs of the Brahmi numeration (see fig. 29–13A).

munication among scholars of diverse origins. In turn, "the Arabs discovered a culture superior to theirs and quickly assimilated the intellectual concepts which the inhabitants of those countries had gradually developed to a high degree. And, in common with the Syrians, Persians, and Jews, the Arabs began constructing a new and distinctive culture" (Youshkevitch). The period of conquest was followed by a very fertile period of cultural assimilation that dazzled the world until the thirteenth century.

The political influence of the "sons of the Arabian desert" began rapidly declining in the second half of the seventh century, when the caliphs of the Ommiad dynasty (661–750) came to power and transferred the capital of the Moslem empire to Damascus. But in the middle of the next century the Ommiads were overthrown by the Abbassids, who gave the Arab-Islamic world a new capital: Baghdad. Founded in 772 by al-Mansur, the second Abbassid caliph (ruled 754–775), Baghdad quickly became a great commercial and intellectual center where the cultural heritage of the conquered nations was enthusiastically received, for "in that city there were not only Moslems but also Christians (mainly Nestorians), Jews and pagans (Parsees), as well as Greeks, Syrians, Egyptians, Armenians, Mesopotamians, Iranians and Hindus" (Becker and Hofmann). It was in Baghdad, during the reign of Caliph al-Mansur and his successors Harun al-Rashid (ruled 786–809) and al-Mamun (ruled 813–833), that the real development of Arab science began. "The Eastern Arabs acquired imperishable merit by preserving precious documents for mankind and carefully collecting all the Greek and Indian scientific works that came to them. Unfortunately we still have access only to a small part of what they bequeathed to us and we can therefore form only an incomplete picture of their work of preservation and their original achievements, which are mainly in algebra and trigonometry" (Becker and Hofmann).

"One of the most brilliant periods in the history of science occurred between the eighth and twelfth centuries. Scientific works were disseminated all over the Moslem world; rich libraries were founded in Baghdad and Cairo, then in Spain." But "in the thirteenth century, with the division of the empire, the Mongol invasion, and the Crusades, the brilliant period of Arab civilization came to an end" (Taton).

Through the Nestorian Christians, to a large extent, the Arabs became acquainted with the works of ancient Greek mathematicians and philosophers (such as Ptolemy, Euclid, Aristotle, Apollonius, Archimedes, Menelaus of Alexandria, Hero, and Diophantus) and translated them into Arabic, often adding original expositions and explanations.

And through trade relations with India, by way of the Persian Gulf and the port of Basra, they also became acquainted with Indian astronomy.

"In year 156 of the Hegira [A.D. 773]," wrote Ali ibn-Yusuf al-Qifti (1172–1248) in his *Dictionary of Scholars*, "there came from India to Baghdad a man deeply learned in the doctrines of his country. This man knew the method of *sindhid* [an Arabic transcription of the Sanskrit *siddhānta*, "astronomical canon"], concerning the movements of the heavenly bodies and equations calculated by means of sines in quarters of a degree. He also knew various ways of determining eclipses and the risings of the signs of the zodiac. He had composed a summary of a work on these subjects, attributed to a prince named Figar. In it, the *kardagas* were calculated by minutes. The caliph ordered that the Indian treatise be translated into Arabic, to help Moslems acquire exact knowledge of the stars. The translation was done by Mohammed ibn-Ibrahim al-Fazzari, the first Moslem to have made a thorough study of astronomy."

Having inherited the scientific works of India and ancient Greece, Arab scholars mingled Greek and Indian methods (and sometimes methods of Babylonian origin) in a remarkable spirit of synthesis, combining the systematic rigor of the Greek mathematicians with the essentially practical aspects of Indian science. They also added important contributions of their own and made progress in arithmetic, algebra, trigonometry, and astronomy. "The Arabs played an outstanding part in the history of science: by preserving the treasures of Greek and Indian science while at the same time giving them new life and an original character, they made possible the European scientific revival in the Middle Ages and the splendid flowering that came later" (Taton).

In arithmetic, the Arabs were influenced by the Greeks, Jews, and Syrians, because they inherited the Greek alphabetic numeration and adapted it to the twenty-eight letters of their alphabet (Chapters 16, 17, 19, and 20). Through the Christians of Syria and Mesopotamia they also inherited the ancient Babylonian place-value system and zero, which they used in their astronomical tables to note sexagesimal unit fractions by means of their own alphabetic numerals.

Still more important was the influence of Indian astronomers, from whom they borrowed, probably beginning in the eighth century, their zero, decimal place-value numeration, and computation methods.

A passage from an Arabic work quoted by Woepcke will give an idea of the enthusiasm that Arab mathematicians and reckoners must have felt when they realized the advantages of this system: "Among the things that they [the Indians] have given us in the way of science, [I must also mention] the [treatise on] numerical calculation reproduced in a more developed form by Abu Ja'far Mohammed ibn-Musa al-Khowarizmi; it is the most succinct and rapid of all calculating methods, and the easiest to understand and learn. It bears witness to the Indians' penetrating intelligence, creative talent and superiority in discernment and inventive genius" (fig. 29–15).

Mohammed ibn-Musa al-Khowarizmi (c. 780–850) was a scholar of Persian origin who lived at the court of the Abbassid caliph al-Mamun shortly after the time when Charlemagne ruled the West. He was among the most illustrious mathematicians of the Arab-Islamic world. It was largely through his work—consisting mainly of two treatises, one on arithmetic and the other on elementary algebra—that the Arabs, and later the peoples of Europe, became acquainted with algebra and the Indian numeration.

His treatise on algebra, *Al-jabr w'al-muqabalah*, was translated into Latin by Gherardo of Cremona (1114–1187) under the title *Liber Maumeti filii Moysi Alchoarismi de Algebra et Almuchabala.* The first word of the Arabic title, *al-jabr*, later came to designate the branch of mathematics known to us as algebra. In the treatise, it and *al-muqabalah* designate the two preliminary operations for solving an equation: *al-jabr* consists in transposing terms in such a way as to have only positive terms in each member of the equation; *al-muqabalah* consists in simplifying the members and canceling their equivalent terms.

His treatise on arithmetic (which has come down to us only in Latin translations) is the earliest known Arabic work in which the Indian place-value numeration and computation methods are specifically explained.

In Europe, al-Khowarizmi's name itself, first Latinized as Algorismi,

Fig. 29–15. Detail of an Arabic manuscript, *Al-Bahir fi 'ilm al-hisab* ("The Luminous Book on Arithmetic"), by As-Samaw'al ibn-Yahya al-Maghribi, showing the "Pascal Triangle" in numerals of the eastern type. It proves that Arab mathematicians knew the expansion of the binomial $(a + b)^n$ for any integral value of n at the beginning of the eleventh century, and therefore probably before Chinese mathematicians (see the Chinese version of the "Pascal Triangle," dating from 1303, in fig. 27–10).

Brought up in the Jewish religion (he was the son of Rabbi Yehuda ben Abbun, a Moroccan), As-Samaw'al ibn-Yahya al-Maghribi was a physician, philosopher, and mathematician who lived in Baghdad, converted to Islam, and died at Maragheh in 1180. In his manuscript he says that he is not the author of the table, having borrowed it from the mathematician Al-Karaji, who was born in the late tenth or early eleventh century. But we do not know whether or not Al-Karaji borrowed it from an earlier source that is still unknown.

then turned into the words "algorism" and "algorithm," designated computation with the Indian written numerations before taking on the more general meaning of computation with any notation.

The Eastern Arabs gradually altered the Indian numerals to bring them closer to the various styles of Arabic writing until finally they took on forms that changed only slightly through the centuries, mainly in the case of the signs for 5 and 0 (fig. 29–16). Called *hindi* numerals by the Arabs, these forms are still commonly used in Egypt, Syria, Turkey, Iraq, Afghanistan, Pakistan, and several Moslem regions of India. Here are the first five *hindi* numerals and their Indian counterparts:

	hindi	*Indian*
1	١	۱
2	٢	২
3	٣	३
4	٤	५
5	۵	৫

The differences between the *hindi* forms and the Indian originals may be at least partly explained by the fact that ancient Moslem scribes wrote vertically downward in columns going from left to right, which were read horizontally from right to left. That is, a page was written in this position:

TOP OF PAGE  BOTTOM OF PAGE

then turned 90 degrees to the right so that the lines could be read from right to left:

TOP OF PAGE

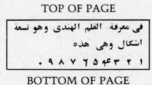

BOTTOM OF PAGE

(This procedure was also used in the Syriac writing of the Syrian Monophysite Church, and by the scribes of the ancient city of Palmyra.)

The most recent changes in the *hindi* numerals seem to be replacement of the old zero sign with a dot and replacement of the old sign for 5, resembling an upside-down B, with a small circle or a sign resembling an upside-down heart. The latter change may have been influenced by the form of the Arabic letter ha (a small circle), which has the value of 5 in the *Abjad* alphabetic numeration.

The change in the zero sign, however, which appears in eleventh-century Arabic manuscripts, is of Indian origin, since a dot was often

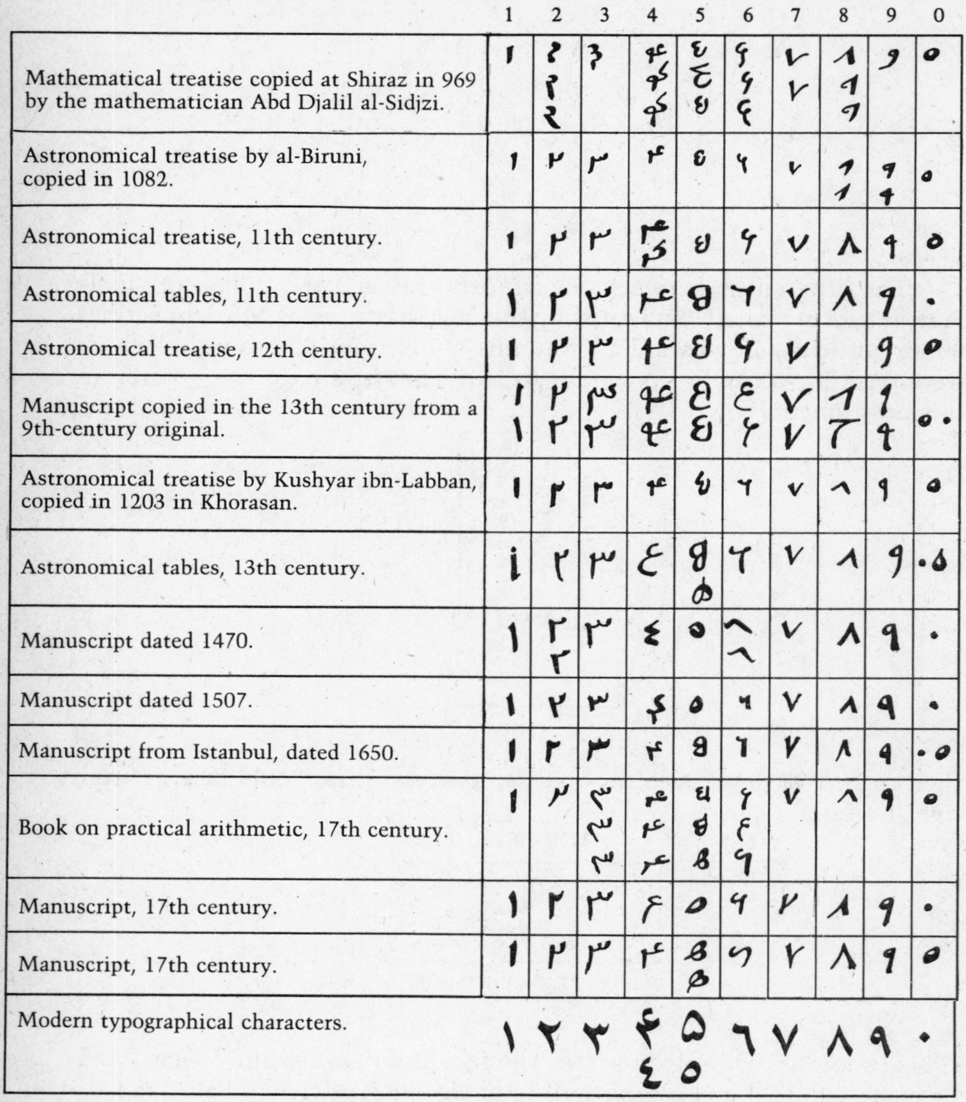

	1	2	3	4	5	6	7	8	9	0
Mathematical treatise copied at Shiraz in 969 by the mathematician Abd Djalil al-Sidjzi.										
Astronomical treatise by al-Biruni, copied in 1082.										
Astronomical treatise, 11th century.										
Astronomical tables, 11th century.										
Astronomical treatise, 12th century.										
Manuscript copied in the 13th century from a 9th-century original.										
Astronomical treatise by Kushyar ibn-Labban, copied in 1203 in Khorasan.										
Astronomical tables, 13th century.										
Manuscript dated 1470.										
Manuscript dated 1507.										
Manuscript from Istanbul, dated 1650.										
Book on practical arithmetic, 17th century.										
Manuscript, 17th century.										
Manuscript, 17th century.										
Modern typographical characters.	١	٢	٣	٤	٥	٦	٧	٨	٩	٠

Fig. 29–16. Numerals of the *hindi* type, used by the Eastern Arabs.

used as the zero sign in Cambodia (figs. 29–6, 29–7), Kashmir, and Punjab (fig. 29–14, Sarada) from the seventh century onward. In a passage quoted earlier in this chapter, Chhüthan Hsi-Ta, an Indian Buddhist living in China in the eighth century, wrote, "Whenever there is an empty space in a column [i.e., a zero], a dot is always placed [to signify it]." This form of the zero sign is also mentioned by al-Biruni (973–1048), an Arab astronomer of Persian descent, in a discussion of the system of Sanskrit symbolic number words. After citing the words *sunya* ("void") and *kha* ("sky") as symbolic designations of zero, he adds, "They signify the dot."

But at first the Arabs evidently used only a small circle as their zero sign. Al-Khowarizmi described it in the early ninth century by comparing it with the Arabic letter ha, which has the same shape.

The circular zero sign was (and still is) used almost exclusively in both northern and southern India, Central and Southeast Asia (Cambodia, Champa, Sumatra, Java, etc.), and by Chinese reckoners who had been influenced by Indian Buddhists (figs. 29–3, 29–6, 29–7, 29–8, 29–13A, 29–14). The dot resulted from the practice, in some regions, of making the circular sign smaller and smaller, until finally the space inside it disappeared.

The Indian zero sign, in either of its forms, is essentially different from the zero sign in the sexagesimal place-value alphabetic notation of Arab astronomers (fig. 26–8). The principle of this notation is of Babylonian origin and its Arabic graphic form is an adaptation of the alphabetic system used by Greek astronomers. Several astronomical treatises, containing expositions of sexagesimal computation and tables compiled according to the base 60, clearly show the differences between the forms and uses of the two zero signs. To indicate the absence of a sexagesimal order, the authors sometimes use *two Indian zeros:*

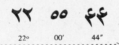

This is contrary to the sexagesimal alphabetic notation, which uses only one sign for that purpose (one of the variants shown in fig. 26–8):

مد ح كب
22°   0'   44"
-------------->

Along with the Indians' decimal place-value numeration, the Arabs also borrowed the main corresponding computation methods. There are good reasons for assuming that, like the Indians, the Arabs must have performed arithmetical operations by writing the first nine numerals on the ground, with a finger or a stick; or, judging from several ancient Arabic works, on a board covered with dust, fine sand, or flour, using a stylus with a pointed end for writing and a flat end for erasing; or on a wax-coated tablet with the same kind of stylus; or perhaps on a clay tablet, following the example of the ancient Babylonians, whose methods of reckoning had probably not been completely forgotten by the Arabs; or with chalk on a slate.

Several twelfth-century Persian poets alluded to such writing materials for computation. In a poem praising Prince Ala al-Dawla Atsuz

(1127–1157), Khaqani wrote, "The seven regions will shiver with quartan fever and the vaulted sky will be covered with dust, like a reckoner's board." And the mystic poet Nizami (died 1203): "From the nine skies [marked] with the nine figures, [God] threw the *hindissi* numerals onto the board of the earth."

Some Arab reckoners must have drawn parallel lines to form columns for the decimal orders they needed in their operations. In this way they could use their first nine numerals according to the place-value principle, leaving an empty column for each missing order:

They could then perform their operations without using the zero sign. This explains why some manuscripts on the Indian computation methods do not mention that sign (figs. 29–18, 29–19, 29–20).

Other Arab reckoners, however, wrote their numerals without columns and used the zero sign, representing the number 608,732,000, for example, in this form:

٦٥٨٧٣٢٥٥٥

6 0 8 7 3 2 0 0 0 ——→

(Although Arabic writing went from right to left, as it still does, it preserved the Indian convention of noting numbers from left to right, beginning with the highest decimal order.)

Some reckoners proceeded by successive corrections, erasing certain intermediate figures and replacing them with others, which reduced the effort required of memory but prevented them from discerning any errors that might occur in intermediate steps. There is an example of this kind of computation in *Kitab fi osul Hisab al-Hind* ("Treatise on Computation with Indian Numerals"), by Abu'l Hassan Kushiyar ibn-Laban al-Gili (971–1029). The manuscript is in the Aya Sofia Library in Istanbul and has been translated and published by A. Mazahéri. In the fourth chapter of the first book, the author writes:

> We want to multiply three hundred and twenty-five by two hundred and forty-three. We write them on the board as follows:
>
> ٣٢٥   325
> ٢٤٣   243

Then we multiply the upper three by the lower two. That makes six, which we place above the upper two and to the left of the upper three:

6 325
243

If this six had happened to have tens, we would have placed them to its left.

Then we multiply the upper three by the lower four. That makes twelve. We place its two above the four and add its one [which represents its ten] to the six [tens] of sixty [-two], which then becomes seventy [-two]:

72325
243

Then we multiply the upper three by the lower three. That makes nine, and we replace the upper three with it:

72925
243

Then we move the lower number one place to the right:

72925
243

And we multiply the two above the lower three by the lower two. That makes four, which, added to the two above the lower two, makes six:

76925
243

Then we multiply the upper two by the lower four. That makes eight, which we add to the nine above the four:

77725
243

Then we multiply the upper two by the lower three. That makes six, with which we replace the upper two, above the lower three:

77765
243

Then we move the lower number one place to the right:

77765
243

Then we multiply the upper five by the lower two. That makes ten, which we add to the tens of the rank above the lower two:

78765
243

Then we multiply the five by the lower four. That makes two [tens], which, added to the tens of [the rank above] the four, makes nine:

78965
243

Finally, we multiply the five by the lower three. That makes fifteen. We do not touch its five. We only add one [ten] to its tens:

78975
243

The [upper] number is the one we wanted to calculate.

Other Arabic treatises on Indian arithmetic, such as those written in the tenth century by al-Uqlidisi and Abu'l Wafa al-Buzjani, show that, while some Arab reckoners used the procedure of erasing numerals in intermediate steps, others crossed them out and wrote intermediate results above them.

For multiplication, the Arabs also used what they called the method of the sieve or the method of the net. In the late Middle Ages it became known in Europe, where it was called multiplication *per gelosia*.

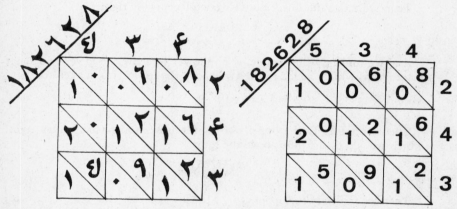

Fig. 29–17. Multiplication *per gelosia* of 534 by 342, with the numerals of the Eastern Arabs. From a sixteenth-century Arabic treatise on written computation.

Figure 29–17 shows an example of this method taken from an Arabic manuscript: the multiplication of 534 by 342. Since the multiplicand and the multiplier have three digits each, we begin drawing a large square containing three columns and three rows of smaller squares. Above (or below) the columns we write the digits of the multiplicand, 534, from left to right, and to the right (or left) of the rows we write the digits of the multiplier, 342, starting at the bottom and going upward. Then we divide each square into two triangles by drawing a line between its upper left and lower right corners. In each square we write the product of the number written above its column and the one written to the right of its row. This product is, of course, less than 100; we write the number of its tens in the left triangle and the number of its units in the right one. If either order is missing, we write a zero in the corresponding triangle or leave it empty. In the upper right square, for example, we write the result of multiplying 4 by 2, with 0 in the left triangle and 8 in the right one. When we have written a product in each square, we add the figures in each diagonal row, beginning with the single figure at the upper right and going downward, carrying a figure from one diagonal row to the next when necessary. In this way we obtain all the digits of the product of 534 and 342: 182,628.

Numerals used by the Western Arabs

"Although the unity of the empire of the caliphs was soon broken, close relations among the different regions inhabited by Moslems continued because of pilgrimages to Mecca, thriving commerce, individual journeys, migrations, and even wars. Therefore Indian arithmetic, already known to the Eastern Arabs, was introduced into the Western Arab regions. Our information on this episode in the history of science is so scanty that we cannot say exactly when it occurred, but we will not be far wrong if we consider it likely that the Arabs of Africa and Spain became acquainted with Indian arithmetic in the ninth century. We still do not know whether they learned it from treatises written by Eastern Arabs and brought into northwestern Africa, or through more direct contact with Indian science, as the Eastern Arabs had done" (Woepcke).

Be that as it may, the Western Arabs expressed the Indian decimal place-value system in numerals quite different from those of the Eastern Arabs. They are called gobar numerals, from the Arabic word for "dust," probably because, like the Indians, the Western Arabs sometimes wrote numerals on a board covered with a layer of dust.

In an eighteenth-century copy of a fourteenth-century manuscript, *Mukhtasar fi 'ilm al-Hisab* ("Summary of Arithmetic"), by Abdel Kadir Ben Ahmed As-Sakhawi, with a commentary by Hosayn Ben Mohammed Almahalli, the numerals of the Eastern Arabs are given as:

9	٨	٧	۴	٥	۴	۳	۲	١
9	8	7	6	5	4	3	2	1

and those of the Western Arabs as:

9	8	7	6	5	4	3	2	1

"The author's meaning," writes the commentator, "is obviously that both ways of forming the signs are of Indian origin, and that is the truth. The learned Chandchuri, in his commentary on the *Murshidah*, referring to the second way of forming the signs, says, 'And it is called Indian, because it was established by the nation of the Indians.' The two are distinguished by their names, however, the first being called *hindi* and the second *ghobari*; the second is called *ghobari* because the ancients had the habit of writing these signs on a wooden table that had been sprinkled with flour."

Then the commentator describes the forms of the gobar numerals in these terms:

ARABIC TEXT	TRANSLATION
وقد نظمها بعضهم فقال	The following verses have been written about these signs:
الف وحاء ثم ج وبعده	An alif (*1*) and a ḥa (ح) then ḥijun (ج) and then
عوّ وبعد العو ع ترسم	An awwun (عو) and after the awwun you write an ayn (ع).
هاء وبعد الهاء شكل ظاهر	A ha (6) and after the ha appears a sign
يبدو كخطّاف اذا هو يرقم	Which, when it is written, resembles a curved spearhead (7).
صفران ثامنها والف بينهما	The eighth [of these signs is formed by] two zeros connected by an alif (8).
والواو تسعها بذلك يختم	And a waw (9) is the ninth, with which [the series] ends.

This must have been a memory aid used in teaching pupils to write the gobar numerals. Similar compositions are found in other manuscripts. They probably helped to stabilize the forms of the numerals.

In the two earliest known documents containing gobar numerals (dating from 874 and 888), their forms are fairly close to those shown above and in figure 29–18.

Examination of these numerals shows that, like those used by the Eastern Arabs, they are altered forms of the Indian numerals associated with the same values. This can be clearly seen by comparing the gobar numerals for 2, 3, 4, 5, 6, 7, and 9 with the signs for the same numbers used at different times and in various regions of northern India and Central Asia (figs. 29–13A, 29–14, 29–18). The differences between the gobar numerals and the others may be explained by a graphic adaptation of the Indian numerals to the style of Arabic writing known as Maghribi, which, at least as early as the tenth century, became widely used in North Africa and Moorish Spain.

It was in their gobar forms that the Indian numerals spread from

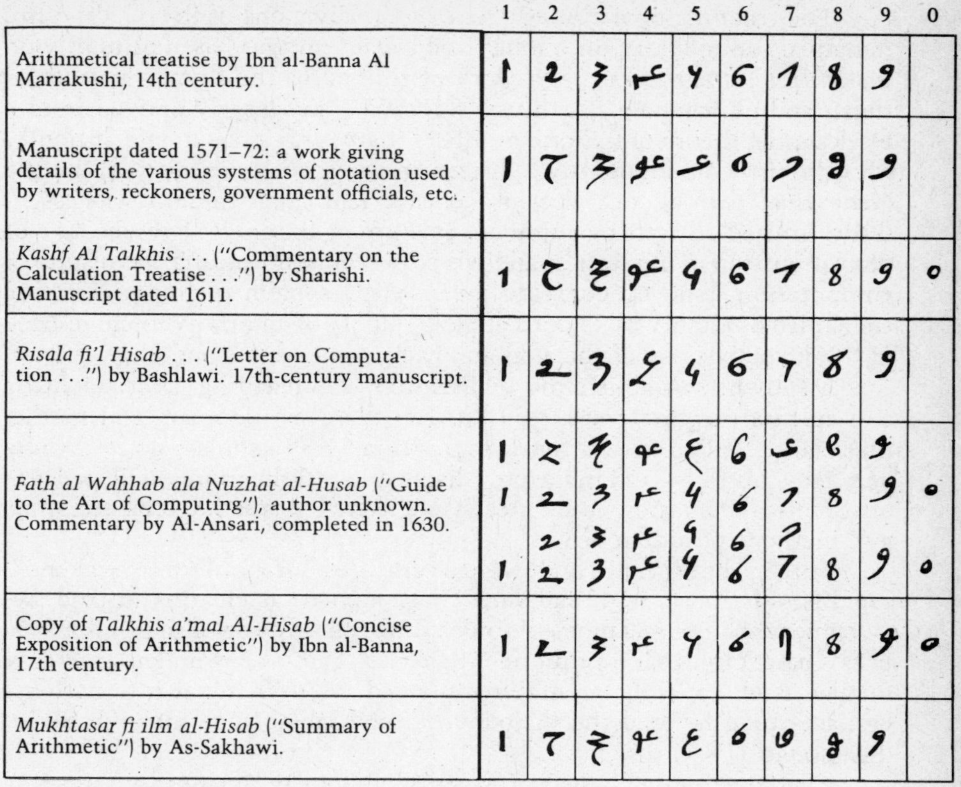

	1	2	3	4	5	6	7	8	9	0
Arithmetical treatise by Ibn al-Banna Al Marrakushi, 14th century.	ו	2	3	ے	٩	6	٦	8	9	
Manuscript dated 1571–72: a work giving details of the various systems of notation used by writers, reckoners, government officials, etc.	ו	٢	3	ﻋﻊ	؎	σ	٦	9	٦	
Kashf Al Talkhis . . . ("Commentary on the Calculation Treatise . . .") by Sharishi. Manuscript dated 1611.	1	٢	3	٩ﻋ	٩	6	٦	8	9	٥
Risala fi'l Hisab . . . ("Letter on Computation . . .") by Bashlawi. 17th-century manuscript.	١	٢	3	٤	٩	6	7	8	9	
Fath al Wahhab ala Nuzhat al-Husab ("Guide to the Art of Computing"), author unknown. Commentary by Al-Ansari, completed in 1630.	١	٢	٣	٩ﻋ	٤	6	ﻋ	٤	٩	
	١	2	3	٣	4	6	٦	8	٩	٥
		2	3	٣	٩	6	٦			
	١	2	3	٣	٩	٥	7	8	٦	٥
Copy of *Talkhis a'mal Al-Hisab* ("Concise Exposition of Arithmetic") by Ibn al-Banna, 17th century.	١	٢	٣	٣	٩	٥	٦	8	٦	٥
Mukhtasar fi ilm al-Hisab ("Summary of Arithmetic") by As-Sakhawi.	١	٢	٣	٩	٤	٥	٤	٦	٦	

Fig. 29–18. Variants of the gobar numerals.

Spain into the rest of Europe, where they gradually took on the forms that are now so familiar to us. For this reason they are still commonly referred to as Arabic numerals but they are also called, more accurately, Hindu-Arabic.

Hindu-Arabic numerals in Europe

In the early Middle Ages, from the fall of the Roman Empire to the end of the ninth century, the Western Christian peoples were plunged in political disorder, economic depression, and intellectual obscurantism. During that period their scientific knowledge, insofar as they can be said to have had any, was elementary. And although it was at the origin of medieval philosophy, the so-called "Carolingian Renaissance"—which began in Benedictine monasteries and, in Charlemagne's time, was marked by educational reform—brought only superficial improvement.

The teaching of arithmetic (one of the divisions of the quadrivium) remained rudimentary for a long time because it was based primarily on *Institutiones Arithmeticae*, a work attributed to the Roman mathematician and philosopher Boethius (480?–524?) and largely derived from a work on the theory of figurate numbers that was famous in late antiquity (though of mediocre quality), written in the second century A.D. by the Neopythagorean Nicomachus of Gerasa. Practical arithmetic was essentially limited to writing numbers in Roman numerals (which did not permit written computation and were used only for noting the results of computation done by concrete means), the system of finger counting transmitted notably by Isidore of Seville (d. 636) and the Venerable Bede (d. 735), and the use of the Roman abacus (Chapters 3 and 8).

Meanwhile Arab-Islamic civilization was reaching great scientific and cultural heights and already had brilliant astronomers and mathematicians. Skilled in written computation, Arab scholars could handle very large numbers and made new discoveries all the more easily because arithmetical operations were greatly facilitated for them by the numerals and methods of Indian origin.

But "in the eleventh and twelfth centuries Europe abruptly wakened: rapid population growth had such consequences as clearing of land, development of cities and monastic orders, and construction of larger churches; prices rose, circulation of money increased, and, as sovereigns gradually quelled feudal anarchy, commerce revived. More frequent international contacts then favored the introduction of Arab science into the West" (Beaujouan 1).

The Frenchman Gerbert of Aurillac was one of the most outstanding scholarly figures of that time. Born in Aquitaine about 945, he became a monk in a monastery at Aurillac. Then, during a stay in Spain from 967 to 970, studying under Bishop Hatto of Vich, he became acquainted with the mathematics, astronomy, and computation methods transmitted by the Arabs. He may have resided in Seville or Cordova, in direct contact with the Western Arabs, but he was more likely in the monastery of Santa Maria de Ripoll, which "offered a striking example of the grafting of Arab elements onto the Isidorian tradition" (Beaujouan 2), the small Catalonian town of Ripoll having served as an intermediary between the Christian and Moslem worlds. From 972 to 982 he directed the diocesan school at Reims; his teachings heavily influenced the schools of his time and revived interest in mathematics in the West. After successively becoming abbot of the Bobbio monastery in Italy, archbishop of Reims, and archbishop of Ravenna, he was elected pope in 999 and took the name of Sylvester II. He died in 1003. He is now regarded as having been the first great scholar to spread the use of Hindu-Arabic numerals (and the astrolabe) in Europe.

The earliest known European manuscript containing the first nine Hindu-Arabic numerals is the Codex Vigilanus, dating from 976 (fig. 29–

19). It was copied by a monk named Vigila at the Albelda monastery in northern Spain. The numerals it shows are very close to the gobar forms, while its Latin letters are in Visigothic script of the type used in northern Spain.

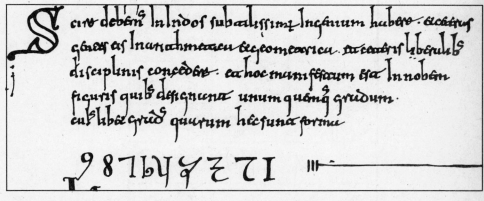

Fig. 29–19. Detail of the Codex Vigilianus, a Spanish manuscript dated 976, containing the earliest known example of the Hindu-Arabic numerals in Europe.

Beginning in the eleventh century, these numerals appear in many other manuscripts from different parts of Europe, in forms that vary with time, place, and individual writing styles.

Contrary to what one might think, however, the diffusion of Hindu-Arabic numerals in Europe did not take place primarily through manuscripts but through oral teaching of a reckoning technique using a new type of counting board that was advocated by Gerbert and his disciples (Bernelinus, Remi of Auxerre, Papias, Heriger, Adalbold, and others). It was an improved version of the old Roman counting board, and the improvement was said to have been made by Gerbert himself.

The Roman counting board (Chapter 8) was marked off in columns associated with decimal orders. Its counters were all the same and each had the value of 1. A given number was represented by placing in each column as many counters as there were units in the corresponding decimal order (fig. 8–5). Arithmetical operations could be performed by a complicated system of moving the counters (fig. 8–6). The device was convenient for addition and subtraction, but more complex operations could be performed on it only with considerable difficulty.

It was simplified in the late tenth century. Instead of identical counters that all had the value of 1, the new system used counters called apices (apex in the singular), each with a value from 1 to 9 indicated on it by a Roman, Greek alphabetic, or, more commonly, Hindu-Arabic numeral.

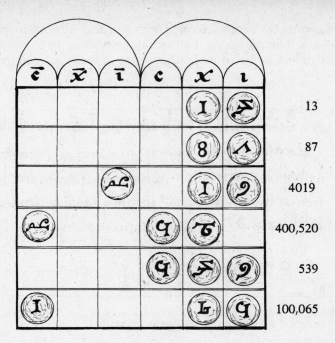

Fig. 29–20. Representation of numbers by means of apices on the improved counting board of Gerbert and his disciples. It had twenty-seven columns joined in groups of three, and the apices had positional values that depended on the columns in which they were placed. The absence of units of a certain decimal order was indicated by leaving the corresponding column empty.

Six identical counters in a column were now replaced with one apex bearing a sign for 6, seven with one apex bearing a sign for 7, and so on. On this type of counting board the numerals acquired place value, and multiplication and division could be performed without the use of a zero sign, since the absence of units of a certain order was indicated by leaving the corresponding column empty (fig. 29–20).

It was mainly by being written on apices that the Hindu-Arabic numerals became widely known in the West during the eleventh and twelfth centuries, though in this initial phase they seem to have been propagated without the zero sign.

After a time, the numerals themselves came to be called apices. They were given the following names, whose origins remain unclear:

1: *igin*		6: *caltis*
2: *andras*		7: *zenis*
3: *ormis*		8: *temenias*
4: *arbas*		9: *celentis*
5: *quimas*		0: *sipos*

The various forms of the Hindu-Arabic numerals used in Europe during this period were for a long time called "Boethian apices" because their earliest known appearance was believed to be in a work entitled *Geometry*, attributed to Boethius. Its author claims that the first nine numerals and their use according to the place-value principle were invented by the Neopythagoreans in connection with use of the counting board. Scholars who accepted this claim made various conjectures in an effort to reconcile it with the known facts about propagation of the Hindu-Arabic numerals in Europe. But modern analysis has proven that the *Geometry* attributed to Boethius is actually the work of an anonymous eleventh-century author and that use of the Hindu-Arabic numerals was unknown to the Neopythagoreans and the Greeks in general. Furthermore, the early European forms of the numerals are very close to the gobar forms and, in some cases, to those used by the Eastern Arabs.

From the tenth century to the twelfth, European manuscripts show great diversity in the forms given to the Hindu-Arabic numerals. This may be at least partly explained by adaptation of the original numerals to the different regional writing styles of medieval Europe. But there may also be another reason for it, as Beaujouan has stated on the basis of a thorough study. While European numerals have relatively stable forms from the thirteenth century onward, they seem to be highly varied during the period of their introduction into the West. However, "the apparently diverse forms taken by each numeral between the tenth and twelfth centuries become closely similar when they are rotated through an angle that varies in different cases" (Beaujouan 3). This is particularly true of the signs for 3, 4, 5, 6, 7, and 9 (fig. 29–21).

"This period of instability coincides with use of the counting board, which was learned much less from books than from practice. The first numerals were therefore propagated in the West on the counters, made of horn, that were then used for reckoning. It is likely that the teachers and pupils in many schools *acquired the habit of using the apices upside down.* Some scribes came to replace the original forms with the ones they were used to, and the mistake soon became deeply rooted because even books taught the signs upside down. It could have been avoided by placing a dot at the bottom of each counter, but that solution was not adopted. Nothing was done except to differentiate the only two signs that could be confused: the six was given an angular form while the nine remained rounded" (Beaujouan 2).

The next period—distinguished by the twelfth-century European revival and the rise of the universities and scholastic science in the thirteenth and fourteenth centuries—marked a decisive stage not only in the propagation of written computation by means of the Hindu-Arabic numerals but also in the stabilization of their forms.

Scientific studies "developed from the twelfth century onward because Arabic works began reaching the West, at first by way of the Moor-

Dates	1	2	3	4	5	6	7	8	9	0
976										
992										
Before 1030										
1077										
11th century										
1049(?)										
11th century										
11th century										
11th or 12th century										
11th century										
11th century										
11th century										
Early 12th century										
Late 11th century										
Late 11th century										
Late 11th century										
12th century										
12th century										
12th century										
12th century										
12th century										
12th century										
Early 13th century										
13th century										

Fig. 29–21. Forms of the apices in the Middle Ages. For more complete information on the subject, see the work by G. F. Hill listed in the bibliography.

ish schools in Spain. Among the translators of those works we must first mention the English monk Adelard of Bath, a great traveler who made journeys to Asia Minor, Egypt, and Spain, and in about 1120 brought back from Cordova the first Latin translation (from an Arabic translation) of Euclid's *Elements.* Adelard also translated al-Khowarizmi's astronomical tables and perhaps his arithmetic. Soon afterward, under the impetus of Archbishop Raimund, a center of learning was formed in Christian Spain, at Toledo. It was there that John of Luna, a converted Jew, wrote his *Liber Algorismi de Practica Arismetrice,* and Gherardo of Cremona (1114–1187), who devoted a great deal of time and effort to translation, found the Arabic editions of Ptolemy's *Almagest"* (Pérès).

In this third phase of the Middle Ages, a decisive change took place in the practice of computation: Gerbert's counting board gradually fell into disuse; arithmetical operations were performed by writing numerals, *including zero,* in sand or dust, and the columns disappeared. Abacism (use of the abacus or counting board) gave way to algorism (written calculation with the Hindu-Arabic numerals), which was more elegant and rapid. For many generations, however, there were still a few reckoners who went on using the old, obsolete methods.

It was also during this time that European numerals began to acquire forms that differed sharply from the apices and seemed to mark a return to the original Arabic forms (fig. 29–22). They were stabilized little by little and finally became our modern numerals. It is therefore in the twelfth century (and not in the fifteenth, nor in the time of Gerbert and his disciples) that we must place the real roots of the forms of the Hindu-Arabic numerals that have been adopted all over the world, for the apparently radical transformations that those numerals underwent at the end of the Middle Ages were simply incorporated into the general tendencies of humanistic writing.

And finally, it was also in the twelfth century that use of the zero sign was propagated in the West, having been previously made unnecessary by the columns of Gerbert's counting board (fig. 29–20).

When it was adopted in Europe various names were given to it, all of them variations of the Arabic word *ṣifr,* which means "zero" (literally "void") and is the Arabic translation of the Sanskrit word *śūnya,* with the same meaning.

In his *Liber Abaci,* the Italian mathematician Leonardo Fibonacci (1180?–1250?) called it *zephirum,* and Italian arithmetic books used that name till the Renaissance. Then, after a few alterations, it became the Italian *zefiro,* and finally *zero,* which appeared for the first time in *De Arithmetica Opusculum,* by Philippi Calandri, printed at Florence in 1491 (fig. 29–23).

But the Arabic *ṣifr* also gave rise to such words as the French *chiffre,* the German *Ziffer,* and the Italian *cifra,* all meaning "numeral." This was a later development, however. Earlier transcriptions of the Arabic

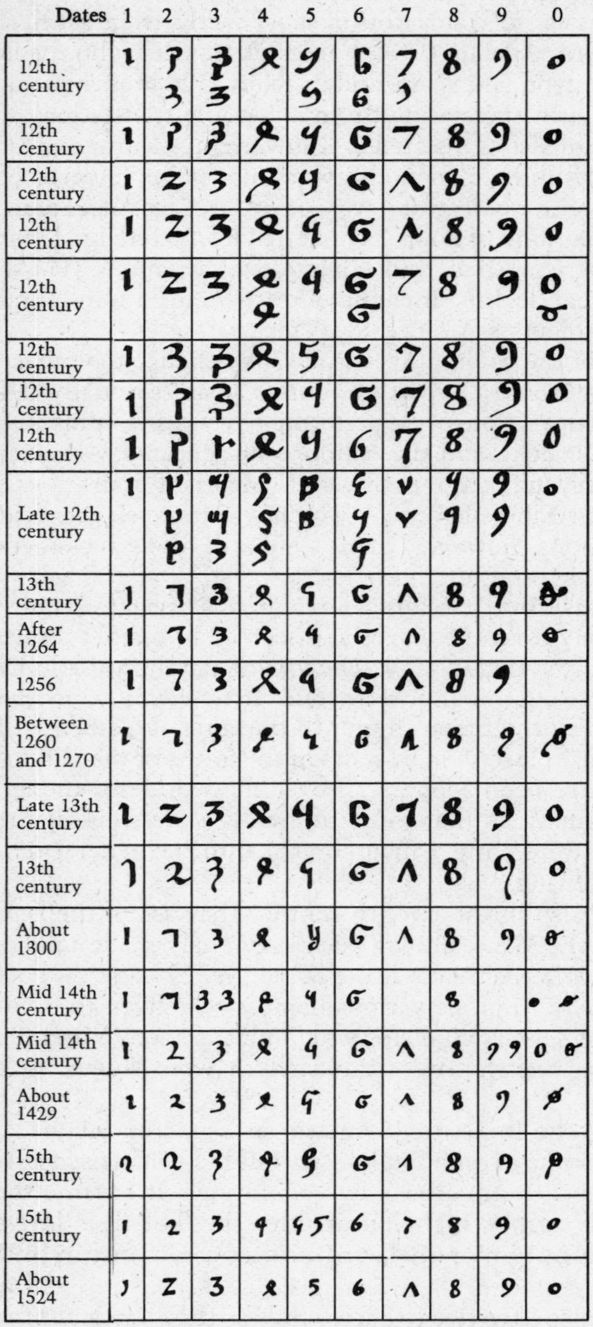

Dates	1	2	3	4	5	6	7	8	9	0
12th century	ı	ꝛ 3	ꝫ 3	℞	y ꝯ	6 6	7 ꝫ	8	ꝯ	o
12th century	ı	ꝛ	ꝫ	℞	ꝯ	6	7	8	ꝯ	o
12th century	ı	z	3	℞	ꝯ	6	⋏	8	ꝯ	o
12th century	ı	z	3	℞	ꝯ	6	⋏	8	ꝯ	o
12th century	ı	z	3	℞ ꝯ	ꝯ	6 6	7	8	ꝯ	o ꝟ
12th century	ı	3	ꝫ	℞	5	6	7	8	ꝯ	o
12th century	ı	ꝛ	ꝫ	℞	ꝯ	6	7	8	ꝯ	o
12th century	ı	ꝛ	ⱳ	℞	ꝯ	6	7	8	ꝯ	o
Late 12th century	ı	ꝟ ꝟ Ꝓ	ꝯ ꝯ 3	ʃ ꝫ ʃ	ꞵ ꞵ	ℰ ꝯ ꝯ	ꝟ ꝟ	ꝯ ꝯ	ꝯ ꝯ	o
13th century	)	ꝺ	3	℞	ꞓ	�G	⋏	8	ꝯ	ꝯ
After 1264	ı	ꝺ	ꝫ	℞	ꝯ	⪗	⋏	8	ꝯ	ꝯ
1256	ı	7	3	℞	ꝯ	⪗	⋏	8	ꝯ	
Between 1260 and 1270	ı	ꝺ	ꝫ	ʄ	Ꝯ	⪗	⋏	8	ℊ	ꝯ
Late 13th century	ı	z	ꝫ	℞	4	⪗	⋏	8	ꝯ	o
13th century	ꝺ	2	ꝫ	℞	ꞓ	⪗	⋏	8	ꝯ	o
About 1300	ı	ꝺ	3	℞	y	⪗	⋏	8	ꝯ	ꝯ
Mid 14th century	ı	ꝺ	33	ʄ	ꝯ	⪗		8		o ꝯ
Mid 14th century	ı	2	3	℞	4	⪗	⋏	8	ꝯꝯ	o ꝯ
About 1429	ı	2	3	℞	ꞓ	⪗	⋏	8	ꝯ	ꝯ
15th century	ꞯ	ꞯ	ꝫ	4	ꝯ	⪗	⋏	8	ꝯ	ꝓ
15th century	ı	2	ꝫ	4	ꝯ5	6	ꝯ	8	ꝯ	o
About 1524	ꝺ	z	3	℞	5	6	⋏	8	ꝯ	o

Fig. 29–22. European numerals, twelfth to sixteenth centuries. See the work by G. F. Hill listed in the bibliography.

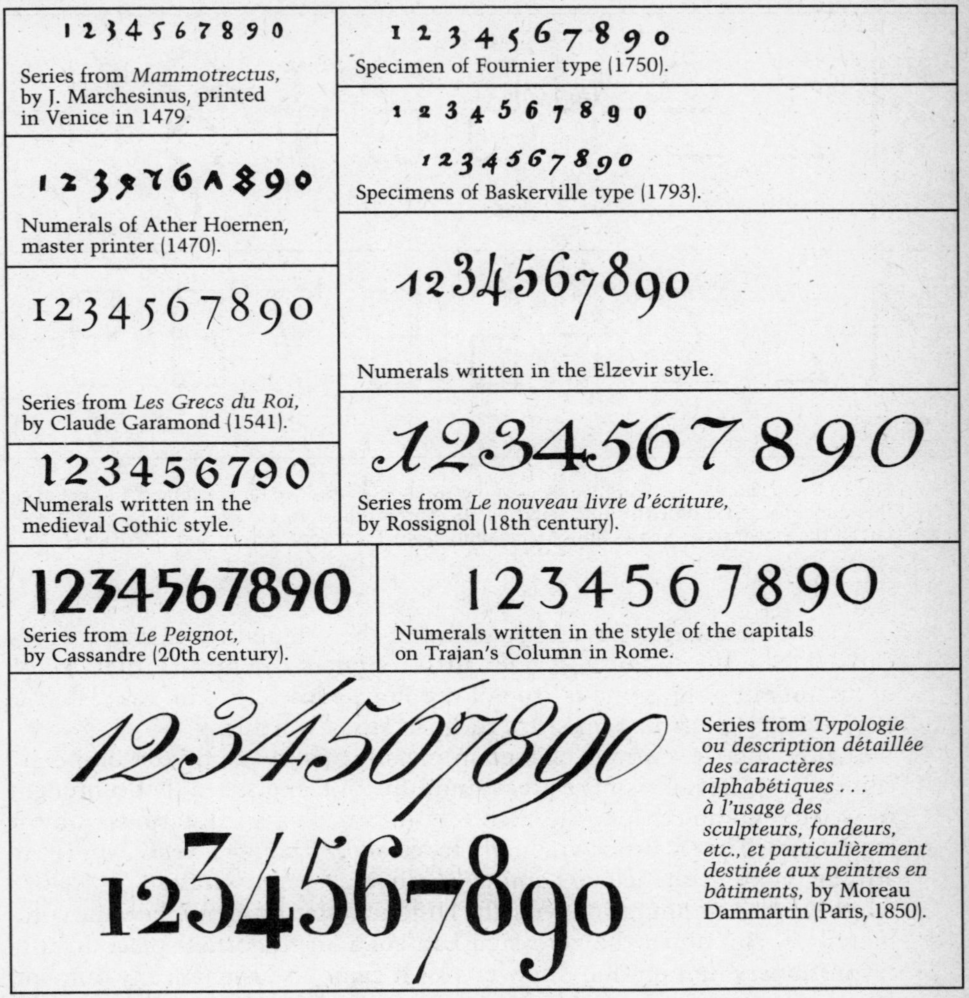

1 2 3 4 5 6 7 8 9 0

Series from *Mammotrectus*, by J. Marchesinus, printed in Venice in 1479.

1 2 3 9 7 6 Λ 8 9 0

Numerals of Ather Hoernen, master printer (1470).

1 2 3 4 5 6 7 8 9 0

Series from *Les Grecs du Roi*, by Claude Garamond (1541).

1 2 3 4 5 6 7 9 0

Numerals written in the medieval Gothic style.

1 2 3 4 5 6 7 8 9 0

Series from *Le Peignot*, by Cassandre (20th century).

1 2 3 4 5 6 7 8 9 0

Specimen of Fournier type (1750).

1 2 3 4 5 6 7 8 9 0

1 2 3 4 5 6 7 8 9 0

Specimens of Baskerville type (1793).

1 2 3 4 5 6 7 8 9 0

Numerals written in the Elzevir style.

1 2 3 4 5 6 7 8 9 0

Series from *Le nouveau livre d'écriture*, by Rossignol (18th century).

1 2 3 4 5 6 7 8 9 0

Numerals written in the style of the capitals on Trajan's Column in Rome.

1 2 3 4 5 6 7 8 9 0

1 2 3 4 5 6 7 8 9 0

Series from *Typologie ou description détaillée des caractères alphabétiques . . . à l'usage des sculpteurs, fondeurs, etc., et particulièrement destinée aux peintres en bâtiments*, by Moreau Dammartin (Paris, 1850).

Fig. 29–23. Handwritten and printed numerals since the late fifteenth century.

word—*sifra, cyfra, tzyphra, cifre, cyfre,* etc.—were all used only with the meaning of "zero."

In thirteenth-century France a worthless man was called a *cyfre d'angorisme* or *cifre en algorisme,* which can be translated literally as "a zero in the place-value numeration." The French word eventually became *chiffre,* but it was not until after 1491 that it acquired its modern meaning of "numeral"; from then on the word *zéro,* of Italian origin, was the only one used with the meaning of "zero" (fig. 29–24).

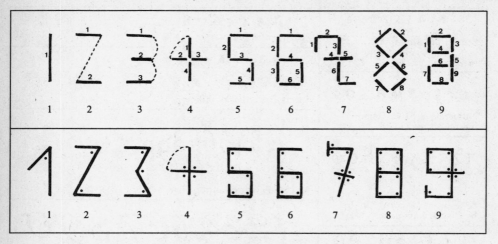

Fig. 29–24. Examples of memory aids used during the Renaissance for teaching the forms of the numerals. (According to a folk tradition still current in Morocco, Syria, and Egypt, the second of these two series shows how Arabic numerals were invented by a North African.)

Other languages have kept traces of the original meaning of words derived from the Arabic *ṣifr*. In English, "cipher" can mean either "zero" or "numeral." The same is true of the Portuguese *cifra*. In Swedish, the expression *Han är just en siffra* means "He is worthless."

How did the curious change in meaning from "zero" to "numeral" come about? Because of the great simplification and speed that it brought to written computation, the zero was regarded as a mysterious, almost magic sign. That attitude gradually faded away (though a vestige of it can still be found in the use of "cipher" to mean "secret writing" or "code") when written computation with the Hindu-Arabic numerals became commonplace. But since the zero sign had such an important place in that revolutionary numeration, the word for it came to designate any numeral of the system, and the French *chiffre*, for example, took on the meaning it still has today.

This is eloquent testimony to public awareness of one of the great discoveries in the history of humanity, a discovery which, made by Indian scholars nearly fifteen centuries ago and then transmitted to the Arabs, became "one of the fundamental contributions of the Middle Ages to the intellectual equipment of Western science" (Beaujouan 1).

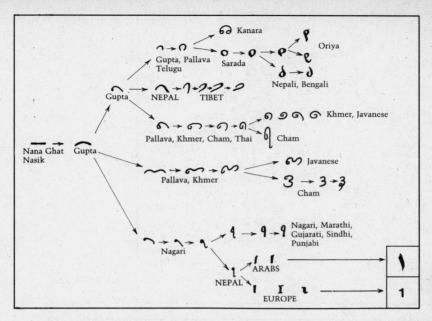

Fig. 29–25A. Origin and evolution of the numeral 1.

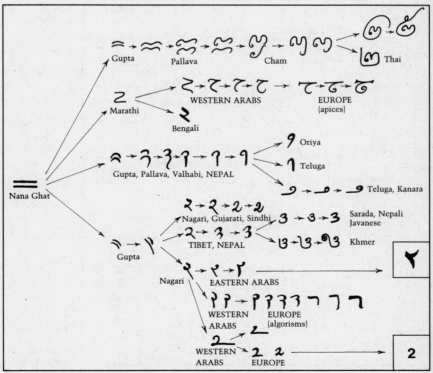

Fig. 29–25B. Origin and evolution of the numeral 2.

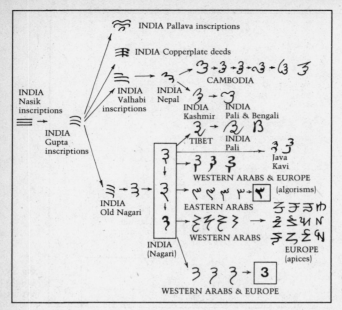

Fig. 29–25C. Origin and evolution of the numeral 3.

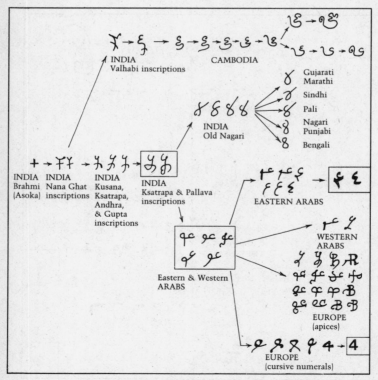

Fig. 29–25D. Origin and evolution of the numeral 4.

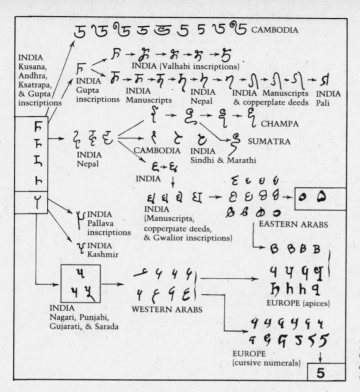

Fig. 29–25E. Origin and evolution of the numeral 5.

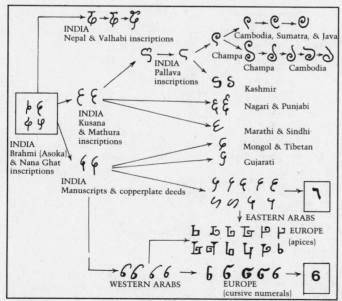

Fig. 29–25F. Origin and evolution of the numeral 6.

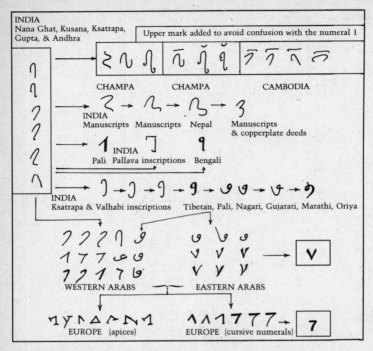

Fig. 29–25G.
Origin and
evolution of
the numeral 7.

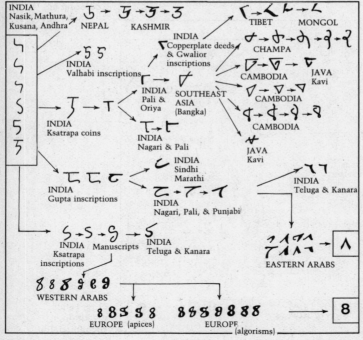

Fig. 29–25H.
Origin and
evolution of
the numeral 8.

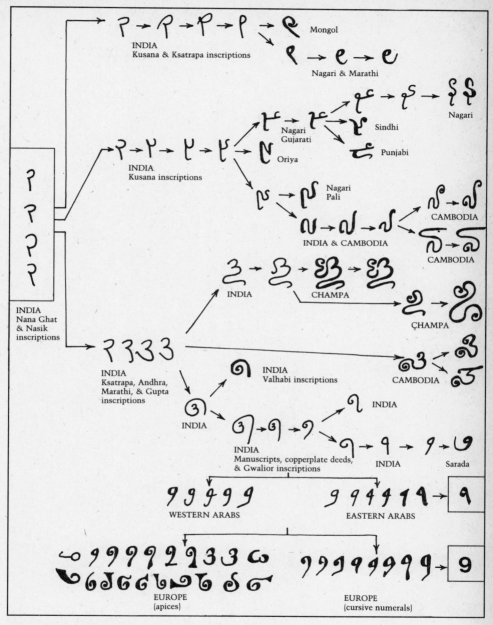

Fig. 29–25I. Origin and evolution of the numeral 9.

Chronology

As used below, the word "appearance" does not have the meaning of "invention" or "discovery." It simply refers to the earliest time for which we have evidence of a given development. Since the evidence we now have is not necessarily the oldest in existence, some of what follows may have to be revised in the light of new archaeological discoveries.

W = History of writing
N = History of numerical notation

30,000–20,000 B.C.
    W: First cave paintings.
    N: First notched bones.

3300–2850 B.C.
    N: Appearance of Sumerian, Proto-Elamite, and Egyptian hieroglyphic numerals.

3100–2900 B.C.
    W: Appearance of the earliest known forms of writing (still close to pictography) in Sumer, Elam, and Egypt.

2700 B.C.
    W: Appearance of cuneiform (wedge-shaped) signs on Sumerian tablets.
    N: Appearance of Sumerian cuneiform numerals.

2600–2500 B.C.
    W: Appearance of Egyptian hieratic writing (a condensed, cursive form of hieroglyphic writing, used concurrently with it).
    N: Appearance of Egyptian hieratic numerals.

Second half of the third millennium B.C.
    W: The Mesopotamian Semites borrow cuneiform characters for recording their own language.

N: The Mesopotamian Semites borrow Sumerian cuneiform numerals and gradually adapt them to the base 10.

2300 B.C.

W: Appearance of Proto-Indian writing at Mohenjo-Daro, in the Indus Valley, Pakistan.

First half of the second millennium B.C.

W: Diffusion of Assyro-Babylonian cuneiform writing in the Near East, where it becomes the writing of diplomacy.
Appearance of writing in Crete (Minoan civilization).

N: The Assyro-Babylonian cuneiform decimal notation (the system in common use) gradually supplants the Sumerian sexagesimal system and spreads through the Near East.

1900–1800 B.C.

N: Appearance of the oldest known place-value written numeration (cuneiform numerals, base 60), used by Babylonian scholars.

Seventeenth century B.C.

W: First known attempt at alphabetic writing: a group of Semites in the service of the Egyptians in the Sinai Peninsula use a few phonetic signs derived from Egyptian hieroglyphs (Serabit el Khadim inscriptions).

Second half of the second millennium B.C.

N: The Egyptian hieratic numeration reaches its full development.

Fourteenth century B.C.

W: Appearance of Hittite hieroglyphic writing.
Appearance, on Ugaritic tablets, of the first known entirely alphabetic writing (alphabet of thirty cuneiform signs).

Late fourteenth century B.C.

W: Appearance of the oldest known specimens of Chinese writing (divinatory inscriptions on bones and tortoise shells) at Xiaodun.

N: Appearance of the oldest known Chinese numerals.

Late twelfth century B.C.

W: First known specimens of the Phoenician alphabet (prefiguration of modern alphabets).

Early first millennium B.C.
> W: Diffusion of alphabetic writing in the Near East and the Eastern
> Mediterranean (Phoenicians, Aramaeans, Hebrews, Greeks, etc.).
> N: The Hebrews borrow Egyptian hieratic numerals.

Late ninth–early eighth centuries B.C.
> W: The Greeks complete the principle of modern alphabetic writing
> by adding vowels to the consonants of Phoenician origin.

Eighth century B.C.
> W: Appearance of Egyptian demotic writing (a cursive derived from
> hieratic, which it supplanted in common use).

Late eighth century B.C.
> N: Oldest known examples of the Aramaic numeration.

Seventh century B.C.
> W: Appearance of Etruscan writing.

Late seventh–early sixth centuries B.C.
> W: Appearance of archaic Latin writing.

Late sixth century B.C.
> N: Oldest known examples of the Phoenician numeration.

Second half of the first millennium B.C.
> N: Diffusion of Aramaic numerals in the Near East (Mesopotamia,
> Syria, Palestine, Egypt, northern Arabia).

Fifth century B.C.
> W: Aramaic writing becomes the system of international corre-
> spondence in the Near East, supplanting the cuneiform writing
> of Assyro-Babylonian origin.
> Appearance of Zapotec writing in Central America.
> N: Appearance of the Greek acrophonic numeration in Attica.
> Appearance of the South Arabic acrophonic numeration in Sa-
> baean inscriptions.

Fourth century B.C.
> N: Diffusion of the Greek acrophonic numeration in the Hellenic
> world.

Late fourth–early third centuries B.C.
> N: The first known specimens of the Greek alphabetic numeration
> appear in Egypt.

Middle of the third century B.C.

W: Appearance of Kharoshthi writing (a cursive derived from Aramaic writing, used in northwestern India, Pakistan, and Afghanistan) in the edicts of King Asoka.

Appearance of Brahmi writing in the edicts of King Asoka. It is the first known strictly Indian writing, and all other Indian scripts are derived from it.

N: Diffusion of the Aramaic numeration in northwestern India, Pakistan, and Afghanistan.

Diffusion of the Greek alphabetic numeration in the Near East and the eastern Mediterranean.

Appearance of the first known zero, in the Babylonian place-value system.

Second century B.C.

W: First known specimens of Square Hebrew, the modern form of Hebrew writing, derived from the Aramaic cursive letters.

Buddhist inscriptions in a cave at Nana Ghat, India, in a form of writing directly derived from Brahmi.

N: More complete appearance of Brahmi numerals in the Nana Ghat inscriptions. These numerals, not yet used according to the place-value principle, are a prefiguration of the modern Indian, Arabic, and Western numerals.

Late second century B.C.

N: Oldest known examples of the use of Hebrew letters as numerals.

Second century B.C.–second century A.D.

W: Reform of Chinese writing and appearance of the *li-shū* style, which gradually developed into the modern style.

N: Appearance of the Chinese place-value decimal numeration, using rod numerals, without a zero.

First or second century A.D.

W: Buddhist inscriptions in caves at Nasik, India, closely resembling the Nana Ghat inscriptions.

Early centuries of the Christian era

W: A cursive branch of Aramaic writing evolves into Arabic writing, whose first known examples date from the fourth century.

N: The Indian numerals 1 to 9 are evidently not yet used according to the place-value principle.

Late third–early fourth centuries

W: Earliest known examples of Mayan writing.

Appearance of Chinese *kăi-shū* writing, the style that is used today.

N: Earliest known Mayan Initial Series inscriptions for expressing dates and periods of time.

Fourth–sixth centuries

W: Appearance of Indian Gupta writing, from which are derived all northern Indian and some Central Asian writings.
Pallava, Chalukya, and Valhabi inscriptions, whose system is the origin of southern Indian writings.

N: Probable time of the appearance of the Mayan astronomer-priests' zero and place-value numeration.

August 28, 458

N: Date of the *Lokavibhaga*, a Jain work on cosmology, in Sanskrit. It is the oldest known example of the use of Sanskrit symbolic number words and shows a clear understanding of the place-value principle and the zero concept.

C. 510

N: Two examples of the use of Sanskrit symbolic number words by the Indian astronomer Aryabhata, who also refers to the zero and the place-value principle.

C. 575

N: Sanskrit symbolic number words are abundantly employed (with a zero) by the Indian astronomer Varahamihira. They then begin to be used almost exclusively by Indian astronomers and mathematicians.

595

N: Date of the oldest known Indian epigraphic document (a deed on copperplate) showing use of the nine numerals according to the place-value principle. This new system, whose signs are derived from the Brahmi notation, is the first written decimal numeration with the same structure as ours.

598

N: Oldest known Sanskrit inscription in Cambodia. Its date, 520 Shaka, is expressed with Sanskrit symbolic number words and a zero, according to the place-value principle.

628–629

N: Use of the decimal place-value written numeration with a zero sign is already well established in India: the astronomer Bhaskara uses not only the place-value system of the Sanskrit symbolic number words but also the Indian numerals, including zero.

Seventh century
> N: Appearance of the Syriac alphabetic numeration.

662
> N: Reference to calculation by means of the Indian numerals in a work by a Syrian bishop, Severus Sebokht.

683
> N: Oldest known Khmer inscription dated by means of Indian numerals.

683–686
> N: Oldest known inscriptions written in Old Malay and dated by means of Indian numerals.

687
> N: Oldest known Sanskrit inscription in Champa with a date expressed in symbolic number words according to the place-value principle.

718–729
> N: An Indian Buddhist astronomer living in China refers to the Indian zero and computation with Indian numerals.

732
> N: Oldest known Sanskrit inscription in Java with a date expressed in symbolic number words according to the place-value principle.

760
> N: Oldest known vernacular inscription in Java (Kavi system) with a date expressed in Indian numerals, including zero.

Eighth century
> N: Appearance of the Arabic *Abjad* alphabetic numeration.
> Appearance of the Indian zero in the Chinese decimal place-value numeration (rod numerals).

Late eighth century
> N: Introduction of the Indian zero and decimal place-value numeration in Islamic countries.

813

     N: Oldest known vernacular inscription in Champa dated by means of Indian numerals.

875–876

     N: Oldest known Indian stone inscriptions expressing numbers by means of the zero and the nine Nagari numerals.

Ninth century

     N: Appearance of the gobar numerals among the Arabs of Spain and northwestern Africa. These numerals prefigure those used in the Middle Ages and those of our own time.

976–992

     N: Two manuscripts from non-Moslem Spain show the nine numerals in a form close to the gobar type; they are the oldest known European examples of the use of Hindu-Arabic numerals.

Tenth–twelfth centuries

     N: European reckoners perform their arithmetical operations on the column abacus of Gerbert and his disciples, using counters marked with the first nine Hindu-Arabic numerals in the forms known as apices.

Twelfth century

     N: Introduction of the zero sign in the West. European reckoners perform their operations by writing the Hindu-Arabic numerals in sand. The forms of the numerals begin to be stabilized.

Thirteenth–fourteenth centuries

     W: Appearance of calculations written with pen and paper, using the Hindu-Arabic numerals, in western Europe.

Fifteenth century

     W: Invention of printing in Europe.

     N: Use of the Hindu-Arabic numerals becomes common in Europe, and their forms are gradually standardized.

Bibliography of Works Quoted or Mentioned in This Book

Alleton, Viviane. *L'écriture chinoise.* Paris: Presses Universitaires de France, 1970.

Amiet, P. *Bas-Reliefs imaginaires de L'Ancien Orient d'après les cachets et les sceaux-cylindres.* Paris: Hôtel de la Monnaie, 1973.

Babelon, J. *Archeologia* no. 22 (May–June 1968): 21.

Balmès, R. *Leçons de philosophie,* vol. 2. Paris: Editions de l'Ecole, 1965.

Beaujouan, Guy.
  1. "La science dans l'Occident médiéval chrétien," *Histoire générale des sciences,* ed. R. Taton. Paris: Presses Universitaires de France, 1957.
  2. "Etude paléographique sur la 'rotation' des chiffres et des *apices* du Xᵉ au XIIᵉ siècle," *Revue d'Histoire des Sciences* 1 (1947): 301–313.
  3. *Recherches sur l'histoire de l'arithmétique du Moyen Age.* Ecole nationale des chartes. Position des thèses de la promotion de 1947.

Becker and Hofmann. *Histoire des mathématiques,* trans. R. Jouan. Paris: Lamarre, 1956.

Billard, Roger. *L'astronomie indienne.* Paris: Publications de l'Ecole française d'Extrême-Orient, 1971.

Bloch, R. "Estrusques et Romains: problèmes et histoire de l'écriture," *L'Ecriture et la psychologie des peuples.* Paris: Armand Colin, 1963.

Bouché-Leclercq, A. *Histoire de la divination dans l'Antiquité.* Paris, 1879.

Brice, W.C. "The Writing and System of Proto-Elamite Account Tablets," *Bulletin of the John Rylands Library* 45 (1962): 15–39.

Bruins, E.-M., and M. Rutten. "Textes mathématiques de Suse," *Mémoires de la Mission archéologique en Iran.* Paris: Paul Geuthner, 1961.

Coedès, G. "A propos de l'origine des chiffres arabes," *Bulletin of the London School of Oriental and African Studies* 6, pt. 2 (1931): 323–328.

Cohen, M. *La Grande Invention de l'écriture et son évolution.* Paris: Klincksieck, 1958.

Colin, S.-S. "Ḥisāb al-Djummal," *Encyclopédie de l'Islam* (2ᵉ édition française).

Cottrell, Leonard, ed. *The Concise Encyclopedia of Archaeology.* New York: Hawthorn Books, 1960.

Couderc, P. *Le Calendrier.* Paris: Presses Universitaires de France, 1970.

Cushing, Frank Hamilton. "Manual Concepts: A Study of the Influence of Hand-Usage on Culture-Growth," *American Anthropologist* 5, no. 4 (October 1892): 289–317.

Dantzig, Tobias. *Number: The Language of Science.* New York: Macmillan Company, 1930.

Daremberg, C., and E. Saglio, eds. *Dictionnaire des antiquités grecques et romaines.* Paris: Hachette, 1881.

Dédron, P., and J. Itard. *Mathématiques et mathématiciens.* Paris: Magnard, 1959.

Devambez, P. "Monnaie," *Dictionnaire de la civilisation grecque.* Paris: Fernand Hazan, 1966.

Dols, P.-J. "La vie chinoise dans la province de Kan-Su," *Anthropos* 12–13 (1917–18): 964 *et seq.*

Drioton, E., and J. Vandier. *L'Egypte,* 4th ed. Paris, 1962.

Duhamel, Georges. *Les Plaisirs et les jeux.* Paris: Mercure de France, 1922.

Evans, Arthur John. *Scripta Minoa,* vol. 1. Oxford: Clarendon Press, 1909.

Février, J.-G. *Histoire de l'écriture.* Paris: Payot, 1959.

Gallenkamp, Charles. *Maya: The Riddle and Rediscovery of a Lost Civilization.* New York: David McKay, 1959.

Gardiner, Alan. *Egyptian Grammar,* 2nd ed. London: Oxford University Press, 1950.

Gendrop, P. *Les Mayas.* Paris: Presses Universitaires de France, 1978.

Gernet, J. "La Chine, aspects et fonctions psychologiques de l'écriture," *L'Ecriture et la psychologie des peuples.* Paris: Armand Colin, 1963.

Gerschel, Lucien.
  1. "La conquête du Nombre: des modalités du compte aux structures de la pensée," *Annales, economies, sociétés, civilisations,* 1962.
  2. "Comment comptaient les anciens Romains?" *Hommages à Léon Herrmann.* Brussels: Latomus, 1960.
  3. "L'Ogam et le nom." *Revue des études celtiques* 10 (1962): 516–557.
  4. "L'Ogam et le nombre," *Revue des études celtiques* 10 (1962): 127–166.

Girard, R. *Le Popul-Vuh.* Paris: Payot, 1972.

Gmür, M. "Schweїzerische Bauermarken und Holzurkunden," *Abhandlungen zum schweizer Recht.* Bern, 1917.

Godart, L., and J.-P. Olivier. *Recueil des inscriptions en linéaire A.* Paris, 1976.

Godron, G. *Revue d'Egyptologie* 8 (1951): 98–100.

Griaule, M. *Dieu d'Eau. Entretiens avec Ogôtemmêli.* Paris: Fayard, 1966.

Guéraud, O., and P. Jouguet. "Un livre d'écolier du IIIᵉ siècle avant J.C.," *Publications de la Société Royale égyptienne de Papyrologie.* Cairo, 1938.

Guitel, G.
  1. *Histoire comparée des numérations écrites.* Paris: Flammarion, 1975.
  2. "Calendrier grégorien, calendrier maya," *Mélanges en l'honneur de Charles Morazé.* Toulouse: Privat, 1979.

Hambis, L. "La monnaie en Asie centrale et en Haute Asie," *Dictionnaire archéologique des techniques,* vol. 2. Paris: Editions de l'Accueil, 1963–64.

Harmand, Jules.
  1. "Les Races indochinoises," *Mémoires de la Société d'Anthropologie de Paris* 2, 2nd series (1882): 338–339.
  2. "Le Laos et les populations sauvages de l'Indochine," *Tour du monde,* 1879–80.

Hawtrey, Seymour H. C., "The Lengua Indians of the Paraguayan Chaco," *Journal of the Anthropological Institute of Great Britain* 31: 280–299.

Higounet, C. *L'Ecriture.* Paris: Presses Universitaires de France, 1969.

Hill, G. F. *The Development of Arabic Numerals in Europe.* Oxford: Clarendon Press, 1915.

Homeyer. *Haus- und Hofmarken.* Berlin, 1870.

Ivanoff, P. *Maya.* Paris: Fernand Nathan, 1975.

Jouglet, René. "Les Paysans," *Europe,* March 1951, p. 61.

Keimer, L. "Bemerkungen zur Schiefertafel von Hierakonopolis (I. Dynastie)," *Aegyptus, Rivista italiana di egittologia e di papyrologia* 7, nos. 3–4 (December 1926).

Lafaye, G. "Micatio," *Dictionnaire des antiquités grecques et romaines,* ed. C. Daremberg and E. Saglio. Paris: Hachette, 1881.

Laroche, E. *Les Hiéroglyphes Hittites.* Paris: Editions du C.N.R.S., 1960.

Lebrun, A. *Cahiers de la délégation archéologique française en Iran,* no. 1 (1971): 163–214.

Lebrun, A., and F. Vallat. "L'origine de l'écriture à Suse," *Cahiers de la délégation archéologique française en Iran,* no. 8 (1978): 11–59.

Lefebvre, G. *Grammaire de l'égyptien classique,* 2nd ed. Cairo: Institut français d'archéologie orientale, 1956.

Lehmann, H. *Les Civilisations précolombiennes.* Paris: Presses Universitaires de France, 1973.

Lemoine, J.-G. "Les Anciens procédés de calcul sur les doigts en Orient et en Occident," *Revue des études islamiques* (1938): 1–58.

Levi della Vida, G. *Der Islam* 10 (1920): 243.

Lévy-Bruhl, Lucien. *How Natives Think,* trans. Lilian A. Clare. New York: Alfred A. Knopf, 1926.

Maspero, Gaston. *Histoire ancienne des peuples de l'Orient classique,* vol. 1. Paris: Hachette, 1895.

Masson, O. "Les Ecritures crétoises et mycéniennes," *L'Ecriture et la psychologie des peuples.* Paris: Armand Colin, 1963.

Mazahéri, A. *Les Origines persanes de l'arithmétique.* Nice, 1975.

Menninger, Karl. *Number Words and Number Symbols,* trans. Paul Broneer. Cambridge, Mass.: MIT Press, 1969.

Métraux, A. *Les Incas.* Paris: Editions du Seuil, 1961.

Milik, Joseph T. "Numérotation des feuilles des rouleaux dans le Scriptorium de Qumrân," *Semitica* 27 (1977): 75–81.

Montluca, J.-F. *Histoire des mathématiques.* Paris: A. Blanchard, 1968.

Morley, Sylvanus Griswold. *The Ancient Maya,* 3rd ed., revised by George W. Brainerd. Palo Alto, Calif.: Stanford University Press, 1956.

Needham, Joseph. *Science and Civilisation in China,* vol. 3. Cambridge: Cambridge University Press, 1959.

Negev, A. "Monnaie," *Dictionnaire archéologique de la Bible.* Paris: Fernand Hazan, 1970.

Neugebauer, O., and A. Sachs. *Mathematical Cuneiform Texts.* (Vol. 29, American Oriental Series.) New Haven: American Oriental Society, 1945.

Niebuhr, Carsten. *Description de l'Arabie* (French translation), edition of 1770, I, p. 91 (edition of 1779: I, p. 145).

Ninni, A. P. *Sui segni prealfabetici usati anche ora nella numerazione scritta dai pescatori Clodieusi.* Venice, 1889.

Peignot, J., and G. Adamoff. *Le Chiffre.* Paris: Editions P. Tisné, 1969.

Perdrizet, P. "Isopséphie," *Revue des études grecques* 17 (1904): 350–360.

Pérès, J. "Les sciences exactes," *Histoire du monde,* vol. 13, ed. E. Cavaignac. Paris: E. de Boccard, 1930.

Perni, P. *Grammaire de la langue chinoise*, vol. 1. Paris: Maisonneuve et E. Leroux, 1873.

Peterson, Frederick. *Ancient Mexico*. New York: Capricorn Books, 1962.

Philippe, A. *Michel Rondet*. Paris, 1949.

Poole, Reginald Stuart. *A Catalogue of Greek Coins of the British Museum*. Bologna, 1963.

Prescott, William H. *History of the Conquest of Peru*. Philadelphia: J. B. Lippincott & Co., 1875.

Prou, M. *Recueil de fac-similés d'écritures du Vᵉ au XVIIᵉ siècle*. Paris, 1904.

Reichlen, P. "Abaque" and "Calcul," *Dictionnaire archéologique des techniques*, vol. 1. Paris: Editions de l'Accueil, 1963.

Ross, Kurt. *Le Codex Mendoza*. Paris: Seghers, 1978.

Ruelle, Charles-Emile. "Arithmetica," *Dictionnaire des Antiquités grecques et romaines*, ed. C. Daremberg and E. Saglio. Paris: Hachette, 1881.

Sainte-Fare Garnot, J. Article in *L'Ecriture et la psychologie des peuples*, pp. 51–71. Paris: Armand Colin, 1963.

Scheil, V. "Documents archaïques en écriture proto-élamite," *Mémoires de la délégation en Perse*, vol. 6. Paris: E. Leroux, 1905.

Schmandt-Besserat, Denise. "The Envelopes That Bear the First Writing," *Technology and Culture* 21, no. 3 (July 1980): 357–385.

Simoni-Abbat, M. *Les Aztèques*. Paris: Editions du Seuil, 1976.

Škarpa, F. "Raboš u Dalmaciji," *Zbornik za Narodni Život i Običaje južnih Slavena* (Zagreb), 29 (1934).

Smirnoff. "Sur une inscription du couvent de Saint-Georges de Khoziba," *Comptes rendus de la Société impériale orthodoxe de la Palestine* 12, no. 1 (1902).

Smith, David Eugene. *History of Mathematics*, vol. 1. New York: Dover Publications, Inc., 1958.

Soustelle, Jacques.
  1. *Les Aztèques*. Paris: Presses Universitaires de France, 1974.
  2. *La Vie quotidienne des Aztèques, à la veille de la conquête espagnole*. Paris: Hachette, 1955.

Stresser-Péan, G. "La science dans l'Amérique précolombienne," *Histoire générale des sciences*, vol. 1, ed. R. Taton. Paris: Presses Universitaires de France, 1957.

Taton, R. *Histoire du calcul*. Paris: Presses Universitaires de France, 1969.

Thompson, J. Eric S. *The Rise and Fall of Maya Civilization*. Norman, Okla.: University of Oklahoma Press, 1954.

Thureau-Dangin, François. *Esquisse d'une histoire du système sexagésimal*. Paris: Paul Geuthner, 1932.

Vercoutter, J. *L'Egypte ancienne*, 7th ed. Paris: Presses Universitaires de France, 1973.

Vissière, A. "Recherches sur l'origine de l'abaque chinois et sur sa dérivation des anciennes fiches à calcul," *Bulletin de géographie historique et descriptive* (1892): 54–80.

Wessely. "Die Zahl neunundneunzig," *Mittheilungen aus der Sammlung der Papyrus Rainer* 1 (1887): 113–116.

Woepcke, F. "Mémoire sur la propagation des chiffres indiens," *Journal asiatique* 1 (1863).

Youshkevitch, A. P. *Les Mathématiques arabes*, trans. M. Cazeneuve and K. Jaouiche. Paris: Vrin, 1976.

Acknowledgments

The history of numerals and numerical systems covers a vast portion of humanity's intellectual development, and like the history of writing, to which it is closely related, it stands, so to speak, at the intersection of the human sciences.

It requires the collaboration not only of mathematicians, logicians, philosophers, and psychologists but also (and perhaps chiefly) of anthropologists, prehistorians, archaeologists, historians of civilization, religion, and science, paleographers, and experts on ancient texts and implements. It is a multidisciplinary endeavor that takes in such a broad and rich field that no one can explore all of it without guidance.

My ambitious undertaking would surely have gone beyond my capabilities if a number of specialists had not generously helped me, guiding me in my bibliographic research, informing me of the latest archaeological discoveries, and providing me with a rich documentation that cannot be found in works of popularization.

For their valuable assistance, advice, observations, and criticisms, I want to express my deepest gratitude to those scholars and friends: Marie-Thérèse d'Alverny, Pierre Amiet, Ruth Antelme, Daniel Arnaud, Louis Bazin, Joëlle Beaucamp, Guy Beaujouan, Pierre Becquelin, Roger Billard, Jean Bottéro, Jean Bousquet, Alain Brieux, Dominique Briquel, André Caquot, Roger Caratini, Françoise de Cenival, Dominique Charpin, Alfred Cordoliani, Bernard Delavault, Jean-Marie Durand, Essam El-Banna, Jean Filliozat, Paul Garelli, Raphaël Giveon, Hattiče Gonnet, Salah Guerdjouma, Antoine Guillaumont, Claude Jacques, Francis Johannes, Jean Jolivet, Gérard Klein, Maurice Lambert, W. G. Lambert, Emmanuel Laroche, Alain Lebrun, Jean Leclant, Henri Lehmann, André Lemaire, Jean Mallon, Emilia Masson, Olivier Masson, Joseph T. Milik, Christian Peyre, Youssef Ragheb, Jacques Raison, Roshdi Rashed, Christian Robin, Maxime Rodinson, Olivier Rouault, François Secret, Gabrielle Sed-Rajna, Marie-Rose Seguy, Mireille Simoni-Abbat, Janine Sourdel, Georgette Soustelle, Maurice Sznycer, Javier Teixidor, Rabbi Charles Touati, Georges Vajda, François Vallat, Léon Vandermeersch, Pascal Vernus, Chi Yu Wu, Jean Yoyotte, Alain Zivie.

I also thank Editions Robert Laffont-Seghers for the advice and encouragement they gave me during the whole time when I was working on this book, as well as for the aid they accorded me from 1977 on; it enabled a solitary researcher to carry out an ambitious project that no university or official organization was willing to subsidize.

And I thank the Centre National des Lettres for making it easier to finish my work by giving me a grant in 1980.